T0140204

Intelligent Systems Reference Library

Volume 76

About this Series

The aim of this series is to publish a Reference Library, including novel advances and developments in all aspects of Intelligent Systems in an easily accessible and well structured form. The series includes reference works, handbooks, compendia, textbooks, well-structured monographs, dictionaries, and encyclopedias. It contains well integrated knowledge and current information in the field of Intelligent Systems. The series covers the theory, applications, and design methods of Intelligent Systems. Virtually all disciplines such as engineering, computer science, avionics, business, e-commerce, environment, healthcare, physics and life science are included.

More information about this series at http://www.springer.com/series/8578

Alejandro Peña-Ayala

Editor

Metacognition: Fundaments, Applications, and Trends

A Profile of the Current State-Of-The-Art

 Springer

Editor
Alejandro Peña-Ayala
World Outreach Light
    to the Nations Ministries
Escuela Superior de Ingeniería
    Mecánica y Eléctrica, Zacatenco
Instituto Politécnico Nacional
Mexico City
Mexico

ISSN 1868-4394   ISSN 1868-4408   (electronic)
ISBN 978-3-319-35591-7   ISBN 978-3-319-11062-2   (eBook)
DOI 10.1007/978-3-319-11062-2

Springer Cham Heidelberg New York Dordrecht London

Springer is part of Springer Science+Business Media (www.springer.com)

Preface

Metacognition is a term used to identify a kind of cognition oriented to monitor and regulate cognition engaged in a given mental activity (e.g., listening, reading, memorizing). Human beings consciously, even unconsciously, acquire and exploit metacognitive knowledge, and develop metacognitive skills every day to perform complex cognitive duties such as: learning, decision-making, and problem solving. Thus, individuals daily deal with metacognitive experiences!

This book offers a glance of recent metacognition labor, which pursues to extend the research, application, and practice reached in the field. The chapters make up a sample of the work currently achieved in countries from three continents, which illustrates a sample of the state-of-the-art. According to the nature of the contributions accepted for this volume, five kinds of topics are identified as follows:

- *Conceptual* introduces a profile of the metacognition to describe its nature, purpose, components, skills, models, and methods. Through this section, readers are stimulated to acquire an overall idea of the metacognition, metacognitive knowledge, metacognitive strategies, and visualize theoretical concepts such as meta-metacognition.

- *Frameworks* outline essential logistics to command the stimuli, performance, and evaluation of metacognitive components and skills. In this section, readers are enabling to enhance metacognitive skills and self-regulatory functions, assess metacognitive skills and self-regulated learning, and profit from executive management metacognition and strategic knowledge metacognition.

- *Studies* share research experiences oriented to respond how metacognition contributes to accomplish successful learning. Along the section, readers are informed of the results obtained from the following academic domains: English listening proficiency, science informal learning, and cognitive self-regulation during the process of learning with stress at the university.

- *Approaches* depict how metacognition is addressed in computer-supported collaborative learning (CSCL) settings and training programs for teaching practices. In this section, readers appreciate the development of social metacognition in CSCL to facilitate collaborative problem solving, content authoring, as well as the recreation of metacognitive experiences to train teachers.

- *Tools* describe and promote intelligent authoring systems (ITS) as a platform to develop learners' diagnosis and inquiry skills. Readers can examine in this section a couple of ITS, where one provides virtual patient cases to practice medical diagnostic; and other provides examples and partial information to trigger learner inquires about historical events.

This volume is the result of the research recently achieved by authors, who are willing to share their models, methodologies, results, and findings to the community of practitioners, pedagogues, psychologists, computer scientists, academics, and students interested in the valuable topic of metacognition!

After the cycle of chapter submission, revision, and tuning according to the Springer quality principles, fourteen works were approved, edited as chapters, and organized according to the prior five topics.

The first part contains *conceptual* topics that are presented in Chaps. 1–3; the second part depicts *frameworks* through Chaps. 4–6; the third part concerns with *studies* that are outlined in Chaps. 7–9; the fourth part is related to *approaches* described in Chaps. 10–12; and finally the fifth part unveils *tools*, which are characterized in Chaps. 13 and 14. A profile of the chapters is given as follows:

1. Chapter 1 offers a review of models that guide metacognitive strategy instruction as well as a hybrid strategy instruction sequence to improve comprehension monitoring and self-regulatory control processes. A distinction of specific instructional strategies that can be used to improve monitoring and control processes is also made.
2. Chapter 2 introduces the cognitive level of meta-metacognition. Specifically, second-order judgments which regulate metacognitive judgments. In addition, a method for analyzing how people make second-order judgments to assess the accuracy of metacognitive judgments is outlined. Some reflections are made concerning individual differences in second-order judgments and the existence of human meta-metacognitive ability.
3. Chapter 3 proposes a conceptual model of the metacognitive activity. It is based on neurological and biological viewpoints. The main idea is to depict metacognition as a living system that behaves as an autopoietic system. Thus, metacognition is a mechanistic system that generates metacognitive products according to its own components, such as processes and skills.
4. Chapter 4 draws a framework supported by tools to stimulate individual and group metacognitive skills, as well as self-regulatory functions and critical thinking. It claims the synergy between tools, teams, and talents, where learners develop skills. Based on those items, a model of metacognition is composed by subjects such as motivation, mindfulness, and emotions to be implemented in a CSCL setting. Authors assert such tools facilitate the development of higher-order critical, reflective, and collaborative metacognitive thinking skills.
5. Chapter 5 depicts an evaluation framework for the application of metacognitive skills and self-regulated learning (SRL) at problem solving. The guide applies several instruments such as calibration, feedback, and rubrics. These instruments assess the analysis process used by the student to solve a problem, as well as the feedback demanded on self-learning and given by the teacher.

6. Chapter 6 describes WebQuest as a learning object for embedding metacognition through inquiry-oriented learning activities delivered through the web. The design of WebQuests includes seven subjects: welcome, introduction, tasks, process, evaluation, conclusion, and teacher page. A series of illustrative examples are given for diverse courses oriented to junior and high school levels.
7. Chapter 7 reports a study about the relationship between metacognitive awareness of listening strategies and listening proficiency about English as a foreign language. The case study concerns high school students who respond to three instruments to assess their English proficiency, metacognitive awareness listening, and academic SRL. Once an inferential statistic analysis is estimated, it was found out that only problem-solving strategies have a major role in listening proficiency of mid self-regulated students!
8. Chapter 8 examines the impact of citizen science programs particularly on student self-regulation with an emphasis on metacognition and motivation. In the study a microanalytic methodology is applied to assess metacognitive and motivational processes of self-regulated learning, which are tailored for novice, transition, and expert users. In addition, a conceptual model and some guidelines for creating effective citizen science programs are outlined.
9. Chapter 9 defines conceptual characteristics and relationships of personal self-regulation, self-regulated learning, and coping strategies, which are used for dealing with stress at learning. The study claims the importance of personal self-regulation (SR) in determining the degree of cognitive SR during the process of learning with stress, as well as the relationship between personal SR and the type of coping strategies, and the relationship between SRL and coping.
10. Chapter 10 aims metacognition as socially shared construct in CSCL settings that contributes to collaborative problem solving. Thus, a series of trials are performed to identify how subjects externalize their thoughts during the process of problem solving with their peers. One of the results indicates that for socially shared metacognition to emerge, it is required that individual group members make their thinking and metacognitive monitoring visible.
11. Chapter 11 explains how to teach students programming by means of games that they collaboratively play, as well as how the metacognition contributes to achieve the goal in a CSCL setting. The approach claims that such an environment facilitates the development of monitor and regulation skills by means of common task development, working in groups, cooperative behavior, positive interdependence, and individual accountability and responsibility.
12. Chapter 12 highlights an approach that supports teachers to develop their metacognitive lifelong learning skills and to reconstruct their conceptual knowledge and procedural strategies when necessary. Authors aim at revealing expressions of metacognition among teachers and examine the changes they designed and applied in their teaching units and teaching processes. The approach concludes that metacognitive knowledge and metacognitive experiences are essential for the development of good, established teachers.
13. Chapter 13 describes how to scaffold metacognitive activities in medical problem solving training by means of BioWorld. Such an ITS instructs novice physicians in developing medical diagnostic reasoning as they receive feedback in

the context of solving virtual cases. The system dynamically assesses the user's performance against expert solution paths. It provides help by means of a set of pre-defined rules based on the context of the learner's activity.

14. Chapter 14 concerns with dysregulated learning, where different classes of failures lead to minimal learning. The work presents the MetaHistoReasoning tool, which is an ITS that traces a domain-specific account of the metacognitive activities involved in learning while performing inquiries about the causes of historical events. The tool induces confusion by failing to mention any information pertaining to the causes of an event. In this way, learners should attain a coherent understanding of the event by seeking and transforming information obtained from sources in accordance with disciplinary-based practices.

I express my gratitude to authors, reviewers, my assistants Adriana Cárdenas and Monserrat Hernández, the Springer editorial team, and the editors Dr. Thomas Ditzinger, Professor Janusz Kacprzyk, and Professor Lakhmi C. Jain for their valuable collaboration to fulfill this work.

I also acknowledge the support given by the Consejo Nacional de Ciencia y Tecnología (CONACYT) and the Instituto Politécnico Nacional (IPN) of Mexico through the grants: CONACYT-SNI-36453, IPN-SIP-20141249, IPN-COFAA-SIBE-DEBEC/647-391/2013, IPN/SIP/DI/DOPI/EDI/848/14.

Last but not least, I acknowledge the strength given by my Father, Brother Jesus, and Helper, as a part of the research projects of World Outreach Light to the Nations Ministries (WOLNM).

July 2014   Alejandro Peña-Ayala

Contents

Part IV   Approaches

10   What Makes Metacognition as Socially Shared in Mathematical Problem Solving?

Part V   Tools

Contributors

Sajad Abedi Faculty of Humanities, English Department, Shahid Rajaee Teacher Training University, Lavizan, Tehran, Iran

Carl Martin Allwood Psychology Department, University of Gothenburg, Gothenburg, Sweden

Eshwar Bachu Department of Computing and Information Technology, The University of the West Indies, St. Augustine, Trinidad and Tobago

Margaret Bernard Department of Computing and Information Technology, The University of the West Indies, St. Augustine, Trinidad and Tobago

Sandra Buratti Psychology Department, University of Gothenburg, Gothenburg, Sweden

Leonor Cárdenas ESIME Zacatenco, Instituto Politécnico Nacional, Mexico, D.F., Mexico

Ronan J. Conway School of Psychology, National University of Ireland, Galway, Galway County, Republic of Ireland

Tenzin Doleck Department of Educational and Counselling Psychology, McGill University, Montreal, QC, Canada

Christopher P. Dwyer School of Psychology, National University of Ireland, Galway, Galway County, Republic of Ireland

Osnat Eldar Oranim Academic College of Education, Tivon, Israel

Jesús de la Fuente Department of Psychology, University of Almería, Almería, Spain

Antonio P. Gutierrez Department of Curriculum, Foundations, and Reading, Georgia Southern University, Statesboro, GA, USA

Owen M. Harney School of Psychology, National University of Ireland, Galway, Galway County, Republic of Ireland

Hope J. Hartman The City College of New York, New York, NY, USA

Suzanne E. Hiller College of Education and Human Development, George Mason University, Fairfax, VA, USA

Michael J. Hogan School of Psychology, National University of Ireland, Galway, Galway County, Republic of Ireland

Tarja-Riitta Hurme Department of Teacher Education, Centre for Learning Research, University of Turku, Turku, Finland; Faculty of Education, Department of Educational Sciences and Teacher Education, Learning and Educational Technology Research Unit, University of Oulu, Oulu, Finland

Amanda Jarrell Department of Educational and Counselling Psychology, McGill University, Montreal, QC, Canada

Sanna Järvelä Faculty of Education, Department of Educational Sciences and Teacher Education, Learning and Educational Technology Research Unit, University of Oulu, Oulu, Finland

Anastasia Kitsantas College of Education and Human Development, George Mason University, Fairfax, VA, USA

Susanne P. Lajoie Department of Educational and Counselling Psychology, McGill University, Montreal, QC, Canada

José Manuel Martínez-Vicente Department of Psychology, University of Almería, Almería, Spain

Kaarina Merenluoto Department of Teacher Education, Centre for Learning Research, University of Turku, Turku, Finland

Shirley Miedijensky Oranim Academic College of Education, Tivon, Israel

Eduardo Montero-García Department of Electromechanical Engineering, Higher Polytechnic School, University of Burgos, Burgos, Spain

Chris Noone School of Psychology, National University of Ireland, Galway, Galway County, Republic of Ireland

Alejandro Peña-Ayala WOLNM, Leyes Reforma, Mexico City, Mexico; ESIME Zacatenco, Instituto Politécnico Nacional, Mexico, D.F., Mexico

Eric G. Poitras Advanced Instructional Systems and Technologies Laboratory, University of Utah Educational Psychology, Salt Lake City, UT, USA

Dave Putwain Faculty of Education, Edge Hill University, Ormskirk, UK

Mehrak Rahimi Faculty of Humanities, English Department, Shahid Rajaee Teacher Training University, Lavizan, Tehran, Iran

María Consuelo Sáiz-Manzanares Faculty of Humanities and Education, Department of Educational Sciences, University of Burgos, Burgos, Spain

Pekka Salonen Department of Teacher Education, Centre for Learning Research, University of Turku, Turku, Finland

Paul Sander Department of Psychology, Cardiff Metropolitan University UWIC, Cardiff, UK

Gregory Schraw Department of Educational Psychology, University of Nevada, Las Vegas, NV, USA

Lucía Zapata Education and Psychology I+D+i. Science and Technology Spin-Off, University of Almería, Almería, Spain

Part I
Conceptual

Chapter 1
Metacognitive Strategy Instruction that Highlights the Role of Monitoring and Control Processes

Gregory Schraw and Antonio P. Gutierrez

Abstract In this chapter we discuss the role metacognitive strategy instruction plays in learning as well as the effects of metacognitive strategy interventions on learning outcomes. We begin with a brief review of the literature regarding metacognitive strategy instruction followed by a succinct explanation of various conceptual models used in metacognitive strategy instruction research. Next, we survey how the body of literature has treated metacognitive strategy instruction, particularly with respect to monitoring and control processes, and propose a hybrid model of strategy instruction. We end the chapter by discussing implications for metacognitive strategy instruction research and practice and identifying gaps in the research literature and areas that merit further investigation.

Keywords Metacognition · Self-regulated learning · Education · Comprehension monitoring · Control processes

Abbreviations

MSI   Metacognitive strategy training

G. Schraw (✉)
Department of Educational Psychology, University of Nevada—Las Vegas,
4505 Maryland Parkway, 453003, Las Vegas, NV 89154-3003, USA
e-mail: gschraw@unlv.nevada.edu

A.P. Gutierrez
Department of Curriculum, Foundations, and Reading,
Georgia Southern University, 8144, Statesboro, GA, USA
e-mail: agutierrez@georgiasouthern.edu

© Springer International Publishing Switzerland 2015
A. Peña-Ayala (ed.), *Metacognition: Fundaments, Applications, and Trends*,
Intelligent Systems Reference Library 76, DOI 10.1007/978-3-319-11062-2_1

3

1.1 Introduction

Effective learners use metacognitive knowledge and strategies to self-regulate their learning [1–7]. Students are effective self-regulators to the extent that they can accurately determine what they know and use relevant knowledge and skills to perform a task and monitor their success.

Unfortunately, many students experience difficulty learning because they lack relevant knowledge and skills, do not know which strategies to use to enhance performance, and find it difficult to sequence a variety of relevant strategies in a manner that enables them to self-regulate their learning [1, 8].

Nevertheless, strategy training is a powerful educational tool that improves learning and performance in academic domains such as elementary and middle school mathematics [9, 10], as well as non-academic skills such as driving and anxiety management [11]. The purpose of this chapter is to review models of metacognitive strategy instruction in the classroom and make recommendations for an instructional sequence designed to maximize the acquisition and use of an integrated strategy repertoire with an emphasis on improving comprehension monitoring and self-regulatory control processes.

The chapter is organized into five sections. The first reviews models from the strategy instruction literature. The second section reviews main findings from the strategy literature and summarizes over 30 strategies that are used commonly to self-regulate one's learning. The third section reviews recent strategy training studies that focus on monitoring and control processes to identify a hybrid strategy instruction sequence that can be implemented in a relatively short timeframe, yet produces significantly improved learning among older students. The fourth section discusses specific instructional strategies from the research literature that can be used to improve monitoring and control processes. Section five discusses implications for instructional practice and future research.

1.2 Models of Strategy Instruction

Strategy instruction has been one of the most active and important lines of educational research over the past 30 years because it is seen as a relatively quick and efficient way to improve student learning [12–16]. One of the primary motivations for strategy instruction is the assumption that limited instructional time with younger children produces life-long benefits once strategies are automatized [4, 15, 17].

Theoretical explanations of strategy instruction typically have been embedded with models of self-regulated learning. Three models first developed in the 1990s have been used to understand the role of different types of strategies and the underlying cognitive mechanisms they invoke to improve learning.

The first model is based on the work of Michael Pressley and colleagues [16] and is known as the *good information processor* model, which includes four main characteristics: (1) a broad repertoire of strategies, (2) metacognitive knowledge about why, when and where to use strategies, (3) task-relevant knowledge, and (4) automaticity in the previous three components.

The second is the four-phase model of self-regulated learning developed by Philip Winne and colleagues [18]. This model emphasizes the interactive relationship among task (e.g., instruction and time factors), cognitive (e.g., domain knowledge, motivation and belief factors), and metacognitive (e.g., monitoring and control processes) conditions during learning.

A third model of special interest to this chapter is the two-level model of metacognitive regulated learning developed by Nelson and Narens [19].

This model proposed a cyclical, two-level model of metacognition that distinguishes between an *object level* characterized by task-relevant knowledge and the use of a repertoire of automated strategies, and a *meta-level* characterized by an explicit mental model of integrated strategy use that controls and regulates learning. The two levels are connected by *monitoring* and *control* processes in which monitoring enables an individual to assess the demands and outcomes of learning in order to inform the construction of a mental model at the meta-level. In turn, updated meta-level information enables the learner to control performance and subsequent monitoring in a top-down manner based on the use of the continually updated mental model [20]. Serra and Metcalfe [21] coined the term *accuracy-control link* to describe the cyclical process of monitoring, updating a mental representation at the meta-level, and control of subsequent learning In general, it is believed that the *accuracy-control link* is the underlying mechanism that enables an unregulated learner to gradually become self-regulated as the individual acquires a repertoire of automated strategies, monitors performance, constructs a model of ongoing learning, and uses that model to control subsequent learning. Although mental models are challenging for learners to construct, there is evidence that scaffolded practice using worked examples and expert modelling enhances their construction and use in control processes [22, 23].

1.3 A Typology of Strategy Use

Using a flexible repertoire of strategies in a systematic manner not only produces learning gains, but also empowers students psychologically by increasing their self-efficacy [20]. Strategy instruction should begin early, be embedded within all content areas, modeled by teachers and self-regulated students, practiced until automatized, and discussed explicitly in the classroom to provide the greatest benefit to students. Pressley and Wharton-McDonald [24] recommend that strategy instruction is needed before, during, and after the main learning episode, which we discuss in detail below. Strategies that occur before learning include *setting goals, making predictions,* determining *how new information relates to prior knowledge,* and *understanding how the new information will be used.* Strategies needed during learning include *identifying important information, confirming* predictions, *monitoring, analyzing,* and *interpreting.* Strategies typically used after learning include *reviewing, organizing,* and *reflecting.* Good strategy users should possess some degree of competence in each of these areas to be truly self-regulated.

Analyses of strategy research over the past 30 years [15, 16] strongly support the following seven claims:

1. *Strategy instruction is typically moderately to highly successful, regardless of the strategy used or instructional method.* This means that students usually benefit from strategy instruction, whether on single strategies or combinations of strategies. Strategy instruction appears to be most beneficial for younger students, as well as for low-achieving students of all ages [25]. One reason may be that younger and lower-achieving students know fewer strategies and therefore have far more room for improvement.

2. *Programs that combine several interrelated strategies are more effective than single-strategy programs* [15, 16]. One reason may be that no single strategy is enough to bring about a substantial change in learning, because most learning is typically complex. A repertoire of four or five general strategies that can be used flexibly, however, can be quite effective [26]. Interested readers are referred to Pressley and Harris [16], *Transactional Strategies Instruction,* for detailed descriptions of a successful multi-strategy program. Key among these strategies are training that promotes self-questioning and teaching students to generate pre-study questions [27], self-explanation [28, 29], and explicit reflection on strategy use [30].

3. *Six principles promote better comprehensive strategy instruction* [13], including (1) teach strategies necessary to learn all core skills in a domain such as vocabulary, comprehension and engagement in reading; (2) teacher planning in advance of instruction is essential, (3) vary the emphasis on each strategy to meet student needs, (4) each strategy requires explicit instruction and modeling, (5) vary the support and amount of responsibility given to students based on ongoing assessment, and (6) use small group discussions preceded by instruction in *effective discussion.*

4. *Strategy instruction programs that emphasize the role of conditional knowledge are especially effective.* One explanation is that conditional knowledge enables students to determine when and where to use the newly acquired strategy. A number of authors provide in-depth discussions of conditional knowledge [6, 31], as well as practical ways to increase explicit conditional knowledge [32].

5. *Programs that span 6 weeks to several months are more effective.* One reason is that students are able to model, practice and automatize strategies, as well as develop conditional knowledge about them through long-term experience [13]. A second reason is that teachers become more adept at teaching and modeling strategies over time.

6. *Newly acquired strategies do not readily transfer to new tasks or unfamiliar domains, but are more likely to do so with integrated strategy instruction.* Transferring newly mastered skills such as strategies between two different domains is very difficult and frequently does not occur spontaneously [33]. It is helpful to teach specifically for transfer of strategies [16]. Two ways of

accomplishing this are to help students make the link between strategies and their application [33, 34] and to have them practice applying strategies in a variety of settings.

7. *Train teachers to be metacognitively aware and to teach metacognitive skills.* Teachers who are more aware of their own metacognition are better able to make strategy instructional decisions in their classroom and better able to teach students about metacognition [35–37].

Another question of interest is what kind of strategies should be prioritized in strategy instruction. Hattie et al. [15] compared rank orderings for approximately 25 learning strategies across three cultures (Japanese, Japanese–Australian, and Australian). Results indicated that a handful of general learning strategies were rated as most important among all cultures.

These included, in order of importance, self-checking, creating a productive physical environment, goal setting and planning, generating questions, reviewing and organizing information after learning, summarizing during learning, seeking teacher assistance, seeking peer assistance, and self-explanation. Not surprisingly, most of the commonly used strategy instruction programs incorporate these skills [16, 29].

Pressley and Afflerbach [38] developed perhaps the most detailed typology of strategy use based on activities that occur before, during and after the main learning episode. Table 1.1 summarizes what we consider to be the most important strategies into an abbreviated typology of essential strategies. Columns in Table 1.1 correspond to strategy use before, during and after the learning episode, while rows parcel strategies into five main types of metacognitive activities described by Schraw and Dennison [39] during the development of the metacognitive awareness inventory. In this framework, planning strategies refer to setting goals and preparing to learn in the most effective manner; organizing refers to implementing strategies that help one manage information; monitoring refers to online assessment of one's learning, performance or strategy use; debugging refers to strategies used to correct performance errors or assumptions about the task or strategy use; and evaluation refers to post hoc analysis of performance and strategy effectiveness.

Table 1.1 includes strategies described in a wide variety of sources in the instructional literature [16, 29, 40]. Two conclusions seem warranted from Table 1.1 about the general structure of strategy instruction. One is that there is a large potential collection of strategies a learner might choose from. These strategies span five functional types of learning activities, including planning, organization of information, comprehension monitoring, debugging learning difficulties, and summative evaluation of what was learned. Presumably, effective strategy instruction interventions should include the most essential strategies from each of these five categories. In our view, the most effective interventions include core strategies from each of these five categories that are taught and modeled in a systematic, sequential manner.

A second conclusion is that most strategies are optimally effective either before during or after learning; thus, it incumbent upon teachers and students to understand the best time and place to use them. To do so, students must acquire both

Table 1.1 A typology of essential strategies

	Before learning	During learning	After learning
Plan	Set learning goals; identify problems; identify important and relevant information; self-question; read instructions; activate prior knowledge; read end of chapter summary first	Re-assess plans	
Organize	Plan time management; create comfortable physical environment; allocate resources; overview material; skim; look ahead; make predictions; select appropriate strategies based on conditional knowledge	Organizing information, make inferences; elaborate, paraphrase; summarize, self-question; selectively focus on important and relevant information; highlight key ideas; makes notes of important information; relate information to what you know; draw pictures or graphs; think aloud and self-explain; use evidence to verify understanding; integrate across a text to get the "big picture" by constructing a meta-level mental model	Use graphic organizers to organize, compare and contrast after studying; write down main ideas; paraphrase themes; construct an interpretation; draw conclusions and verify their credibility
Monitor		Check understanding; assess attainment of learning goals; self-test; stop to reflect	Check final comprehension against learning goals
Debug		Reread; ask for help from teachers and peers; look back; review main ideas; change strategies; re-evaluate assumptions	Review; reread; seek additional information
Evaluate		Look back and reread; review main ideas; review learning goals; assess final learning and performance	Look back and reread; review main ideas; summarize what you've learned; assess the adequacy of conclusions

the strategy and *conditional knowledge* about situated strategy use as well, where conditional knowledge refers to knowledge about when, where and why to use the strategy. Schraw [32] developed a strategy evaluation matrix and regulatory check-list that can be used in conjunction with strategy training to increase conditional knowledge.

1.4 A Hybrid Strategy Instruction Model

Scaffolded strategy instruction embedded within a content domain such as math or reading benefits learners when it includes multiple strategies matched to student needs and is sequenced over time to promote automaticity. Not all strategy instruc-tion is equally effective, however, especially with younger learners and those with disabilities [41]. For example, Jitendra et al. [42] identified both conceptual and methodological problems common to strategy instruction programs with younger, learning disabled students, though these problems may be endemic to strategy interventions with older students too.

The main conceptual problems were the lack of a clear theoretical framework for program planning and poor strategy selection and sequencing. The main meth-odological problems were inadequate teacher training, poor treatment fidelity evi-dence, a limited array of outcome variables that tended to assess shallow rather than deep learning, poor instrument reliability, and failure to report relevant effect sizes.

We believe that some or all of these problems can be overcome by conducting a thorough review of the literature in order to identify an integrated strategy design that sequences strategies to promote better monitoring and control.

Indeed, a number of recent studies have included explicit monitoring train-ing within strategy instruction. For example, Hacker et al. [43] reported positive effects for reflection that facilitated monitoring and control processes.

In addition, two recent studies reported a positive effect for self-explanation instruction on metacognition and monitoring during teacher-led [29] and com-puter-assisted [44, 45] instruction. Self-explanation and reflection also may help students construct a meta-level model that facilitates subsequent control [30].

Several studies have investigated the effect of integrated strategy instruction on monitoring among college undergraduates. Nietfeld and Schraw [46] found that college students who received strategy instruction showed higher learning and more accurate monitoring. This study investigated the effects of strategy training when solving probability problems.

Participants received a 2-h instructional sequence with four components, including understanding and solving problems involving percentages, the addition rule (adding together separate probabilities of mutually exclusive outcomes to find the probability that any one outcome will occur), the multiplication rule (multiply-ing separate probabilities of independent outcomes to find the probability that the outcomes will occur together), and conditional probability.

The participants in addition reviewed five specific strategies discussed during the training, including:

1 Draw a picture.
2 Look for key words. The word *or* signifies addition; the word *and* signifies multiplication.
3 Ask yourself whether events are independent or dependent.
4 Ask yourself whether there is replacement or no replacement.
5 Compute a probability by constructing a ratio comparing the sample space and total outcome space. To do so, identify the total number of possible events and use this as the denominator. Then identify the number of observed events and use this as the numerator.

More recently, Nietfeld et al. [47] found that distributed monitoring training with feedback produced higher performance, confidence and accuracy among college students, while Huff and Nietfeld [48] found that monitoring instruction with 5th grade students improved performance and accuracy as well.

Gutierrez and Schraw [49] likewise found that integrated strategy instruction improved learning and monitoring accuracy independent of external incentives to perform well among college students.

We contend that well-planned strategy training can help middle school and older students to learn to monitor, construct meta-level representations, and utilize control processes with greater efficiency. We refer to a hybrid sequence as *metacognitive strategy instruction* (MSI) because it is designed explicitly to enhance the processes described in the Nelson and Narens [19] model of metacognition.

Table 1.2 shows an extension of the strategy intervention used by Gutierrez and Schraw [49] that aligns strategy instruction to monitoring and control processes and also includes instruction on constructing meta-level representations that connect the object level to the meta-level via the *accuracy-control link* [21].

We suggest that the following instructional elements be included in metacognitive strategy instruction programs:

1. Select a repertoire of 6–8 strategies that will benefit the student population of interest the most.
2. Embed these strategies into specific academic courses such as reading or science on a daily basis for six to 12 weeks at a minimum.
3. Provide explicit conditional knowledge about each of the strategies through teacher modeling that is sequenced to include student modeling as well [50]. Modeling and student discussion should focus on the why, when and where of strategy use, and ideally, should use an external record keeper such as the *strategy evaluation matrix* described by Schraw [32].
4. As discussed in detail below, teach monitoring and control strategies explicitly to older students (i.e., middle school and above) to develop monitoring awareness and proficiency through repeated practice and feedback. Monitoring instruction should focus on judgments of one's learning and strategies to debug and repair

Table 1.2 Summary of metacognitive strategies and their relation to calibration and theory

Strategies	Expected effect on monitoring and control processes	Rationale for enhanced monitoring and control processes
Review main objectives of the text and focus on main ideas and overall meaning	Monitoring	Enhance learning and monitoring through clarifying misunderstandings and tying details to main ideas
Read and summarize material in your own words to make it meaningful; use elaboration and create your own examples	Monitoring and control	Enhances learning and monitoring by transforming knowledge into something personally meaningful. Create meta-level model
Reread questions and responses and reflect on what the question is asking; evaluate the question paying attention to relevant concepts	Monitoring and control	Monitor ongoing learning. Construct model of to-be-learned information; select relevant control strategies
Highlight key ideas; integrate concepts and themes	Control	Highlighting and underlining can assist one to focus on main ideas and construct meta-level model
Promote teacher and peer modeling of strategies and meta-level understanding through discussion	Monitoring and control	Interactive discussion and expert guidance provide examples of skilled monitoring, strategy selection, and implementation of control processes
Relate information by using diagrams, tables, pictures, graphs, etc. to help organize information	Control	Integrated meta-level model facilitates selection of self-regulation strategies
Model the construction of a meta-level mental model through instruction, teacher feedback, and peer discussion	Control	A mental model serves as a basis to monitor and self-regulate ongoing control processes
Evaluate learning and performance	Monitoring	Reflect on mental model at meta-level; check comprehension, conclusions and performance outcomes against model

learning problems [51, 52]. Control instruction should focus on identifying learning goals and selection of appropriate strategies to achieve those goals [50].

5. Teach students to construct a meta-level model of their task-specific learning to promote deeper understanding. We recommend using modeling, self-explanation, making inferences, reflection, and evaluative discussion to do so [22, 29].

6. Demonstrate how experts (e.g., teachers) navigate the accuracy-control loop to self-regulate their learning [35, 52]. Thinking aloud before, during and after learning and problem solving is an excellent way to do so.

1.4.1 Instructional Strategies for Teaching Monitoring and Control Processes

Unfortunately, few strategy instruction programs focus on teaching monitoring and control skills explicitly. Those that do typically report a significant increase in learning, monitoring accuracy, and better control as measured by greater efficiency and deeper learning [1, 43, 46, 47]. We summarize below 10 strategies to improve monitoring and control processes that have appeared in the literature.

1 Use a systematic approach to monitoring such as the *discrepancy-reduction model of self-regulated learning* proposed by Thiede et al. [53] that helps students set achievable learning goals and monitor their performance in order to reduce the gap between goals and learning [54, 55].

2 Make predictions about one's learning in terms of what to learn and how well to learn it. Use real-time monitoring to determine whether predictions were accurate [56].

3 Use the monitoring heuristics described in Serra and Metcalfe [21, pp. 281–289]. These include assessing ones current knowledge, using information familiarity as a cue for learning and monitoring [57], evaluating the extent to which one has accurately summarized information, and assessing one's knowledge of the test [58].

4 Distribute one's studying across time rather than massing it in a single condensed study episode [2, 59].

5 Space monitoring judgments over time [47]. This includes making pre-reading predictions, online judgments of learning, and post-reading judgments of comprehension. Monitoring research has reported a robust *delayed judgment effect* in which monitoring one's comprehension and test performance after a brief delay increases monitoring accuracy [60, 61].

6 Engage in regular monitoring practice with real-time feedback [43, 62, 63].

7 Use the regulatory checklist developed by Schraw [32] in which learners are prompted to assess planning, monitoring and evaluation of learning. Failure to respond positively to any of the 12 questions on the checklist reminds learners to pause and reflect on their learning.

8 Improve retrieval of information through rehearsal and repeated study of important information [2, 64].

9 Think aloud during learning and test-taking, using self-explanation of monitoring and control processes [29, 43].

10 Think aloud after performance using post-learning reflection logs to review and evaluate performance and comprehension [30].

1.5 Discussion and Implications for Practice and Research

We highlight four suggestions for better strategy instruction designed explicitly to improve monitoring and control process. One is to train all students to acquire and automatize a repertoire of strategies that span the five main categories in Table 1.1

using best instructional practices [13, 16, 40]. A second is to train teachers to be better strategy users in order to teach and model strategy use and conditional knowledge about strategies [7, 32].

A third is to include explicit monitoring and control training with feedback for older students based on the 10 strategy instruction guidelines above [35]. A fourth is to help students learn to construct mental models of the learning task in order to better understand their comprehension and to self-regulate their learning [21–23, 29].

Several questions require additional research as well. One concerns the overall instructional plan for older learners in terms of what kind, how many, and how best to sequence strategy instruction. Thus far, most studies have been conducted with younger students and have not included explicit monitoring and control instruction. It is important to note that metacognitive strategy training interventions have generally yielded effect sizes ranging from small (Cohen's $d = 0.41$; $\eta^2 = 0.04$) to moderate ($\eta^2 = 0.08$ to $\eta^2 = 0.12$) to more robust effects ($\eta^2 = 0.15$). A second is to establish a better assessment approach for determining when students are academically and developmentally ready for monitoring and control instruction.

Some students undoubtedly benefit more from this instruction than others; however, there are no studies that we found that examined the degree to which aptitude (e.g., intelligence, working memory, or argumentation skills) by instructional interactions are robust indicators of "readiness."

A third question concerns the effects of MSI on different levels of understanding, including simple recall and recognition, understanding main ideas, constructing integrative inferences, generating and verifying predictions, and constructing a mental model of the task that is used as both a retrieval guide and a means to monitor and control one's learning.

1.6 Conclusions

Several conclusions can be drawn from the information we have presented in this chapter. The first is that results of previous investigations on metacognitive strategy training (e.g., [46, 49]) suggest that metacognitive strategy training interventions, even when they are brief and compact, enhance the monitoring and control processes depicted in the *accuracy-control link hypothesis* [21].

The strategies learners internalize during the training aid them in being more alert and attentive to the task by deliberately slowing down cognitive processing, which improves reflection and awareness of one's internal cognitive processes [49]. This, in turn, allows learners to more accurately and effectively gather information regarding the task, such as what the task demands truly are.

Consequently, they are able to more precisely feel what they know and do not know about the task, and thus, are better able to respond to changing task demands and adjust their cognitive resources accordingly to yield better learning outcomes. This clearly shows how strategy training influences learners' monitoring and control capabilities.

A second conclusion is that enhanced monitoring and control improves learners' information-gathering, and thus, metacognitive awareness. As a result of an improved monitoring-control process, learners are able to maximize their conditional knowledge (e.g., applying appropriate and effective strategies while learning predicated on task demands).

References

1. Bol, L., Hacker, D.J.: Calibration research: where do we go from here? Front. Psychol. **3**, 1–6 (2012)
2. Bjork, R.A., Dunlosky, J., Kornell, N.: Self-regulated learning: beliefs, techniques and illusions. Annu. Rev. Psychol. **64**, 417–447 (2013)
3. Ekflides, A.: Interactions of metacognition with motivation and affect in self-regulated learning: the MASRL model. Educ. Psychol. **46**(1), 6–25 (2011)
4. McCormick, C.B.: Metacognition and Learning. Handbook of Psychology: Educational Psychology, pp. 79–102 (2003)
5. Winne, P.H.: Students' calibration of knowledge and learning processes: Implications for designing powerful software learning environments. Int. J. Educ. Res. **41**(6), 466–488 (2004)
6. Zeidner, M., Boekaerts, M., Pintrich, P.R.: Self-regulation: directions and challenges for future research. In: Boekaerts, M., Pintrich, P.R., Zeidner, M. (eds.) Handbook of Self-regulation, pp. 749–768. Academic Press, San Diego (2000)
7. Zohar, A., David, A.: Paving a clear path in a thick forest: a conceptual analysis of a metacognitive component. Metacogn. & Learn. **4**(3), 177–195 (2009)
8. Grimes, P.W.: The overconfident principles of economics students: an examination of metacognitive skill. J. Econ. Educ. **33**(1), 15–30 (2002)
9. Carr, M., Taasoobshirazi, G., Stroud, R., Royer, J.M.: Combined fluency and cognitive strategies instruction improves mathematics achievement in early elementary school. Contemp. Educ. Psychol. **36**(4), 323–333 (2011)
10. Montague, M., Krawec, J., Enders, C., Dietz, S.: The effects of cognitive strategy instruction on math problem solving of middle-school students of varying ability. J. Educ. Psychol. **106**(2), 469–481 (2014)
11. Soliman, A.M., Mathna, E.K.: Metacognitive strategy training improves driving situation awareness. Soc. Behav. Pers. **37**(9), 1161–1170 (2009)
12. de Bruin, A.B.H., van Gog, T.: Improving self-monitoring and self-regulation: from cognitive psychology to the classroom. Learn. Instr. **22**(4), 245–252 (2012)
13. Dougherty-Stahl, K.A.: Synthesized comprehension instruction in primary classrooms: a story of successes and challenges. Read. Writ. Q. Overcoming Learn. Difficulties **25**(4), 334–355 (2009)
14. Ekflides, A.: Metacognition: defining its facets and levels of functioning in relation to self-regulation and co-regulation. Eur. Psychol. **13**(4), 277–287 (2008)
15. Hattie, J., Biggs, J., Purdie, N.: Effects of learning skills interventions on student learning: a meta-analysis. Rev. Educ. Res. **66**(2), 99–136 (1996)
16. Pressley, M., Harris, K.R.: Cognitive strategy instruction: from basic research to classroom instruction. In: Alexander, P.A., Winne, P.H. (eds.) Handbook of Educational Psychology, 2nd edn. pp. 265–286. Academic Press, San Diego (2006)
17. Palincsar, A.S.: Scaffolded instruction of listening comprehension with first graders at risk for academic difficulty. In: McKeough, A.M., Lupart, J.L. (eds.) Toward the Practice of Theory-based Instruction, pp. 50–65. Lawrence Erlbaum, Hillsdale, NJ (1991)
18. Winne, P.H., Nesbit, J.C.: Supporting self-regulated learning with cognitive tools. In: Hacker, D.J., Dunlosky, J., Graesser, A.C. (eds.) Handbook of Metacognition in Education, pp. 259–277. Routledge, New York, USA (2009)

19. Nelson, T.O.: Metamemory: a theoretical framework and new findings. Psychol. Learn. Motiv. **26**, 125–173 (1990)
20. Dunlosky, J., Metcalfe, J.: Metacognition. Sage Publications, Thousand Oaks (2009)
21. Serra, M.J., Metcalfe, J.: Effective implementation of metacognition. In: Hacker, D.J., Dunlosky, J., Graesser, A.C. (eds.) The Handbook of Metacognition in Education, pp. 278–298. Erlbaum, Mahwah (2009)
22. Kintsch, W.: Comprehension: A Paradigm for Cognition. Cambridge University Press, Cambridge (1998)
23. Miwa, K., Morita, J., Nakaike, R., Terai, H.: Learning through intermediate problems in creating cognitive models. Inter. Learn. Environ. **22**(3), 326–350 (2014)
24. Pressley, M., Wharton-McDonald, R.: Skilled comprehension and its development through instruction. Sch. Psychol. Rev. **26**(3), 448–466 (1997)
25. Swanson, H.L., Lussier, C., Orosco, M.: Effects of cognitive strategy interventions and cognitive moderators on word problem solving in children at risk for problem solving difficulties. Learn. Disabil. Res. Pract. **28**(4), 170–183 (2013)
26. Jitendra, A.K., Griffin, C.C., Haria, P., Leh, J., Adams, A., Kaduvettoor, A.: A comparison of single and multiple strategy instruction on third-grade students' mathematical problem solving. J. Educ. Psychol. **99**(1), 115–127 (2007)
27. Rosenshine, B., Meister, C., Chapman, S.: Teaching students to generate questions: a review of the intervention studies. Rev. Educ. Res. **66**(2), 181–221 (1996)
28. Griffin, T.D., Wiley, J., Thiede, K.W.: Individual differences, rereading, and self-explanation: concurrent processes and cue validity as constraints on metacomprehension accuracy. Mem. Cogn. **36**(1), 93–103 (2008)
29. McNamara, D.S., Magliano, J.P.: Self-explanation and metacognition. In: Hacker, D.J., Dunlosky, J., Graesser, A.C. (eds.) Handbook of Metacognition in Education, pp. 60–81. Erlbaum, Mahwah (2009)
30. Belcher, D.A.: Self-monitoring tools and student academic success: when perception matches reality. J. Coll. Sci. Teach. **41**(5), 26–32 (2012)
31. White, B., Frederiksen, E.: A theoretical framework and approach for fostering metacognitive development. Educ. Psychol. **40**(4), 211–223 (2005)
32. Schraw, G.: Promoting general metacognitive awareness. Instr. Sci. **26**(1), 113–125 (1998)
33. Dunlosky, J., Rawson, K.A., Middleton, E.L.: What constrains the accuracy of metacomprehension judgments? Testing the transfer-appropriate-monitoring and accessibility hypothesis. J. Mem. Lang. **52**(4), 551–565 (2005)
34. Duffy, G.: The case for direct explanation of strategies. In: Block, C.C., Pressley, M. (eds.) Comprehension Instruction: Research-based Best Practices, pp. 28–41. Guilford Press, New York (2002)
35. Duffy, G.G., Miller, S., Parsons, S., Meloth, M.: Teachers as metacognitive professionals. In: Hacker, D.J., Dunlosky, J., Graesser, A.C. (eds.) Handbook of Metacognition in Education, pp. 240–256. Erlbaum, Mahwah (2009)
36. Zohar, A.: The nature and development of teachers' metastrategic knowledge in the context of teaching higher-order Thinking. J. Learn. Sci. **15**(3), 331–377 (2006)
37. Zohar, A., Barzilai, S.: A review of research on metacognition in science education: Current and future directions. Stud. Sci. Educ. **49**(2), 121–169 (2013)
38. Pressley, M., Afflerbach, P.: Verbal Protocols of Reading: The Nature of Constructively Responsive Reading. Erlbaum, Hillsdale (1995)
39. Schraw, G., Dennison, R.S.: Assessing metacognitive awareness. Contemp. Educ. Psychol. **19**(4), 460–475 (1994)
40. Harris, K.R., Graham, S., Brindle, M., Sandmel, K.: Metacognition and children's writing. In: Hacker, D.J., Dunlosky, J., Graesser, A.C. (eds.) Handbook of Metacognition in Education, pp. 131–153 (2009)
41. Kearns, D.M., Fuchs, D.: Does cognitively focused instruction improve the academic performance of low-achieving students? Except. Child. **79**(3), 263–290 (2013)

42. Jitendra, A.K., Burgess, C., Shigajria, M.: Cognitive strategy instruction for improving expository text comprehension of students with learning disabilities: The quality of evidence. Except. Child. **77**(2), 135–159 (2011)
43. Hacker, D.J., Bol, L., Bahbahani, K.: Explaining calibration accuracy in classroom contexts: the effects of incentives, reflection, and explanatory style. Metacogn. Learn. **3**(2), 101–121 (2008)
44. Azevedo, R., Witherspoon, A.M.: Self-regulated use of hypermedia. In: Hacker, D.J., Dunlosky, J., Graesser, A.C. (eds.) Handbook of Metacognition in Education, pp. 319–339. Routledge, New York (2009)
45. Roediger III, H.L., Karpicke, J.D.: Test-enhanced learning: Taking memory tests improves long-term retention. Psychol. Sci. **17**(3), 249–255 (2006)
46. Nietfeld, J.L., Schraw, G.: The effect of knowledge and strategy training on monitoring accuracy. J. Educ. Res. **95**(3), 131–142 (2002)
47. Nietfeld, J.L., Cao, L., Osborne, J.W.: The effect of distributed monitoring exercises and feedback on performance and monitoring accuracy. Metacogn. Learn. **1**(2), 159–179 (2006)
48. Huff, J.D., Nietfeld, J.L.: Using strategy instruction and confidence judgments to improve metacognitive monitoring. Metacogn. Learn. **4**(2), 161–176 (2009)
49. Gutierrez, A.P., Schraw, G.: The effect of strategy training and incentives on calibration accuracy. J. Exp. Educ (2014)
50. Schmitt, M.C., Sha, S.: The developmental nature of meta-cognition and the relationship between knowledge and control over time. J. Res. Reading **32**(2), 254–271 (2009)
51. Shiu, L., Chen, Q.: Self and external monitoring of reading comprehension. J. Educ. Psychol. **105**(1), 78–88 (2013)
52. Volet, S.E.: Modelling and coaching of relevant metacognitive strategies for enhancing university students' learning. Learning and Instruction. **1**(4), 319–336 (1991)
53. Thiede, K.W., Anderson, M.C.M., Therriault, D.: Accuracy of metacognitive monitoring affects learning of texts. J. Educ. Psychol. **95**(1), 66–73 (2003)
54. Hacker, D.J., Keener, M.C., Kircher, J.C.: Writing is applied metacognition. In: Hacker, D.J., Dunlosky, J., Graesser, A.C. (eds.) Handbook of Metacognition in Education, pp. 154–172. Erlbaum, Mahwah (2009)
55. Hadwin, A.F., Webster, E.A.: Calibration in goal setting: examining the nature of judgments of confidence. Learn. Instr. **24**, 37–47 (2013)
56. Hacker, D.J., Bol, L., Horgan, D.D., Rakow, E.A.: Test prediction and performance in a classroom context. J. Educ. Psychol. **92**(1), 160–170 (2000)
57. Koriat, A., Levy-Sadot, R.: The combined contributions of cue familiarity and accessibility heuristics to feelings of knowing. J. Exp. Psychol. Learn. Mem. Cogn. **27**(1), 34–53 (2001)
58. Koriat, A.: The self-consistency model of subjective confidence. Psychol. Rev. **119**(1), 80–113 (2012)
59. Son, L.K.: Spacing one's study: evidence for a metacognitive control strategy. J. Exp. Psychol. Learn. Mem. Cogn. **30**(3), 601–604 (2004)
60. Mitchum, A.L., Kelley, C.M.: Solve the problem first: constructive solution strategies can influence the accuracy of retrospective confidence judgments. J. Exp. Psychol. Learn. Mem. Cogn. **36**(3), 699–710 (2010)
61. Thiede, K.W., Griffin, D., Wiley, J., Redford, J.S.: Metacognitive monitoring during and after reading. In: Hacker, D.J., Dunlosky, J., Graesser, A.C. (eds.) Handbook of Metacognition in Education, pp. 85–106. Erlbaum, Mahwah (2009)
62. Bol, L., Hacker, D.J., O'Shea, P., Allen, D.: The influence of overt practice, achievement level, and explanatory style on calibration accuracy, and performance. J. Exp. Educ. **73**(4), 269–290 (2005)
63. Hartwig, M.K., Was, C.A., Isaacson, R.M., Dunlosky, J.: General knowledge monitoring as a predictor of in-class exam performance. Br. J. Educ. Psychol. **82**(3), 456–468 (2012)
64. Karpicke, J.D.: Metacognitive control and strategy selection: Deciding to practice retrieval during learning. J. Exp. Psychol. Gen. **138**(4), 469–487 (2009)

Chapter 2
Regulating Metacognitive Processes— Support for a Meta-metacognitive Ability

Sandra Buratti and Carl Martin Allwood

Abstract Second-order judgments aim to regulate metacognitive judgments or at least to assess the accuracy of metacognitive judgments (first-order judgments). For this reason, second-order judgments can be seen as a form of meta-metacognition. In this chapter, we clarify the concept of meta-metacognition and how it relates to first-order metacognitive judgments. Furthermore, we explain why the concept of second-order judgments is an important addition to the research literature on metacognition and why it is an important concept in the context of learning and memory. We also present a new generalizable method for eliciting and measuring the accuracy (realism) of second-order judgments in the context of confidence judgments of semantic and episodic memory performance and suggest how this method can be computer implemented. An asset of this method is that it allows for fine-grained analyses of the strategies that people use when they make second-order judgments without reverting to think-aloud reports.

Keywords Confidence judgments · Second-order judgment · Regulation · Episodic memory · Semantic memory

Abbreviations

JOL   Judgments of learning
M Mean

S. Buratti (✉) · C.M. Allwood
Psychology Department, University of Gothenburg,
Box 100, 40530 Gothenburg, Sweden
e-mail: sandra.buratti@psy.gu.se

C.M. Allwood
e-mail: cma@psy.gu.se

© Springer International Publishing Switzerland 2015
A. Peña-Ayala (ed.), *Metacognition: Fundaments, Applications, and Trends*,
Intelligent Systems Reference Library 76, DOI 10.1007/978-3-319-11062-2_2

2.1 Introduction

In recent years, a new line of research within the field of metacognition has
emerged. This research concerns second-order judgments of first-order metacognitive judgments [for short reviews, see [1, 2]. By the term 'second-order metacognitive judgment', we mean a judgment that aims to regulate a first-order
metacognitive judgment (e.g., the accuracy of confidence judgment, judgments of
learning, etc.) or that at least aims to assess the accuracy of a first-order judgment.
A second-order judgment thus has a first-order metacognitive judgment as its target and can therefore be seen as meta-metacognition.

The assessment of the accuracy of metacognitive judgments is important in
many different contexts, such as in learning. For example, these judgments guide
children in allocating study time, choosing the best search strategies during
problem solving, and finding the best answer to a learning task [3]. It is clearly
important to the learner that this guidance is correct. Another important context
is forensic psychology, especially eyewitness psychology, in which eyewitnesses
often are asked to give metacognitive judgments of their memory performances
[4]. Thus, it is of importance that the witness has an accurate judgment of the
trustworthiness of their memory reports.

A substantial amount of the research on second-order judgments has focused on a
specific type of metacognitive judgment, namely confidence judgments of different
kinds of performances. Confidence judgments may, according to the model presented
by Koriat and Goldsmith [5], have a central role when a person decides whether to
externally report information that he or she has retrieved covertly in memory.

According to Koriat and Goldsmith's model, when people covertly retrieve
information in their memory, at approximately the same time they also generate a
spontaneous feeling of confidence associated with the correctness of the retrieved
memory. The model next assumes that the person's decision whether to externally
report the retrieved memory or not is based on a comparison between the level of
the person's spontaneous confidence and the cost of reporting an incorrect memory in the specific context the person is in. For example, in a court of justice, a
person is likely to use a stricter report criterion than when making conversation
with a friend at a cafeteria.

Confidence judgments are important in many settings, partly because of the central role confidence judgments play when people decide whether to report retrieved
memories, but also for other reasons. For example, confidence judgments of semantic memory information are important for optimizing learning outcomes [6, 7].

The confidence judgments a student makes about a certain performance are part
of a self-monitoring process that can result in different types of action. For example, if a student takes an oral exam and does not feel confident about a certain
answer, the student might decide not to report the answer openly.

If, on the other hand, the student is confident about the answer, he is more
likely to report the answer to the test question. Confidence judgments often are
also made in decision-making contexts. For example, people make confidence
judgments in their professions [8]. Thus, physicians may judge how likely it is that

their patient will have a heart attack based on the patient's symptoms. Judges in court make decisions about whether to release an offender or not, based on how confident they are that the offender will not commit new offenses.

Finally, confidence judgments are often made about episodic memory performance, such as by witnesses in the justice system. In the context of the justice system, research shows that the professionals involved in the process, such as the police, prosecutors, attorneys, judges, and jurors, often judge the credibility of witness testimony by how confident the witness appears [9–11]. Thus, such professionals and actors have been found to believe confident witnesses more than less confident witnesses [e.g., 12]. For this reason among others, it is important that the witness has an accurate understanding of the accuracy of their confidence.

A problem is that studies have shown that the accuracy of (first-order) confidence judgments in a number of different contexts is poor [13]. In many judgment situations, people are more confident in their performance than they are correct in that performance. This phenomenon is known as the "overconfidence effect". Although settings exist in which people display underconfidence [14–16], some research supports a persistent overconfidence phenomenon in many types of situations [17].

For example, the overconfidence effect is common for both episodic and semantic memory. It is therefore important to investigate whether people have the ability to make their confidence judgments more realistic. Numerous models and theories have been presented that have explained the overconfidence effect [13, 17]. This overconfidence phenomenon is troubling in different contexts. For the student in the example above, it is problematic because overconfidence may hinder him or her from correctly allocating study time to information that he or she has not learnt properly yet.

In line with this, Dunlosky and Rawson [18] asked undergraduate students to learn six definitions, each with different subcomponents, and after each trial asked students to report their confidence in their recalled definition at three qualitative levels. After a student had reported the highest confidence for their recalled definition in three trials in a row, it was not practiced anymore.

The results showed that students who exhibited the most overconfidence performed at a low level at a final recall session 2 days after the first experimental session. In the forensic system, overconfidence also is problematic [19]. In real life, the testimony of overly confident witnesses has often led to the conviction of innocent people [20]. Likewise, research has shown that people's confidence judgments in most of the researched professions are miscalibrated in the sense that they are overconfident.

In brief, confidence judgments can have an important impact on the person making the confidence judgments as well as on other people facing the consequences of the judgments. It is therefore very important that people's confidence judgments are as *realistic* as possible (i.e., as close as possible to what is actually the case with respect to the person's performance). More specifically stated, the realism of confidence judgments depends on their relation to the correctness of the actual performance. Confidence realism is in some research traditions also called *confidence accuracy*.

In this chapter, we first discuss the notion of metacognition and then explain our conception of second-order judgments that we relate to the concept of *meta-metacognition* and show that there are different types of meta-metacognitive judgments. Next, we review this new line of research of second-order metacognitive judgments and then present a new method that we have developed for research on second-order judgments of confidence judgments of both semantic and episodic memory. After this, we present our own research using this method in which we investigated if people have the ability to make successful second-order judgments of already given first-order confidence judgments.

In this context, we describe the different strategies used when making second-order judgments in the two investigated memory contexts. The method can easily be adopted to assess second-order judgments of different kinds of metacognitive judgments and can, as we describe, be implemented in computer-learning contexts. Finally, we suggest directions for future research within this field.

2.2 What Is Metacognition?

To clarify what we mean by meta-metacognitive ability, we first address what we mean by metacognition. Flavell provided one of the earliest definitions of metacognition in the 1970s. According to Flavell, metacognition is "one's knowledge concerning one's own cognitive processes or anything related to them" [21, p. 232].

Flavell [22] proposed that metacognitive processes do not necessarily differ from cognitive processes, although the target of metacognitive processes is different because the metacognitive processes target cognition itself. Over time, the definition of metacognition has been expanded. For example, Koriat [23, p. 261] noted that metacognition more broadly can be seen "to refer to cognition about cognition in general, as well as self-regulation processes that take cognitive processes as their object". Thus, metacognition is commonly seen to include the regulation of cognitive processes and may, for example, also include knowledge of other people's knowledge. Metacognition has become a highly multidimensional concept, and its definitions and components vary extensively among researchers [for a review, see 24].

One of the most well-known models of metacognition is the two-level model presented by Nelson and Narens [3, 25]. Nelson and Narens [3] noted that this model is abstract and referred for their terminology to texts written by the mathematician David Hilbert in the 1920s and by the philosopher Rudolf Carnap in the 1930s.

We discuss this model next because it includes an early framework for the discussion of the concept of meta-metacognition, without mentioning this term. In this model, what is called *the meta level* controls *the object level* and receives monitoring information from the object level, that is, the cognition level. Through *control processes*, the meta level modifies the object level, but not vice versa, and the control exerted by the meta level on the object level is said by Nelson and Narens to be analogous to speaking into a telephone handset. *Monitoring* is said by Nelson and Narens to be "logically (even if not always psychologically) independent of the control component" (p. 127). They also state that monitoring means that "the meta-level is *informed* by the object level" (p. 127, Nelson and Narens' italics). Seen from the perspective of the monitoring level, this concept is said to be analogous to a person listening to a handset. Control can have three effects on the object level: (1) initiating an action; (2) continuing an action; or (3) terminating an action. Information from the object level may change the model at the meta level of the situation at hand.

To further explain Nelson and Narens' model, the following example is offered: For a student taking a test to answer a question concerning some topic, the student needs to self-direct his or her search for the answer and thus select a search strategy. This selection of search strategy and the termination of the search are control processes. The confidence the student expresses in the answer is based on information from the object level, communicated by a monitoring process. The confidence level will be heeded when the person determines whether the answer is at a sufficiently satisfactory level to be presented during the test (compare Koriat and Goldsmiths' model [5]), or if a new search for a better answer candidate should be initiated. Nelson and Narens' model has been highly influential within the field of educational metacognitive research.

However, it is, as noted above, somewhat abstract and seems primarily not to have been intended as psychological but rather as a more formal abstract model. In line with this, Nelson and Narens addressed few of the specific processes pertaining to the model. The somewhat abstract formulation is illustrated in the context of their claim that the two-level model they presented could easily be generalized to more than two levels, in such a way that the meta level may be the object level of a higher meta level. Thus, depending on context, it seems that something that is on the meta level on one occasion (e.g., a confidence judgment) can be on the object level on a following occasion.

2.3 How Does Metacognition Relate to Meta-Metacognition?

Although Nelson and Narens [3, 25] did not use the term meta-metacognition, they, as described above, proposed that their two-level model could be generalized to include more levels, in such a way that the meta level may be the object level of a higher meta level. In general, just as it is useful to separate metacognition from cognition, it is also relevant to separate meta-metacognition from metacognition.

This suggestion was presented by Renkl et al. [26] in a paper where they proposed the importance of having knowledge of one's own metacognitive knowledge. Similar to Renkl et al., Roberts and Erdos [27] proposed that there could be different levels of metacognitive awareness where cognition of metacognitive processes was defined as a possible form of meta-metacognition. In a somewhat different version of meta-metacognition, Efklides and Misailidi [28] suggested that "This kind of metacognition, however, that has as its object the cognition of others could represent a meta-metacognition level, that is a social level of cognition" [28, p. 16].

In line with most of these ideas, we suggest that, just as metacognition can be referred to as "any knowledge or cognitive activity that takes as its object, or regulates any aspect of any cognitive enterprise" [29, p. 150], any activity that targets the regulation of a metacognitive judgment can be referred to as a form of meta-metacognition [e.g., 30]; more in general, meta-metacognitive judgments are judgments that include the evaluation of the accuracy of metacognitive judgments.

This definition illustrates that second-order judgments provide a possibility for individuals to improve their metacognitive performance without relying on external sources, such as social and environmental sources, focused on by Kim, Park, Moore, and Varma [31].

The difference between meta-metacognitive and metacognitive and cognitive processes is primarily the target of the processes. Meta-metacognitive processes target a metacognitive performance product.

Too little is presently known, however, to say more about exactly in which manner meta-metacognitive processes differ in nature from metacognitive processes, except formally as discussed above. In this chapter, we present a method for examining and regulating second-order judgments, which we see as a form a meta-metacognition because they target metacognitive judgments.

There are also, however, other forms of meta-metacognition, such as having knowledge about metacognitive knowledge, as proposed by Renkl et al. [26]. Moreover, as discussed below, some forms of meta-metacognition may be of a more mixed form, or different from what we have suggested above [28, 32]. Efklides and Misailidi [28] suggested using the term 'meta-metacognition' for metacognition that "has as its object the cognition of others" [28, p. 16]. This activity, although obviously of great interest, might perhaps better be called "social metacognition" or similar because it explicitly excludes meta-metacognition of *one's own* metacognitive performance.

Yet other usages of the term meta-metacognition exist that do not involve the regulation of metacognitive judgments but that include evaluation of the accuracy of a previous metacognitive judgment as a part of making a confidence judgment about a decision that also may include other considerations [e.g., 33]. This phenomenon is explained more in detail further below.

2.4 Recent Research on Second-Order Metacognitive Judgments

Recently researchers have investigated the making of second-order judgments [e.g., 1, 30, 33–40]. In general, the results of these studies show that people can make successful second-order judgments of different types of metacognitive judgments such as confidence judgments, confidence intervals, judgments of learning (JOLs), and prediction of exam scores.

Dunlosky et al. [37] investigated second-order judgments in the context of JOLs (i.e., a person's ratings of the likelihood that they will be able to recall a recently studied item). The authors defined second-order metacognitive judgments as "an individual's confidence in the JOLs themselves" (p. 335) and reported that their participants could make successful second-order assessment of their JOLs. More specifically, Dunlosky et al. asked their participants to rate the accuracy of their JOLs, and the results showed that the participants were more confident in JOLs that predicted recognition success than they were in JOLs that predicted recognition failure.

A further finding showed that the participants had higher confidence for their delayed JOLs, that is, JOLs made some minutes after their item study session, compared to their immediate JOLs. Because the delayed JOLs were a better prediction of the participants' performance, this result is an indication of good realism of the second-order confidence judgments. Using a similar method, Serra and Dunlosky [40] compared the second-order confidence judgments of older adults (66 years) with those of younger adults (20 years).

For immediate JOLs, the older adults' second-order judgments were somewhat lower than those of the younger participants, but for delayed JOLs, the second-order judgments were about the same for the two age groups. The authors speculate that the reason may be that there were fewer salient cues (specifically: retrieval fluency) available in the context of immediate JOLs than after a delay and that the older participants for this reason were less willing to give high second-order judgments after the immediate JOLs compared with the delayed JOLs.

Meta-metacognitive judgments were also studied by Cesarini et al. [36], although the authors did not describe their research as lying within this framework. The participants were first asked to answer 10 numerical questions relating to economics by providing values for each question that constituted the upper and lower limits of a 90 % confidence interval for the correct answer. Next, they were asked to estimate how many of their 10 intervals included the correct answer. After this, all participants who had not estimated that 9 out of their 10 intervals included the correct answer were instructed to revise their intervals so that they thought that 9 of their intervals contained the correct answer. For the control group, the number of questions that included the correct answer was on average about 4.5 after the first round and nearly 6 questions after the second round. The adjustments in the confidence intervals thus resulted in better assessments because the goal was to construct intervals that included the correct answer for 9 of the 10 questions.

Miller and Geraci [39] investigated students' predictions of their exam scores. Students were first asked to make judgments of their number of correct answers to the exam (called *global judgments*) and then to make a confidence rating of the correctness of their predictions. Low-performing students were found to be more overconfident in their first-order predictions of their total exam performance than high-performing students, (i.e., they predicted higher scores than they had). However, the more interesting result was that the low-performing students were more accurate than the high-performing students in their second-order judgments of their exam predictions. That is, they were better at evaluating the accuracy of their first-order global judgments compared with the high-performing students.

Händel and Fritzsche [38] studied the same issue and also investigated the effect of different types of five-step confidence scales on the success of the second-order judgments. In contrast to the study by Miller and Geraci [39], their results showed that the *high-performing* students made more successful second-order judgments of the correctness of their global judgments of their exam scores. It is not clear why this discrepancy in results between these two studies was found, and the authors did not relate their findings to those reported by Miller and Geraci.

In addition, the results presented by Händel and Fritzsche showed that bipolar non-numeric and non-verbal second-order confidence scales were more successful than unipolar versions of the same scale types and that the bipolar smiley scale (more or less sad–happy five-level smiley scale) was the best of the investigated scales.

In a recent study, Arnold [33] investigated a form of second-order judgments in a face-recognition task. In the task, the participants first saw 48 faces and then, after a filler task, performed a recognition test in which 24 of the already presented faces were mixed with 24 new faces. For each face, participants reported whether the face was *new* (they had never seen the stimuli before) or *old* (the stimuli had appeared during the study phase). The participants had to choose one of the options and were then, directly after the report for a face, asked to confidence rate how sure they were that the face was new/old. After this, the participants were asked if they wanted to report or withhold their answers, with the possibility of winning or losing points depending on the correctness of the reported answer.

The participants were then asked to rate their confidence in that they had made the right decision to go for points/withhold answer. In this way, Arnold separated the confidence a person had in an answer to the recognition task from the confidence he or she had in a decision about whether to withhold or report the answer and showed that these can differ. It is noteworthy that the confidence judgments of the decision to report probably did not solely pertain to the previous metacognitive performance because the decision likely would also have included figuring out the consequences of the specific reward structure the researcher had associated with the decision to report (compare with the Koriat & Goldsmith model [5]). Thus, the type of meta-metacognition investigated by Arnold may not be a pure form but rather a mixed or diluted form.

A majority of the studies investigating second-order judgments have focused on confidence as the means whereby a second-order judgment evaluates a first-order judgment of some kind (e.g., a confidence judgment, a global judgment, a confidence interval judgment, or a JOL). This relationship may be in line with how second-order judgments function in everyday life.

Such judgments may be triggered for various reasons; thus, for some reason, we may ask ourselves how confident we are in our metacognitive judgment—is it accurate (realistic) or does it need to be adjusted? One reason for such triggering may be that someone, for example, in court or a researcher, asks us to consider how realistic our confidence is. Another reason may be that we ourselves, more or less automatically, note that some information that we heed contradicts our (for example) high confidence. This observation may lower our confidence in our first-order confidence and may make us reevaluate our confidence.

This thought is analogous to the process described in Koriat and Goldsmith's model [5] where, as described above, confidence judgments are used to regulate which information a person wants to report or withhold. If the person is so confident in a report that this confidence exceeds an accuracy criterion (which depends on the context), then the person will choose to report the information. If the confidence does not exceed the accuracy criterion, the person will not report the information. Second-order judgments of metacognitive judgments can be argued to function in a similar way. The second-order judgment is, or at least includes, an evaluation of the first-order confidence judgment and can be the basis for a person's decision to adjust a metacognitive judgment or keep it as it is. As noted above, the first-order metacognitive judgment does not, of course, need to be a confidence judgment but could be any type of metacognitive judgment, such as a JOL.

2.5 New Method for Studying Second-Order Judgments

Recently, we developed a new method for assessing second-order judgments, and with this method, we have shown that people have the ability to increase the realism in their confidence judgments for both episodic [1, 30, 34] and semantic memory tasks [35].

This method can be used to test a whole group of participants at the same time and, in brief, consists of an instruction about *confidence realism* and two tasks called the Confidence and the Adjustment tasks. The instructions on *confidence realism* explain the concepts of *realism* and *confidence judgments* in a way that makes them understandable to the participants. They address issues like *What is a confidence judgment?* and *How do confidence judgments relate to the actual memory report?*

The instructions also explain *What does it mean to show overconfidence, perfect realism, or underconfidence?* The instructions are 10 min long and help the person understand the concepts on a rather deep level. To check if the participants have understood the concepts explained, they are then provided with a test concerning these concepts.

After this, the participants continue to the Confidence task. In this task, the person gives a memory report of some kind and then provides confidence judgments for separate parts of this memory report. Depending on the type of memory tested, preparations for the Confidence task differ. For episodic memory, questions concerning some stimuli experienced by the participants are answered, and for this reason, the participants can, for example, be shown a short video clip about some event at the beginning of the test session. For semantic memory, general knowledge questions can be used, and no preparations before the confidence task thus may be needed. The format of the memory report can also differ. In our studies, we have mostly used directed recall questions for which the participants are not given any answer alternatives but are told to provide their own answer.

When the participants have given their full memory report, for each memory question answered, they are asked to rate how confident they are in their answer on a confidence scale that ranges from 0 % ("I am absolutely sure that my answer is incorrect") to 100 % ("I am absolutely sure that my answer is correct").

As an introduction to the Confidence task, the participants are told to attempt to answer as many of the memory report questions correctly as possible and to be as realistic as possible in their confidence judgments. If they do not know the answer to a memory report question, they are told to make a guess. To have an incentive for the task, the participants can be told that the person with the most correct answers to the memory report questions will receive a reward (e.g., a movie ticket).

After the Confidence task, participants continue to the Adjustment task for which they are instructed to go back and change the confidence judgments they believe are unrealistic by making a new confidence judgment. The participants are also told that they are not allowed to change their responses to the memory report questions they answered during the Confidence task, only their confidence judgments. As an incentive, they can be told that the person with the best realism after this task will receive a reward (e.g., a movie ticket).

This method has been used for confidence judgments in episodic and semantic memory tasks, and can, as noted above, easily be adopted for other metacognitive judgments. For example, it can be used with JOLs. In that case, the participants are first asked to provide JOLs in some learning task for several items and then asked to go back and change the JOLs they find unreliable.

So far, we have used only recall questions with this method. However, it can also be used with recognition questions for which two or more answer alternatives are provided. In such cases, the confidence scale used should be altered to account for the possibility to be correct by chance, taking into account the number of options to choose from when answering.

2.6 The Making of Successful Second-Order Confidence Judgments

The above method has been used to investigate whether people can successfully improve their first-order confidence judgments by means of second-order judgments. In two studies, we showed that people can do this for an episodic memory task when being asked to go back and adjust their confidence judgments [1, 34]. In these studies, the participants first saw video clips concerning forensically relevant events (a theft in a park or a kidnapping).

After watching the video clip, they received instructions regarding the concept *realism of confidence* and performed a test that showed that they understood the concept. After this step, they completed the Confidence task, answering 40–50 directed recall questions (depending on the study) concerning the video clip and rating their confidence in their answers on a scale ranging from 0 % ("I am absolutely sure that my answer is incorrect") to 100 % ("I am absolutely sure that my answer is correct").

They then proceeded to the Adjustment task. The improvements in confidence realism after the Adjustment task showed that the participants on average significantly decreased their overconfidence, although the effect was small. These improvements were measured by using *absolute bias* and *calibration* scores, which are traditional realism of confidence measures [41]. Absolute bias is the absolute value of the bias score, and bias is calculated by subtracting the proportion correct from the average confidence level the person expresses in the task. Calibration is the average squared deviance between proportion correct and confidence for each separate confidence class (for example, for each of the 11 confidence classes: 0–9 %, 10–19 %, 20–29 %, ..., 90–99 %, and 100 %). The results in a study by Buratti and Allwood [34] showed that the participants in the control condition improved their absolute bias score from $M = 0.14$ in the Confidence task to $M = 0.09$ in the Adjustment task.

In another study, using the same method, Buratti, Allwood and Kleitman showed that people also can adjust their confidence for semantic memory reports [35]. In that study, the participants answered 40 knowledge questions and rated how confident they were in their answers. The questions were directed recall questions. The results showed that the participants could successfully adjust their confidence judgments but that they accomplished this only by increasing the confidence for memory items that were correct. The confidence for correct items increased from $M = 0.88$ to $M = 0.90$.

The effect in that study was small, similar to studies of episodic memory [1, 34], but this outcome could result from the fact that the realism in this study was already high after the Confidence task. The value for absolute bias was 0.06 after the Confidence task, which is quite close to zero (the level for perfect realism).

This value is in line with earlier research in which less overconfidence is often found for recall questions than for recognition. Future studies should therefore investigate contexts in which first-order overconfidence is high in order to investigate contexts that are more sensitive to the participants' attempts to improve the realism of their confidence.

2.6.1 Strategies Used When Making Second-Order Judgments

When investigating if people have a meta-metacognitive ability, we found some surprising differences in strategies used for increasing realism of confidence. First, when people were asked to go back and try to increase the realism of their confidence judgments, they did not merely use a simple heuristic method in which they on average simply lowered their confidence across all items [1, 34].

In general, it should be noted that to increase the realism for correct items, the participants should increase the level of their confidence. The reverse is the case for incorrect items. Our analyses showed that selectively the participants were able to identify items in need of adjustment. For example, in the Buratti and Allwood study [1], analyses showed that the confidence judgments identified as in need of change were associated with lower realism than the confidence judgments that were not identified. That is, for correct items, the confidence level for chosen items ($M = 0.61$) was lower than the confidence level for items not chosen to be modified ($M = 0.84$). The reverse was true for incorrect items, namely that the items chosen for modification ($M = 0.65$) had a higher confidence level than the items not chosen for modification ($M = 0.51$).

Furthermore, in the Buratti and Allwood study [34], analyses also showed that the new confidence judgments (after change) were more realistic than the identified confidence judgments. For example, in the control condition, the participants' level of absolute bias was $M = 0.27$ for the chosen confidence judgments and $M = 0.14$ after the modification in the Adjustment task. This finding indicates that the participants chose the confidence judgments with the worst realism and increased the realism for these particular items.

Further, we found that depending on the memory task, the strategies used for increasing realism differed. When we analyzed the data separately for items the participants had answered correctly and items that they had answered incorrectly, we found for episodic memory that the participants were more successful with the incorrect items and lowered the confidence significantly for these items [1, 34]. For example, in the control condition in the Buratti and Allwood study [1], the participants decreased their confidence for incorrect items in the Confidence task from $M = 0.52$ to $M = 0.47$ in the Adjustment task. However, the confidence level for the correct items was the same across the Confidence and the Adjustment tasks.

In contrast, as mentioned earlier, the participants in the study investigating semantic memory increased the confidence for correct items, leaving the confidence level for incorrect items untouched [35]. These differences between episodic and semantic memory tasks could be explained by a divergence in encoding possibilities and feedback for episodic and semantic memory, respectively. Such variations have been suggested to explain why research has found differences in confidence judgments made for episodic and semantic memory [42, 43].

For example, the differences found in the regulation of first-order confidence judgments between episodic and semantic memory tasks could be due to variations in the form of feedback received. It may be that semantic memory information in general, compared with episodic information, is subjected to more confirmatory feedback. Over a lifetime, a person is expected to receive more confirmatory feedback that an answer is correct and therefore have more possibilities to encode correct answers. For example, learning that Rome is the capital of Italy is a fact that a person is likely to receive positive feedback on in several different contexts such as in school or when planning a holiday. Given this, it should be easier to know when a given answer in a general knowledge test is correct—i.e., when a low confidence judgment for a correct answer should be increased—than when a high confidence judgment for an incorrect answer should be lowered. This scenario would explain why participants are more likely to increase the confidence for correct semantic memory items than they are to lower it for incorrect semantic memory items.

Episodic memory tasks lack the encoding and feedback possibilities that are common for semantic memory, which may result in a focus on the answers that are incorrect. The reason may be that people target answers that stand out in consciousness because they seem unlikely and that the person therefore may believe are incorrect. These experiences of answers as "unlikely" or "odd" events can been seen as a form of cognitive interruption because the information attended to in the answers does not follow schema-related expectations [44].

Likewise, such experiences may be associated with a decrease in the *Feeling of Rightness* (sense of correctness) suggested by Thompson et al. [45] to accompany answers or performance in problem solving. In line with this reasoning, Allwood [46], in a study using the think-aloud protocol, found that a fruitful way for first-year university students to detect errors that they had made when solving statistical problems was to follow up a noted difference between their *expectations* about some property of the generated answer and the answer they had devised.

2.6.2 Cues for Making Second-Order Confidence Judgments

In our studies we also investigated which cues were used when making second-order confidence judgments. Different types of cues have been assumed to affect confidence judgments [47]. Koriat et al. divided such cues into theory-based and experience-based cues, in which theory-based cues are related to beliefs one might have about one's competence and knowledge whereas experience-based cues are related to features of the person's phenomenological experience during the making of the confidence judgments. In one of our studies, we investigated two such possible experienced-based cues in the context of episodic memory, namely processing fluency and phenomenological memory quality [34]. Processing fluency is the subjective feeling of ease a person experiences when performing a cognitive task such as recalling a certain item [48]. Studies have shown that confidence judgments are to a large extent based on processing fluency so that high processing fluency is associated with high confidence judgments in semantic knowledge tasks [49, 50] as well as in eyewitness situations [51, 52].

Because processing fluency plays such an important part in making first-order confidence judgments, we hypothesized that processing fluency may also be an important cue when making second-order judgments. Thus, in Buratti and Allwood [34], the participants in the processing fluency condition, directly after each memory report question in the Confidence task, also made a rating of how fluently the answer came to mind and in another condition rated the memory quality of the item.

The results in the Buratti and Allwood study [34] showed that the confidence judgments participants tended to modify in the Adjustment task were associated with lower levels of experienced fluency ($M = 3.10$) than the confidence judgments that were not modified ($M = 4.51$). Given that incorrectly remembered items are likely to be associated with lower fluency than correctly remembered items, this finding is in line with the fact that the participants, for episodic memory, mostly opted to change their confidence for incorrect answers. It is also in line with the idea that items that did not cohere with the participants' implicit or explicit expectations may have been associated with lower fluency and a lower Feeling of Rightness.

As mentioned above, we also investigated the experience-based cue of phenomenological memory quality [34]. Two types of phenomenological memory qualities are *remember* and *know* [53]. A memory has a "remember" quality if the person experiencing the memory recollects concrete details about the event or subject at hand. If, on the other hand, the person has only a vague feeling of familiarity during the retrieval, than this memory is associated with the quality "know".

In a study [54] investigating the realism of confidence in an eyewitness situation, a higher degree of realism was found for answers associated with the memory quality "remember" than answers associated with the memory quality "know" [54]. In our study [34], we showed that phenomenological memory quality could also be used as a possible cue when making second-order judgments. Answers associated with the memory quality "know" had worse realism (absolute bias $M = 0.24$) than the "remember" items (absolute bias $M = 0.13$), and the "know" items were chosen to be modified more often (52 % of the "know" items) than answers associated with the memory quality "remember" (20 % of the "remember" items).

2.6.3 Individual Differences in Second-Order Judgment

Our studies also analyzed individual differences associated with the making of successful second-order judgments. One such individual difference is cognitive ability, or more specifically, short-term memory. People who can hold more information activated might be better at regulating their metacognitive judgments than people with low ability in this regard.

However, our results showed no correlation between performance on the digit span task and how successful individuals were at making second-order judgments [30]. In one study, we also investigated the link between several personality and cognitive styles and second-order judgments [35]. Personality factors such as the Big five factors [55] and narcissism [56] were investigated together with cognitive styles such as Need for cognition [57] and Need for closure [58].

The only factor found to predict the confidence level in incorrect items for both first and second-order judgments was Openness [35]. People high in Openness tend to be open to possibilities and solutions. They have an active imagination and have a high intellectual curiosity [55]. They often perform well on cognitive tasks, which may lead them to not question and doubt their abilities, possibly providing the reason why they were significantly more confident for incorrect items than people low in Openness.

The Extraversion/Narcissism factor predicted confidence level in correct items for the first-order judgment only. This outcome is in line with earlier research showing that people high in extraversion [59, 60] and narcissism [61] tend to show overconfidence when performing different types of tasks. The weak relationship between personality and cognitive style variables on the one hand and second-order judgments on the other is not surprising because only weak relationships between personality and first-order judgments have been found [e.g., 62].

We also investigated if there was a relationship between willingness to make second-order judgments that involved changing one's first-order confidence judgments and different personality traits. We believed that people high in self-doubt would be more willing to question their ability and thus more prone to making second-order judgments that involved changing their first-order confidence judgments.

We also believed that people high in narcissism would not be as willing to doubt their ability, which would lead this group to make fewer second-order judgments of the just-described kind. However, no such relations were found between willingness to make second-order judgments and different personality traits [35]. The reason for this lack is unclear and should be investigated in future research.

2.7 Does Research Really Support that We Have a Meta-Metacognitive Ability?

What distinguishes a first-order judgment from a second-order judgment? Is the making of a second-order judgment merely redoing a metacognitive judgment? The process by which people regulate their first-order confidence judgments can be divided into two parts. The first part is an *identification* of the confidence judgments with poor realism that are in need of modification.

In this stage, the person needs to consider not only the confidence judgment per se but also the relation between the memory report that forms the target for the first-order judgment and the confidence judgment. This step may also include reconsidering the correctness of the memory report itself. In brief, the cues from the memory report and the level of the confidence judgment for each item need to match; thus, the whole metacognitive judgment needs to be evaluated. This questioning and evaluation of metacognitive judgments that happen during the identification part are, we believe, partly a process beyond just making another first-order judgment.

The second part in the regulation of first-order confidence judgments is the actual *adjustment* during which a new metacognitive judgment is made, and this process need not differ from a first-order metacognitive judgment [35].

We do not know yet which processes specifically underlie the regulation of confidence judgments. Critics might claim that the improvement in realism after the Adjustment task merely is an effect of redoing the confidence judgment.

One reason for this claim is that studies investigating error models, i.e., models assuming that all judgments are associated with measurement error, have reported that when people make multiple confidence judgments of the same item, their random error decreases, causing the realism to improve [63, 64].

Against this conclusion as the full explanation of the improvement in realism in the Adjustment task is the study of Allwood et al. [65] in which subjects in one condition were told to simply redo their confidence judgments. The results showed that there was no improvement in realism for this group. Another study investigating the dialectical bootstrapping model found that when participants were told to merely redo their estimations for different questions, they did not improve these estimations [66]. Rather, the improvement occurred in the condition in which the participants were told to actually question their first estimation. This outcome supports the idea that there needs to be a questioning of the first-order judgment that calls for a deeper or different processing than merely redoing a first-order judgment.

2.8 Illustration of Computer Implementation of the New Method for Improving Second-Order Judgments in a Learning Context

The new method presented earlier in the chapter could easily be implemented in a computer program, for use in classrooms, for example. Below, an illustration is provided for how this can be done, but various aspects of this illustration can obviously be altered to fit the specific needs of the situation. In the computer implementation, the instructions may be provided in writing on the screen.

If the session where the method is used is planned to involve new learning material, this material is presented first by the teacher in the classroom or on the computer screen. This material could be, for example, language information, some type of text (e.g., concerning events in history), or a short film on some topic. If the goal is to test material that has been presented to the students earlier, the students can start directly at the first step of the method.

This step presents the instructions on the concept of *confidence realism* in a lively way on the screen. Following this presentation, the test evaluating whether participants have understood the instructions on confidence realism could be given in a multi-alternative answer format so that the results can be automatically corrected by the computer program. If a student does not do well on this test, the computer could present further instructions and tests until the student has reached a suitable level of understanding. After this step, the program would allow the teacher to prepare for different types of memory reports from the students, either directed questions, with or without answer alternatives, or free recall instructions.

In the latter case, the student would be asked to enter his or her free recall in the form of one short sentence at a time delimited by use of the Return key. Each answer to a directed recall question and each free recall sentence delimited by a press on the Return key is seen as an "item".

When the memory report task is finished, the student is instructed to confidence rate each answered memory report item on a confidence scale presented just below the reported item. This scale would be fitted to the number of answer alternatives provided in the context of the memory report items. For example, if two answer alternatives were provided, the scale would start at 50 % ("I'm guessing") and run to 100 % ("I'm sure the answer is correct"). When no answer alternatives are provided, the scale could, for the sake of simplicity, go from 0 % ("I'm guessing") to 100 % (I'm sure the answer is correct").

When the students have entered their confidence ratings on the scales, either a pause with another activity could commence or the next phase, the Adjustment phase, can start directly. In the Adjustment phase, the students are instructed to look over their confidence ratings (shown on the screen together with their associated memory reports) and change those ratings that they consider to be lacking in realism. When the students have finished the Adjustment task, they press the "Finished" key.

If the memory questions answered by the student are in recognition form, that is, with two or more provided answer alternatives, the computer can easily be programmed to correct the memory questions and then compute measures of the realism of the students' first- and second-order confidence ratings. When memory report questions without answer alternatives or free recall reports are used, much more advanced programming would be needed to correct the memory reports, and such programming may not be practically possible for many learning materials.

In situations where programming that automatically corrects the students memory reports has been done, however, the computer can next provide the students with feedback on how the realism of their new confidence ratings compares with the realism of their first confidence ratings. Such feedback on first-order confidence tasks has been suggested and tested by Leclercq [67], and similar means of providing feedback can be used in the present context.

For example, the feedback could be provided by means of two curves in a calibration diagram showing the different confidence levels of the confidence scale (e.g., from 0–100 %, in increments of 10 %) on the x-axis and the proportion correct items in each confidence class on the y-axis. As the students should be told when they are instructed on the concept of confidence realism, the diagonal in the calibration diagram shows perfect confidence realism, where, for example, 60 % of the items at the confidence level 60 % are correct. One of the two curves would show the student's first confidence levels and the second curve, in another color, would show their confidence ratings after they have been (possibly partly) changed in the Adjustment task. Multiple rounds of adjustments, each separated by a day or two or more, may be useful for the students.

2.9 Future Research

Because the field of second-order judgments is young, many different venues exist for future research projects. An important aspect that should be explored is how meta-metacognitive processes differ from metacognitive ones. This exploration might be a difficult task because of still-unanswered questions concerning what separates metacognitive from cognitive processes. It might be that meta-metacognitive processes do not differ in nature from metacognition or cognition, but it is yet too early to say.

On a similar note, it would be interesting to further investigate which cues underlie meta-metacognitive judgments. So far, as mentioned above, one study has investigated the use of two examples of experienced-based cues [34]. It would also be interesting to research, however, how theory-based cues may affect second-order judgments. For example, an individual might believe that he or she is often too confident when it comes to their memory about events and thus will be more prone to making second-order judgments of their confidence judgments than people who do not have this theory about their confidence level in relations to their performance.

As described above, differences in strategies between doing successful second-order judgments for episodic and semantic memory have been found [1, 30, 34, 35]. It would be interesting to further investigate the actual reason for these differences in strategies. As suggested above, the divergence could be because of differences in feedback and encoding possibilities for the different types of memory reports.

It would be of interest to investigate whether people can make successful second-order judgments of performances in which individuals show *underconfidence*. Do people have the ability to increase their realism in these cases, as well? It would also be interesting to investigate which strategies the individuals use in these contexts.

Future research endeavors could additionally investigate if individuals using the method proposed earlier in this chapter can regulate other types of first-order judgments than just confidence judgments. Few of the second-order judgments have focused on the regulating aspect of second-order judgments and have instead focused on the assessment of first-order judgments. It would be interesting to investigate whether our method can help students to regulate their first-order JOLs, for example.

Finally, it would be of great theoretical and practical interest to combine research on meta-metacognition in the more individual sense described in this chapter with social versions of meta-metacognition [31, 32]. How could, for example, instructions to small groups of individuals be constructed to enhance their ability to improve the realism of their earlier first-order confidence judgments?

In this context, a study by Fraundorf and Benjamin [68] is of relevance. These authors investigated the possibility of using several first-order estimates of 12 numerical-estimate questions provided by the participants to optimize their performance, and the researchers also in one experiment incorporated the answers from other individuals. Although this study did not investigate second-order judgments because the participants were not asked to target first-order metacognitive judgments, it is still interesting because of the similarities between this type of multiple estimates study and second-order judgment studies in that both types investigate individuals' ability to improve their performance by making several judgments.

Likewise, research on integrating the different individual judgments of people in a larger group is relevant in this context [e.g., 69]. In all of these various contexts, ideas about how the association frame, or decision frame, of the individual or group deliberating can be broadened are important [e.g., 70, 71].

References

1. Buratti, S., Allwood, C.M.: The effects of advice and "try more" instructions on improving realism of confidence. Acta Psychol. **144**(1), 136–144 (2013)
2. Serra, M.J., England, B.D.: Magnitude and accuracy differences between judgments of remembering and forgetting. Q. J. Exp. Psychol. **65**(11), 2231–2257 (2012)
3. Nelson, T.O., Narens, L.: Metamemory: a theoretical framework and new findings. In: Bower, G.H. (ed.) The Psychology of Learning and Motivation, pp. 1–45. Academic Press, New York (1990)
4. Allwood, C.M.: Eyewitness Confidence. In: Granhag, P.A. (ed.) Forensic Psychology in Context: Nordic and International Approaches, pp. 281–303. Willan Publishing, Devon (2010)
5. Koriat, A., Goldsmith, M.: Monitoring and control processes in the strategic regulation of memory accuracy. Psychol. Rev. **103**(3), 490–517 (1996)
6. Kleitman, S., Gibson, J.: Metacognitive beliefs, self-confidence and primary learning environment of sixth grade students. Learn. Individ. Differ. **21**(6), 728–735 (2011)
7. Wang, M.C., Haertel, G.D., Walberg, H.J.: What influences learning? a content analysis of review literature. J. Educ. Res. **84**(1), 30–43 (1990)
8. Allwood, C.M., Granhag, P.A.: Feelings of confidence and the realism of confidence judgments in everyday life. In: Juslin, P., Montgomery, H. (eds.) Judgment and decision making: Neo-Brunswikian and process-tracing approaches, pp. 123–146. Lawrence Erlbaum Associates, Mahwah (1999)
9. Cutler, B.L., Penrod, S.D., Stuve, T.E.: Juror decision making in eyewitness identification cases. Law Hum. Behav. **12**(1), 41–55 (1988)
10. Lindsay, R.C., Wells, G.L., Rumpel, C.M.: Can people detect eyewitness-identification accuracy within and across situations? J. Appl. Psychol. **66**(1), 79–89 (1981)
11. Wells, G.L., Ferguson, T.J., Lindsay, R.C.: The tractability of eyewitness confidence and its implications for triers of fact. J. Appl. Psychol. **66**(6), 688–696 (1981)
12. Boyce, M., Beaudry, J.L., Lindsay, R.C.L.: Belief of eyewitness identification evidence. In: Lindsay, R.C., Ross, D.F., Don Read, J., Toglia, M.P. (eds.) Handbook of Eyewitness Psychology, vol. 2 Memory for People, pp. 501–525. Lawrence Erlbaum, Mahwah (2007)
13. Griffin, D., Brenner, L.: Perspectives on probability judgment calibration. In: Koehler, D.J., Harvey, N. (eds.) Blackwell Handbook of Judgment and Decision Making, pp. 177–198. Blackwell Publishing, Malden (2004)
14. Baranski, J.V., Petrusic, W.M.: Realism of confidence in sensory discrimination. Percept. Psychophys. **61**(7), 1369–1383 (1999)
15. Björkman, M., Juslin, P., Winman, A.: Realism of confidence in sensory discrimination: the underconfidence phenomenon. Percept. Psychophys. **54**(1), 75–81 (1993)
16. Stankov, L.: Calibration curves, scatterplots and the distinction between general knowledge and perceptual tasks. Learn. Individ. Differ. **10**(1), 29–50 (1998)
17. McClelland, A.G.R., Bolger, F.: The calibration of subjective probability: Theories and models 1980–94. In: Wright, G., Ayton, P. (eds.) Subjective Probability, pp. 453–482. Wiley, Oxford (1994)
18. Dunlosky, J., Rawson, K.A.: Overconfidence produces underachievement: inaccurate self evaluations undermine students' learning and retention. Learn. Instruct. **22**, 271–280 (2012)
19. Brewer, N., Wells, G.L.: Eyewitness identification. Curr. Dir. Psychol. Sci. **20**(1), 24–27 (2011)
20. Wells, G.L., Small, M., Penrod, S., Malpass, R.S., Fulero, S.M., Brimacombe, C.A.E.: Eyewitness identification procedures: recommendations for lineups and photospreads. Law Hum. Behav. **22**(6), 603–647 (1998)

21. Flavell, J.H.: Metacognitive aspects of problem-solving. In: Resnick, L. (ed.) The Nature of Intelligence, pp. 231–236. Erlbaum Associates, Hillsdale (1976)
22. Flavell, J.H.: Metacognition and cognitive monitoring: A new area of cognitive-developmental inquiry. Am. Psychol. **34**, 906–911 (1979)
23. Koriat, A.: Metacognition research: An interim report. In: Perfect, T.J., Schwartz, B.L. (eds.) Applied Metacognition. Cambridge University Press, Cambridge (2002)
24. Lai, E.R.: Metacognition: a literature review. Technical Report. Pearson Research Report http://www.pearsonassessments.com/hai/images/tmrs/Metacognition_Literature_Review_Final.pdf (2011)
25. Nelson, T.O., Narens, L.: Why investigate metacognition? In: Metcalfe, J., Shimamura, A.P. (eds.) Metacognition: knowing about knowing, pp. 1–25. MIT Press, Cambridge (2011)
26. Renkl, A., Mandl, H., Gruber, H.: Inert knowledge: ANALYSES and remedies. Educ. Psychol. **31**(2), 115–121 (1996)
27. Roberts, M.J., Erdos, G.: Strategy selection and metacognition. Educ. Psychol. **13**, 259–266 (1993)
28. Efklides, A., Misailidi, P.: Introduction: the present and the future in metacognition. In: Efklides, A., Misailidi, P. (eds.) Trends and Prospects in Metacognition Research. Springer, New York (2010)
29. Flavell, J.H., Miller, S.A., Miller, P.H.: Cognitive development, 3rd edn. Prentice-Hall, Englewood Cliffs (1993)
30. Buratti, S., Allwood, C.M.: The accuracy of meta-metacognitive judgments: Regulating the realism of confidence. Cogn. Process. **13**(3), 243–253 (2012)
31. Kim, Y.R., Park, M.S., Moore, T.J., Varma, S.: Multiple levels of metacognition and their elicitation through complex problem-solving tasks. J. Mathematical Behavior **32**, 377–396 (2013)
32. Efklides, A.: Metacognition: defining its facets and levels of functioning in relation to self-regulation and co-regulation. Eur. Psychol. **13**(4), 277–287 (2008)
33. Arnold, M.M.: Monitoring and meta-metacognition in the own-race bias. Acta Psychol. **144**, 380–389 (2013)
34. Buratti, S., Allwood, C.M.: Improved realism of confidence for an episodic memory event. Judgment Decis. Making **7**(5), 590–601 (2012)
35. Buratti, S., Allwood, C.M., Kleitman, S.: First- and second-order metacognitive judgments of semantic memory reports: THE influence of personality traits and cognitive styles. Metacognition Learn. **8**(1), 79–102 (2013)
36. Cesarini, D., Sandewall, Ö., Johannesson, M.: Confidence interval estimation tasks and the economics of overconfidence. J. Econ. Behav. Organ. **61**(3), 453–470 (2006)
37. Dunlosky, J., Serra, M.J., Matvey, G., Rawson, K.: Second-order judgments about Judgments of learning. J. Gen. Psychol. **132**(4), 335–346 (2005)
38. Händel, M., Fritzsche, E.S: Students' confidence in their performance judgments: a comparison of different response scales. Educational Psychology: Int. J. Exp. Educ. Psychol. (In Press) (2014)
39. Miller, T.M., Geraci, L.: Unskilled but aware: reinterpreting overconfidence in low-performing students. J. Exp. Psychol. Learn. Mem. Cogn. **37**(2), 502–506 (2011)
40. Serra, M.J., Dunlosky, J.: Do older adults show less confidence in their monitoring of learning? Exp. Aging Res. **34**, 379–391 (2008)
41. Yates, J.F.: Subjective probability accuracy analysis. In: Wright, G.A., Ayton, P. (eds.) Subjective Probability, pp. 381–410. John Wiley and Sons, Oxford (1994)
42. Perfect, T.J.: The role of self-rated ability in the accuracy of confidence judgements in eyewitness memory and general knowledge. Appl. Cogn. Psychol. **18**(2), 157–168 (2004)
43. Perfect, T.J., Watson, E.L., Wagstaff, G.F.: Accuracy of confidence ratings associated with general knowledge and eyewitness memory. J. Appl. Psychol. **78**(1), 144–147 (1993)
44. Touroutoglou, A., Efklides, A.: Cognitive interruption as an object of metacognitive monitoring: feeling of difficulty and surprise. In: Efklides, A., Misailidi, P. (eds.) Trends and Prospects in Metacognition Research, pp. 171–208. Springer, New York (2010)

45. Thompson, V.A., Prowse Turner, J.A., Pennycook, G.: Intuition, reason, and metacognition. Cogn. Psychol. **63**, 107–140 (2011)
46. Allwood, C.M.: Error detection processes in sta-tisti-cal problem solving. Cogn. Science **8**, 413–437 (1984)
47. Koriat, A., Nussinson, R., Bless, H., Shaked, N.: Information-based and experience-based metacognitive judgments: evidence from subjective confidence. In: Dunlosky, J., Bjork, R.A. (eds.) A Handbook of Memory and Metamemory, pp. 117–136. Lawrence Erlbaum, Mahwah (2008)
48. Alter, A.L., Oppenheimer, D.M.: Uniting the tribes of fluency to form a metacognitive nation. Pers. Soc. Psychol. Rev. **13**(3), 219–235 (2009)
49. Kelley, C.M., Lindsay, D.S.: Remembering mistaken for knowing: Ease of retrieval as a basis for confidence in answers to general knowledge questions. J. Mem. Lang. **32**(1), 1–24 (1993)
50. Koriat, A.: How do we know that we know? The accessibility model of the feeling of knowing. Psychol. Rev. **100**(4), 609–639 (1993)
51. Robinson, M.D., Johnson, J.T., Herndon, F.: Reaction time and assessments of cognitive effort as predictors of eyewitness memory accuracy and confidence. J. Appl. Psychol. **82**(3), 416–425 (1997)
52. Robinson, M.D., Johnson, J.T., Robertson, D.A.: Process versus content in eyewitness meta-memory monitoring. J. Exp. Psychol. Appl. **6**(3), 207–221 (2000)
53. Tulving, E.: Memory and consciousness. Can. Psychol. /Psychologie Canadienne **26**(1), 1–12 (1985)
54. Seemungal, F.V., Stevenage, S.V.: Using state of awareness judgements to improve eyewitness confidence-accuracy judgements. In: Chambres, P., Izaute, M., Marescaux, P.J. (eds.) Metacognition: Process, Function and Use, pp. 219–231. Kluwer Academic Publishers, Dordrecht (2002)
55. Goldberg, L.R.: The development of markers for the big five structure. Psychol. Assess. **4**, 26–42 (1992)
56. Raskin, R., Terry, H.: A principal components analysis of the narcissistic personality inventory and further evidence of its construct validity. J. Pers. Soc. Psychol. **54**, 890–902 (1988)
57. Cacioppo, J.T., Petty, R.E.: The need for cognition. J. Pers. Soc. Psychol. **42**(1), 116–131 (1982)
58. Webster, D.M., Kruglanski, A.W.: Individual differences in need for cognitive closure. J. Pers. Soc. Psychol. **67**(6), 1049–1062 (1994)
59. Pallier, G., Wilkinson, R., Danthir, V., Kleitman, S., Knezevic, G., Stankov, L., Roberts, R.D.: The role of individual differences in the accuracy of confidence judgments. J. Gen. Psychol. **129**(3), 257–299 (2002)
60. Schaefer, P.S., Williams, C.C., Goodie, A.S., Campbell, W.K.: Overconfidence and the big five. J. Res. Pers. **38**(5), 473–480 (2004)
61. Campbell, W.K., Goodie, A.S., Foster, J.D.: Narcissism, confidence, and risk attitude. J. Behav. Dec. Making **17**(4), 297–311 (2004)
62. Dahl, M., Allwood, C.M., Rennemark, M., Hagberg, B.: The relation between personality and the realism in confidence judgements in older adults. Eur. J. Ageing **7**(4), 283–291 (2010)
63. Johnson, T.R., Budescu, D.V., Wallsten, T.S.: Averaging probability judgments: Monte Carlo analyses of asymptotic diagnostic value. J. Behav. Dec. Making **14**(2), 123–140 (2001)
64. Wallsten, T.S., Diederich, A.: Understanding pooled subjective probability estimates. Math. Soc. Sci. **41**, 1–18 (2001)
65. Allwood, C.M., Granhag, P.A., Johansson, M.: Increased realism in eyewitness confidence judgments: The effect of dyadic collaboration. Appl. Cogn. Psychol. **17**, 545–561 (2003)
66. Herzog, S.M., Hertwig, R.: The wisdom of many in one mind: Improving individual judgments with dialectical bootstrapping. Psychol. Sci. **20**(2), 231–237 (2009)
67. Leclercq, D.: Validity, reliability, and acuity of self-assessment in educational testing. In: Leclercq, D., Bruno, J. (eds.) Item Banking: Interactive Testing and Self-Assessment, pp. 114–131. Springer Verlag, Berlin (1993)

68. Fraundorf, S.H., Benjamin, A.S.: Knowing the crowd within: metacognitive limits on combining multiple judgments. J. Mem. Lang. **71**, 17–38 (2014)
69. Surowiecki, J.: The Wisdom of the Crowds. Abacus, London, UK (2004)
70. Larrick, R.P.: Debiasing. In: Koehler, D.J., Harvey, N. (eds.) Blackwell Handbook of Judgment and Decision Making, pp. 316–337. Blackwell Publishing, Malden (2004)
71. Larrick, R.P.: Broaden the decision frame to make effective decisions. In: Locke, E.A. (ed.) Handbook of Principles of Organizational Behavior, 2nd edn, pp. 461–480. Wiley, Chichester (2009)

Chapter 3
A Conceptual Model of the Metacognitive Activity

Alejandro Peña-Ayala and Leonor Cárdenas

Abstract This chapter makes a call for contributing to shape a theoretical and well sounded baseline concerning metacognition. It begins recognizing the fuzzy boundaries of the metacognition field and tailors a profile through a wide collection of related works. Particularly, this research focuses on an essential subject: metacognition models. Thus, a sample of proposals for describing the nature, components, and performance of the metacognition is summarized, and a proposal called Conceptual Model of the Metacognitive Activity (CMMA) is introduced. The CMMA is a conceptual model that depicts the metacognitive activity with the purpose of providing a functional view of how metacognition interacts with object-oriented cognition. Such a model takes into account basic aspects of neurology and biology sciences. Additionally, the autopoiesis property is considered to describe the autonomy and performance of the metacognition. Moreover, an analysis of metacognitive models is outlined and a comparison between them and the CMMA is made in order to shape an overall idea of what metacognition is, and the contribution of the CMMA. In this way, valuable topics are provided to encourage research oriented to build the metacognition basis.

Keywords Metacognition · Metacognition model · Metacognitive activity · Autopoiesis · Conceptual model of the Metacognitive Activity

A. Peña-Ayala (✉)
WOLNM: 31 Julio 1859 No. 1099-B, Leyes Reforma, 09310 Mexico City, Mexico
e-mail: apenaa@wolnm.org; apenaa@ipn.mx

A. Peña-Ayala · L. Cárdenas
ESIME Zacatenco, Instituto Politécnico Nacional, Building Z-4, 2nd Floor,
Lab 6, Miguel Othón de Mendizábal S/N, 07320 Mexico, D.F., Mexico
e-mail: adriposgrado@gmail.com

© Springer International Publishing Switzerland 2015
A. Peña-Ayala (ed.), *Metacognition: Fundaments, Applications, and Trends*,
Intelligent Systems Reference Library 76, DOI 10.1007/978-3-319-11062-2_3

Abbreviations

AS	Autopoietic system
AO	Autopoietic organization
CMMA	Conceptual Model of the Metacognitive Activity
FOC	First-order cognition
LS	Living system
MS	Mechanistic system
NNS	Neuronal nervous system
SR	Self-regulation
SRL	Self-regulated learning
SWOT	Strengths, weaknesses, opportunities, and threats

3.1 Introduction

Metacognition, as a mental phenomenon that happens in the brain of human beings as well some species of animals [1], has been a research subject of cognitive developmental psychology. Although, metacognition basically means: *Cognition about cognition* or knowing about knowing [2], it concerns a variety of epistemological processes. Papaleontiou-Louca qualifies metacognition as: A kind of "second-order cognition" to highlight: thinking about thinking, knowing about knowing, regulating about regulation, and so on [3]. She infers: If "first-order cognition" (FOC) concerns understanding, memorizing, and so forth then, metacognition implies being aware of one's own comprehension, memory... Thus, diverse cognitions about cognitions might be named: metacomprehension, metamemory... with metacognition remaining the superordinate term.

Even though the essential concept given to the metacognition term is quite simple, its broad meaning and nature have been qualified as "fuzzy" [4]. In this regard, Veenman claims: One of problems with metacognition is the "fuzziness" of the concept and its constituents, as well as the proliferation of terminologies and disagreement about the metacognition ingredients and their interrelationships [5]. Furthermore, Zohar and Dori assert: "The 'fuzziness' in the metacognition definitions makes difficult to discuss several studies together in an integrated and clear way [6]. Moreover, Whitebread et al. [7] observe: "...metacognitive skillfulness is a rather fuzzy concept. It can be considered as a person's propensity to use these "basic skills" in everyday situations...".

Efklides and Misailidi [8] complain that: "...the distinction between cognition and metacognition is often hard to be made, and the diversity of metacognitive phenomena suggests that there is no single mechanism that can explain them all. Acevedo and Aleven [9] claim: "...there is a great need for theoretical clarity, including better definitions and descriptions of the metacognition components. Beran et al. [10] explain why they called their book "Foundations of

Metacognition": "Given that the term metacognition has acquired several different meanings in literature, a general definition of the term is no longer feasible".

This chapter is an essay on how the metacognitive community can contribute to develop a formal, holistic, and systemic theoretical baseline to ground metacognition. In pursuit of such a call, this chapter offers a conceptual view of the metacognition field that focuses on its background and development, as well as models that depict metacognition, including one that proposes sketching the metacognitive activity. So, the remaining of the chapter is organized as follows: In the second section the background and a sample of metacognition facets are outlined. In the third section several works explain the nature, composition, and activity of the metacognition. In the fourth section a conceptual model to sketch the metacognitive activity is introduced. The aim is to consider essential neuronal and biological aspects of the metacognitive activity. In the fifth section a discussion of the described models is given to sketch an integral view of the metacognition. The last section identifies strengths, weaknesses, opportunities, and threats (SWOT) of the metacognition, as well as the future work to be carried out to ground CMMA.

3.2 A Glance at Metacognition

In order to highlight essential elements to be considered for designing an integral baseline for the metacognition field, a conceptual view is outlined in this section. Thus, prior to recalling the appearance of the metacognition term, some research that makes up its roots is stated. Afterwards, a series of works published since the statement of the metacognition term up to the present is provided. Later on, several collections of works concerning different metacognition facets are identified to shape the nature of the field. Finally, several publications that unveil two of the main metacognitive actors are introduced, one for children and the other for non-human beings.

3.2.1 Previous Works

The metacognition historical roots are deep because they lie in works such as: "Principles of Psychology" published by William James in 1890 [11], the study on memory and the feeling-of-knowing experience made by Hart [12], and the Piaget School, where Flavell [13] made an incursion and contributed with his book "The Developmental Psychology of Jean Piaget" published in 1963.

However, the study of metacognition achieved widespread prominence in the 1970s through the research fulfilled on developmental changes in children's cognition about memory, understanding, communication, and problem solving by Tulving and Madigan [14], Flavell [15], Fischhoff [16], Brown [17], Wellman [18], Lachman [19], and other scientists.

3.2.2 The Birth of a Research Line

In 1976, Flavell coined the *metacognition* term to entitle his paper: "Metacognitive aspects of problem solving" [20]. In such a work, he provides a concept to show what metacognition is: "Any kind of cognitive transaction with the human or non-human environment, a variety of information processing activities may go on. Metacognition refers, among other things, to the active monitoring and consequent regulation and orchestration of these processes in relation to the cognitive objects or data on which they bear, usually in service of some concrete goal or objective".

Originally, Flavell recognized that metacognition consists of both monitoring and regulation aspects. In order to characterize such a viewpoint, Flavell [21] tailored a Formal Model of Metacognitive Monitoring in 1979. His model embraces four classes of phenomena: metacognitive knowledge, metacognitive experiences, tasks or goals, and strategies or activities. In addition, the interrelationships between these phenomena are drawn to explain how they interact to monitor and regulate cognitive activity.

In that moment a research line was born in the field of developmental and cognitive psychology, where scientists of diverse disciplines contribute to extend the theory, baseline, methods, and applications as is summarized in the followings two subsections.

3.2.3 A Chronicle of Metacognition Research Development

With the purpose of recognizing the evolution of the metacognition field, a sample of related works is introduced in this subsection. The works are organized according to the decade of their publication to reveal how the metacognition labor has evolved since the term was coined up to the present.

Once the metacognition term emerged, many partisans of the developmental psychology, cognitive sciences, neural sciences, pedagogy, education, and computer sciences contributed to extended the former achievements of the 1970s. For instance, during the 1980s a sort of relevant works were oriented to: study comprehension monitoring [22], cognitive knowledge and executive control [23], learning, remembering, and understanding [24], performance [25], metacognitive skills [26], reading comprehension [27], strategies [28], motivation [29], and cooperative learning [30].

As for the 1990s, the metacognition research enhanced their lines and explored diverse subjects, such as: instruction [31], self-esteem [32], metamemory [33], metacognition models [34], metacognitive judgments [35], development of metacognition in children [36], frontal lobe supports to metacognitive activity [37], metacognitive theories [38], implicit memory [39], and prefrontal cortex supports to control and monitor memory processes [40].

Concerning the present century, an explosion of works concerning metacognition has been published; particularly books, edited books, specialized journals, conferences, and societies, such as: Metacognition: Process, Function... [41], Thinking and Seeing Visual Metacognition... [42], ...Self-reflective Consciousness [43], Handbook of Metamemory and Memory [44], Metacognition [45], New science of Learning Cognition... [46], Trends and Prospects in Metacognition... [47], Metareasoning... [48], Metacognition in Science Education [49], Foundations of Metacognition [50], ...Handbook of Metacognition... [51], Cognitive Development (since 1986) [52], Journal of Cognitive Neuroscience (since 1989) [53], Trends in Cognitive Sciences journal [54], Journal of Cognition and Development [55], Metacognition and Learning journal (since 1997) [56], Journal of Applied Research in Memory and Cognition [57], International Conference on Metacognition [58], International Association for Metacognition [59].

3.2.4 A Conceptual Shape of the Metacognition Through Its Works

In order to sketch an image of the metacognition, several of its facets are instantiated through a collection of recent works. The traits considered to make up the metacognition shape are the following: disciplines involved in the study, related paradigms, metasubjects, metacognitive facets, self-skills, and support to cognitive processes.

Some disciplines that study metacognition are the following: cognitive psychology [60], developmental psychology [61], educational psychology [62], neuroscience [63], cognitive neuroscience [64], science of learning [46], cognitive sciences [65], science education [49], autonomous artificial life form [66], memory [67] (e.g., amnesic [68], blank in the mind [69]), mental health [70], social psychology [71], social sciences [72], self-regulation (SR) [73], and computer-assisted learning [74].

A sample of work lines that make an explicit reference, or at least implicit in some sense, to metacognition as a "peer", "similar", "related", or "subordinate" construct, are the following: affect [4], cognitive processing [5], self-regulated learning (SRL) [75], executive control [76], critical thinking [77], theory of mind [78, 79], unawareness and uncertainty [80], cognitive load [81], and motivation [82].

The research in metacognition often demands a specialized study of a given subject that is labeled with the prefix "meta" to depict a kind of relationship, collaboration, or subordination such as: meta-metacognition [83], metacomprehension [84], metastrategic knowledge [85], meta-affect and meta-affective (compound term to adjective skill, experience, and knowledge) [86, 87], metamemory [67–69, 88], metarepresentation [89], meta-analysis [90], and meta-attention [91].

In addition, several functions, methods, strategies and techniques have been qualified by the adjective "metacognitive" to highlight its particular nature, such as: strategies [92], accuracy [93], judgments [94], prompts [95], instruction [96], tools [96, 97], inquiry [98], behavior [99], feelings [100], measures [101], scaffolding [102], and feedback [103].

Essentially, knowledge [104], regulation [104, 105], and experiences [106] have been considered the main "components" of the metacognition. However, some works recognize others components or functionalities, such as: skills [107, 108], control [109], monitoring [110, 111], reflection [112], and awareness [113].

Metacognition is involved with the reflective property of its components and other constructs to emphasize that the subject is aware and responsible for the performance of his/her mental activity, such as: metacognitive self-regulation [114, 115], self-esteem [115], self-efficacy [115, 116], self-monitoring [117, 118], self-confidence [118], self-explaining [102, 119], self-knowledge [120], self-perceived [121], self-correction [122], self-assessment [123], and self-management [123].

In spite of many works that relate metacognition with learning [124] and knowledge acquisition [125], metacognition is also involved in essential cognitive functions, such as: reading [126], understanding [127], questioning [128], pronunciation [129], spelling [130], decision-making [131], problem-solving [132], help-seeking [133], collaboration (e.g., co-regulation in learning) [134], and reasoning [135].

3.2.5 Metacognition Research on Children and Animals

Most of the metacognition research is oriented to young and adult people, who are involved in formal settings and long-life learning. However, some scientists are interested in exploring how metacognition is manifested in children and non-human beings. So, this subsection is reserved to highlight a sample of both research targets.

Metacognition is intricately linked to the human mind, including cognitive control, self-awareness, and consciousness. For this reason, it is acknowledged as one of the humans' most sophisticated cognitive capacities, and it is widely accepted that humans are capable of metacognitive processing. Thus, a question is raised: When and how metacognition emerges and is developed? In order to respond to these questions, a collection of works provides an answer that asserts: metacognition evolves naturally (conscious and non-conscious) in informal and formal settings.

Some of the works that study metacognition in childhood are the following: Allwood [94] explores metacognitive judgments in children of 8–9-year-olds and 12–13-year-olds; Whitebread et al. [7] examine a broad variety of metacognition studies in different ranges of young children; Lyons and Ghetti [136] summarize different studies of metacognition in early childhood from 12 to 18 month old babies and from 3 to 5 years old preschool children; Misailidi [79] pursues to bridge the gap between metacognition and theory of mind based on a series of studies made with young children.

More related works about children metacognition are presented as follows: Renkl et al. [119] measure self-confidence and academic achievements in Primary-school children, while Kolić-Vehovec et al. [137] report developmental trends in metacognition and reading comprehension reached by children during elementary and high school (9–17 years) in Croatia; Csíkos and Steklács [138] highlight a similar research in Hungary with 10–11 years old children, whereas Sodian et al. [139] study metacognition in infants and young children. Esken [140] unveils forms of metacognition in children and Roebers et al. [121] examine associations between executive functioning, metacognition, and self-perceived competence in first grade children.

Other interesting works are introduced as follows: Kloo and Rohwer [141] compare the development of earlier and later forms of metacognitive abilities and Bryce and Whitebread [142] inquiry: whether developmental changes in children aged 5–7 years, reflect quantitative or qualitative improvements, and how meta-cognitive skills change with age and task-specific ability.

In addition, Krebs and Roebers [143] examine the influence of retrieval processes on 9–10 and 11–12 years old children's metacognitive monitoring and controlling; whilst, Barfurth et al. [144] examine metacognition in children and adolescents to consider the link between childhood giftedness and adult expertise, as well as understanding ways very able children and adults think and solve problems.

Historically, according to Morgan [145]: Homo sapiens alone were regarded as metacognitive, while animals were considered to have little by way of mental lives, and they were considered much more bound in their behavior to the stimuli that they encountered and the outcomes that they experienced [146]. Thus, a skepticism posture is placed on those who study non-human beings if claims of animal behavior should be considered as the result of metacognitive processes. A sample of works related to metacognition in animals is introduced as follows:

Beran et al. [146] offer evidence that counters that belief and some theoretical objections against the possibility that monkeys' performances reflect metacognitive abilities. Couchman et al. [147] reveal evidence for animal metaminds; Crystal [148] highlights several models of metacognition in animals; Fujita et al. [149] make the question: are birds metacognitive?; Call [150] seeks information in animals; Carruthers and Ritchie [151] study how metacognition emerges before demonstrations of affect and uncertainty in animals.

3.3 A Sample of Models to Describe Metacognition

A second subject worthy to be taken into account for setting a theoretical baseline for the metacognition is the theoretical models. A model represents a conceptual viewpoint that describes the nature, components, and the way they interact in order to explain the metacognitive phenomenon. Therefore, a survey of fifteen works that characterize the metacognition is stated in this section.

3.3.1 Classic Models of the Metacognition

Research in metacognition is grounded on a wide variety of theoretical models, which provide essential concepts to describe its nature. Some of them are briefly described to highlight relevant attributes of the metacognition.

- Flavell's Metacognitive Monitoring Model: identifies four phenomena involved in metacognitive monitoring: knowledge, experiences, goals-tasks, strategies [21, 152]. Where, knowledge represents facts and beliefs that the individual holds about the factors that bias cognitive activities. Knowledge is characterized as person, task, and strategies variables. As for experiences, they are subjective internal responses of an individual to his/her cognitive performance. With respect to goals-tasks, they depict the results to be achieved by a task; whereas strategies are ordered processes set to control one's own cognitive activities and to assert that a cognitive goal has been fulfilled.
- Brown's [153] Knowledge and Regulation of Cognition Model: reveals two closely related categories: knowledge of cognition and regulation of cognition. The former means "*knowing that*" and represents activities that involve conscious reflection on ones cognitive abilities and activities. The later corresponds to activities triggered by self-regulatory mechanisms during an ongoing attempt to learn or solve problems. Such activities are unstable.
- Nelson and Narens's [154] Hierarchical Model: splits cognitive process into *meta-level* and *object-level*. The former sketches a cognitive model of the latter, which is updated as a result of the monitoring flow coming from the object-level. The meta-level reacts to such stimuli by producing control flows oriented to initiate, alter, or terminate mental actions being achieved at object-level.
- Norman and Shallice's [155] Executive-Object Model: embraces two levels named *executive system* and *instance*. The first depicts a view of the perceptual and cognitive functions existent at the instance level; whilst the second contains schemas that are basic units of action and thought. The model asserts the executive system modulates the instance level schemas according to an individual's intentions.
- Shimamura's [156] model: is based on the model proposed by Nelson and Narens [154], and Norman and Shallice [155] to map the meta-level and the object-level onto a hierarchical brain structure. Where posterior cortex supports object level to carry out task performance and prefrontal cortex performs the meta-level that is conceptualized as both monitoring and controlling the object level.

3.3.2 Declarative and Procedural Models of the Metacognition

Besides the "classic" models of metacognition, there are additional models of metacognitive skills that can be typified as either describing components or scheduling the processes involved in metacognitive behavior regardless of timing.

Descriptive Models of the Metacognition. Descriptive models highlight components, facets, and functionalities of the metacognition according to a particular viewpoint. In addition, a kind of conceptual, functional, or hierarchical relationship is linked in order to explain how they are organized and interrelated. A sample of this category is briefly presented as follows.

- Kuhn's [157] model: depicts metacognition as: "Any cognition that has cognition...as its object". It encompasses three components: metacognitive knowing (i.e., refers to one's base of declarative knowledge), metastrategic knowing (i.e., involves procedural knowledge), and epistemological knowing (i.e., individual's broader understanding of what knowledge and knowing are in general).
- Alexander and Schwanenflugel's [158] model: identifies three components of metacognition: declarative metacognitive knowledge (i.e., individual's knowledge about the contents of the mind), cognitive monitoring (i.e., individual's ability to read one's own mental states), and regulation of strategies (i.e., ability to strategically use metacognitive knowledge to achieve goals).
- Tobias and Everson's [159] Componential Model: considers monitoring prior learning as a prerequisite for metacognitive process. The model focuses on the ability to monitor, evaluate learning, select strategies, and make plans for one's learning, as well as the control of these processes. The knowledge monitoring is the ability to know: what you know and knowing what you do not know.
- Schraw et al. [84] framework: sketches a hierarchy to split metacomprehension into metacognition and metamemory. Because metacognition refers to knowledge about cognition and cognitive processes, it recognizes them as its essential components, which are also respectively named metacognitive knowledge and metacognitive skills. Where, the former holds three sorts of knowledge: declarative, procedural, and conditional. Whilst, the latter considers two additional skills (information management and debugging) to the classical ones (planning, monitoring, and evaluation).
- Zohar's [85] Metastrategic Knowledge Model: reveals general knowledge about higher-order thinking strategies. It maps the traits of the models proposed by Flavell [21, 152], Schraw [84], and Kuhn [157] prior stated. Where metastrategic knowledge corresponds to three kinds of knowledge about: persons, tasks, and strategies.
- Efklides's [4] model: defines two functions of the metacognition: monitoring and control. The first is manifested by metacognitive knowledge and experiences; whilst the second is expressed by metacognitive skills. The metacognitive knowledge is associated to facets: ideas, beliefs, theories of goals,

task, person, cognitive function... Metacognitive experiences concerns with the facets: feelings of familiarity, difficulty, knowing... and judgments of learning, estimate of effect... Metacognitive skills are related to the facets: conscious for effort and time allocation.

Procedural Models of the Metacognition. Procedural models characterize metacognition as a sequence of stages or processes, which evolves during the time the individual matures. Some of these models are outlined next:

- Veenman's [160] model: extends the model of Nelson and Narens [154] to depict a dynamic bidirectional flow. Where metacognition is seen as both bottom-up and top-down processes. As for the former, anomalies in task performance trigger monitoring activities, which in turn activate control processes on the meta-level. Whereas for the latter, apart from being triggered by task errors, the top-down process is also triggered as an acquired program of self-instructions, whenever the individual is faced with performing a task he/she is familiar with a certain extent. Such a program of self-instructions could be represented by a production system of condition-action rules [161].

- Zelazo conscious awareness model: traces an information processing account through the next stages: (1) at birth, children reveal minimal consciousness because they are aware only of the input stimuli; (2) around the first year, infants unveil recursive consciousness due to they are able to bring back to mind stimuli which are no longer in the environment; (3) around the second year, children are self-conscious as they are accustomed to reflect about the stimuli brought back to their mind; (4) during the proceeding years additional levels of conscious awareness are reached as children progress through further iterative recursions, bringing to mind and reflecting upon the contents of their mental activity from lower levels of consciousness [136, 162].

- Flavell's awareness of uncertainty model: considers children deal with uncertainty through four stages: (1) at birth, babies may not have any experience of uncertainty; (2) young children may have subjective experiences of uncertainty, but fail to be consciously aware of it; (3) children may be consciously aware of the subjective experience of uncertainty, but may not attribute it as such; (4) later, children are consciously aware of their subjective experiences of uncertainty and recognize them as uncertainty [136, 163].

- Efklides's [164] metacognitive and affective model of self-regulated learning: joins metacognition and motivation/affect as two levels of functioning in SRL, *person* and *task x person*. At the first level, person interactions between trait-like characteristics (e.g., cognitive ability, metacognitive knowledge and skills, self-concept, perceptions of control, attitudes, emotions, and motivation in the form of expectancy-value beliefs and achievement goal orientation) are supposed to happen and such person traits guide top-down SR. At the task x person level (i.e., the level at which SRL events occur) metacognitive experiences (e.g., feeling of difficulty) and online affective states play a major role in task motivation and bottom-up self-regulation. The stimuli represented by a cognitive task activate person and task x person levels, as well as reciprocal relationships between them and their respective components. This means, person traits interact with each other at person level; whereas at task x person level, reciprocal relationships between cognition and metacognition–affect, as well as metacognition–affect and SR of affect/effort are triggered.

3.4 A Conceptual Model of the Metacognitive Activity

In order to participate in grounding the metacognition basis, a Conceptual Model of the Metacognitive Activity (CMMA) is proposed in this section. The CMMA includes concepts of the neurology and biology inspired in [165–169]. They provide a theoretical context of neurological structures and biological systems that support the development of cognition, and specifically metacognition. Thus, diverse concepts, premises, and hypothesis are stated to describe the nature of cognition and metacognition.

3.4.1 A View of the Nervous System

Cognitive activity is immanent to living beings. Daily, the *nervous system* of human beings performs cognitive activity to fulfill mental and physical tasks (e.g., thinking, eating...), besides automatic and routine tasks (e.g., breathing, sleeping...). Therefore, the cognitive activity, and more specifically the metacognitive activity, is a daily process of the nervous system.

Multi-cellular animals, such as human beings, have a neuronal nervous system (NNS); whereas unicellular living systems have a molecular nervous system. NNS is organized as a *closed network of neurons* that tailors changing activity relationships. The NNS exists in structural intersection with a larger system, the *organism*, and at the *sensory* and *effector* areas that are used to interact in a medium that is a dynamic totality. Based on the work achieved by Maturana, some operational consequences concerning the manner that the nervous system is constituted and several properties of the NNS are outlined in this section [165–169].

Organism and NNS exist operationally in different non intersecting domains. The organism operates in the domain in which the living system (LS) exists. The NNS operates in the domain in which it is found as a closed neuronal network of changing relations of activities. The interrelation between both domains occurs at the sensory and effector items where organism and NNS are in structural intersection.

The NNS does not interact with the medium, neither act on representations of the medium. Its structure is not fixed; it continuously changes due to the following reasons: (1) the structure of the NNS follows a path of change that is contingent to the flow of the interactions of the organism in the realization and conservation

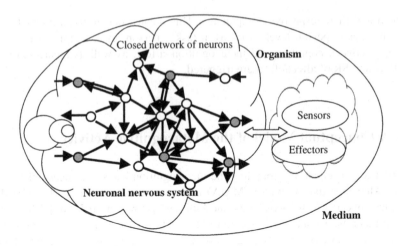

Fig. 3.1 A conceptual view of the neuronal nervous system according to the CMMA

of its life; (2) the structural changes triggered in neuronal items that intersect with internal and external sensory/effectors of the organism as a result of the organism interactions in the external medium; (3) by structural changes triggered in the neuronal items by hormones secreted by the organism endocrine cells; (4) by recursive structural changes triggered in its neuronal components as a result of their participation in its operation as a closed network of changing activity relationships.

The NNS intersects structurally with the sensors and effectors of the latter's sensory and effector surfaces. So, the sensors and effectors of a multi-cellular organism have a dual character and operate both as components of the organism and the NNS. As regards the former, sensors and effectors operate in the interactions of the organism in its existence domain as its sensors and its effectors. For the latter, sensors and effectors operate in their closed dynamics of changing activity relationships, as well as other neuronal elements. A conceptual view of the NNS is drawn in Fig. 3.1 to illustrate the CMMA.

3.4.2 A View of the Biological Context

Human and non-human beings are biological LS, whose cognitive and metacognitive activities are fulfilled through neuronal structures and neuronal activity. Neurons are cells that establish synaptic connections to shape networks of neurons. The neuronal structures facilitate the interchange of chemical and electrical flows as a way to perform neuronal activity. Thus, in order to complement the CMMA baseline, it is pertinent to take into account a biological view to consider metacognition as a LS, specifically an *autopoietic system* (AS), based on the following concepts:

- LS: it is a structure determined system, whose structure defines how the LS is made at any instant. All that happens to the LS at any instant depends on its structure. Any agent impinging on the LS only triggers structural changes determined in it [165]. So a LS is a kind of molecular machine that operates as closed networks of molecular productions. The molecules produced through their interactions generate the same molecular network that produced them, specifying at any instant its extension. Such a phenomenon reveals a recursive property.
- Living organization: is a property to support the belief that all LS must share a common organization. A living organization can only be characterized unambiguously by specifying the network of interactions of components which constitute the LS as a whole, that is, as a "*unity*". Varela et al. [166]. inquiry: "What is the necessary and sufficient organization for a given LS to be a living unity?"
- Unity: is treated from two views: (1) as an analyzable whole endowed with constitutive properties which define it as a unity; (2) as a complex system that is realized as a whole through its components and their mutual relationships. Where a complex system is defined as a unity by: The relationships between its components (which realize the system as a whole) and its properties, which determine the way the unity is defined [166].
- Classes of unities: they are LS determined by the same organization in spite of having different kinds of components, as long as these components have the properties which realize the required relationships. Some classes of unities are labeled as *mechanistic systems* (MS) because their organization is specifiable only in terms of relationships between processes generated by the interactions of their components. This is the case of LS whose living organization is considered to be a MS.
- MS: is a system whose components' properties are capable of satisfying certain relationships that determine the unity of the interactions and transformations of these components. It means the system under study only behaves as it does because all its components contribute. So a mechanistic viewpoint necessarily needs a decomposition of the system into components and their interrelations [167]. There are some MS defined as unities by a specific organization called *autopoietic organization* (AO). Such MS is different from others in the sense the product of its operation as a system is necessarily always the system itself. If the network of processes that constitutes this kind of MS is disrupted, the system disintegrates.
- AO: is defined as a unity by a network of productions of components which satisfies two constrains: (1) components recursively participate in the same network of productions of components that produced them; (2) tailor the network of productions as a unity in the space in which the components exist. This is the case of a cell, which is a network of chemical reactions that produce molecules that realize the cell; and such reactions interact and recursively participate in the same network of reactions which produced them [166].
- AS: is a mechanistic system that exhibits the AO. Biological evidence shows that LS belong to the class of AS [168]. The LS is a molecular AS open to the flow of matter and energy. The AS is a LS closed in its states dynamics

in the sense that it is alive only while all its structural changes conserve their autopoiesis [169]. The AS holds an *autonomy* property of an AO, where the realization of the AO is the product of its operation. As long as an AS exists, its organization is invariant. If the network of productions of components which defines the organization is disrupted, the unity disintegrates. So an AS has a domain in which it compensates perturbations by the realization of its autopoiesis, and it remains a unity.

3.4.3 Nature of the Cognition

Besides the neurological and biological essentials, the CMMA also considers the nature of the cognition. In this section a series of concepts is stated to provide key definitions of several constructs. In addition, the learning ability is chosen to provide an example of the conceptual view of cognition, which is similar to Fig. 3.1.

Essential Concepts about Cognition. The brain is the seat of cognition. Cognition literally means: "to know". Cognition has to do with how a person perceives, understands, and behaves in the world. Cognition claims the acquisition, development, and exploitation of a wide sort of knowledge and cognitive activities.

As for knowledge, it can be seen as memories formed from the manipulation and assimilation of raw input (i.e., information perceived via our senses of sight, hearing, taste, touch, and smell), as well as the result of our daily cognitive activity (e.g., thoughts, reasoning, recalls, learning, mental experiences, cognitive outcomes…).

A large part of cognition involves the organization of our knowledge into associations or categories. These might range from facts (e.g., things one might find in a place) to beliefs (e.g., suppositions about how some people behave). Simple symbols (e.g., dollar currency $) are used to group more complex learned associations such as those between noses, lips, eyes, and smiles. Using knowledge to direct and adapt action towards goals is the foundation of the *cognitive activity*.

In regards cognitive activity, it represents the manifestation of cognition achieved in the brain. People perform cognitive activity to fulfill nearly every human action, from the simplest task to the most complex. Cognitive activity is embodied into *cognitive abilities* to guide their organization, practice and control to accomplish specific cognitive purposes. For instance, answering the phone involves at least: perception (hearing the ring tone), decision making (answering or not), motor skill (lifting the receiver), language skills (talking and understanding language), and social skills (interpreting tone of voice and interacting properly with another human being).

Cognitive abilities (e.g., brain or mental functions) are neural processes, which are represented and performed in the brain. They constitute the ownership of the means to achieve something, or the faculty for practicing a natural or mastered skill needed to do something. They are based on specific constellations of brain structures (e.g., memory skills rely mainly on parts of the temporal lobes, next to the temples, and parts of the frontal lobes, behind the forehead). Cognitive abilities are neural processes, which are represented and performed in the brain. The cognitive abilities are overlapping and not always clearly distinct. Keeping this in mind, the concept of cognition is broken down into some of its more widely cognitive abilities, such as: perception, attention, reasoning, speaking, planning, learning…

Representation of Cognition according to the CMMA. In order to show how the CMMA characterizes the cognition, the *learning* ability is picked to apply the neurological and biological previously stated baseline. Therefore, a series of concepts and a graphical representation are outlined as follows.

Learning is a ordinary cognitive ability that demands a dynamic structure determined by neural networks. It triggers the interchange of information, depicted as chemical and electrical flows, with the NNS, the sensors and effectors, and the organism. The mechanism of *synaptic plasticity*, stated in the Hebbian theory, reveals that an increase in synaptic efficacy arises from the pre-synaptic cell's repeated and persistent stimulation of the postsynaptic cell [170]. This mechanism claims the adaptation of neurons in the brain during learning. The theory claims: "Cells that fire together, wire together". It attempts to depict associative learning, in which simultaneous activation of cells leads to pronounced increases in synaptic strength between those cells.

According to the CMMA, learning can be conceptually characterized as is shown in Fig. 3.2. The model presents a *closed network of components* such as cognitive activities and knowledge. They are sketched as a network of black and white nodes to respectively depict cognitive activities and knowledge. They hold relationships for transferring stimuli between them. The direct relationships are drawn as thin links, the recursive flows are pictured as thick links, and the feedback flows are depicted as dotted lines. The arrowheads of the links reveal direct sense of stimuli transference.

The closed network of components is organized as a structure determined system oriented to accomplish *ordinary cognitive* processes. The processes progressively master new and current knowledge, as well as new and current cognitive skills. The structure determined system operates and evolves throughout several cycles.

During a given cycle, several kinds of outcomes are produced, such as: new or transformed components (e.g., cognitive *activities* and *knowledge*), ordinary cognitive *processes*, new knowledge, and *skills*, where these products are respectively identified in Fig. 3.2 by A, K, P, and S ovals, which appear inside the smallest cloud. In consequence, feedback and recursive flows transfer the outcomes as inputs to trigger the next cycle. In this way, the closed network evolves like a spiral along the time. The closed network and the set of outcomes hold bidirectional flows to transfer stimuli with the sensors and effectors. They are the mediators between the NNS, shaped as the largest cloud in Fig. 3.2, and the organism, sketched as an oval.

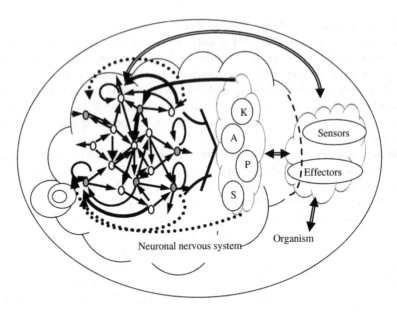

Fig. 3.2 A conceptual view of the learning cognitive ability according to the CMMA

3.4.4 Nature of the Metacognition

Once the neurological, biological, and cognitive baseline has been stated, it is time to explain how the CMMA represents the metacognition. Firstly, key concepts are defined to shape a theoretical reference. Later, a sample of how to depict the metacognition to support learning is outlined in accordance with the previous representations.

Essential Concepts about Metacognition. Most of the cognitive activity, such as all the cognitive abilities introduced in the prior subsection, is considered as FOC due to people daily practicing to accomplish a mental or a physical goal (e.g., acquiring new knowledge, watching an advertisement…), and to interact with others and their environment (e.g., speaking with somebody…). However, there is a category of cognitive activity that pursues to trigger, supervise, evaluate, exert, and take over FOC activity, such a category is called: *metacognition*. Essentially, *metacognition is cognition* and aims at gaining, stimulating, and practicing several kinds of knowledge and activities, which are labeled as *metacognitive*.

For instance, *metacognitive knowledge* is knowledge of cognition. It refers to what individuals know about cognition and their own cognition strengths and weaknesses. Moreover, it accounts for experiences, strategies, and conditions under what some kind of activity is preferred more than others. Furthermore, it shapes a model of the current FOC activity, which is the target of the metacognitive activity.

As regards *metacognitive activity*, it is also the manifestation of neural processes and is supported by brain structures (e.g., metacognitive monitoring and control have been viewed as a function of the prefrontal cortex, which monitors sensory signals from other cortical regions and through feedback loops implements control [44]). Metacognitive activity reveals the practice of metacognitive strategies, processes and skills currently performed in the brain.

Metacognitive activity is guided by *metacognitive strategies*. Such strategies are sequential processes devoted to monitor and control FOC activities and to ensure the fulfillment of a cognitive goal. The organization of metacognitive activities, processes, and strategies to accomplish a metacognitive purpose is named "metacognitive skill".

The repertory of *metacognitive skills* is quite extensive, and nearly includes all the ordinary cognitive abilities. Furthermore, the prefix *self* produces a special version of the former skill term. For instance, metacognitive regulation is the monitoring of one's cognition and includes planning activities, awareness of comprehension and task performance, and evaluation of the efficacy of monitoring processes and strategies [171]. Whereas, metacognitive self-regulation is defined as: self-regulated thoughts, feelings, and behaviors that are oriented to attaining goals. It follows three stages: forethought, performance, and self-reflection.

Characterization of Metacognition according to the CMMA. Based on the prior neurological, biological, and cognitive concepts and practices of representation, this subsection explains how the CMMA characterizes metacognitive activity.

Metacognitive activity is considered a neurological, biological, and LS. The view provides an idea of the nature of the metacognition and how the activity is accomplished according to the support of the NNS. Based on the arguments presented in Sect. 3.4.2, it infers that metacognition is a kind of AS, illustrated in Fig. 3.3.

The meaning of the representation stated by the CMMA for the metacognition in Fig. 3.3 takes into account the prior description given for the common shapes (e.g., direct, feedback, and recursive relations are respectively sketched through thin, thick, and dotted lines...) in Fig. 3.2. In addition, the following observations are pointed out:

The square depicts the *organism*; the oval sketches the *NNS*; the big cloud at the left shows the *metacognition*; the middle cloud corresponds to the *FOC activities*; the smallest cloud concerns the *sensor* and *effectors* items.

The metacognitive image holds three elements: a *closed neuronal network* is illustrated at the left (e.g., where the black and white nodes represent metacognitive activities and metacognitive knowledge respectively); at the center is shown a cloud with ovals to label metacognitive *processes* (e.g., *P1, P2*...); at the right, the wide arrows show metacognitive *strategies* (e.g., *S1, S2*...).

As for the FOC picture, it shapes a *closed neuronal network* (e.g., where the black and white nodes depict cognitive activities and cognitive knowledge); whilst several cognitive activities labeled as *A1, A2*... are outlined like ovals at the right.

Metacognition is performed by the structure and activity of the NNS. The NNS structures are characterized as a closed network of neuronal elements that

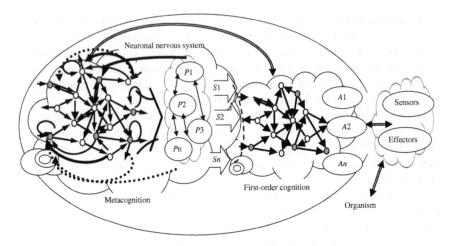

Fig. 3.3 A conceptual view of the metacognition according to the CMMA

establishes changing activities relationships. The NNS provides a mediated structural intersection with the organism by means of the participation of FOC activities. They transfer and receive communication with sensory and effector items. Such items make up the structural intersection between organism and NNS. The organism operates in the domain in which human being exists; whereas, the metacognition operates in the domain in which it exists as a closed neuronal network of changing activity relationships.

Metacognition, as a manifestation of cognition with a given purpose, monitors and controls the performance of FOC activities. It is considered a kind of *LS* that holds a living organization. Metacognition is thought as a closed network of interactions of basic components (e.g., metacognitive knowledge and metacognitive activities).

The closed network establishes direct, feedback, and recursive relationships between the basic components. As a result, several kinds of metacognitive processes are fulfilled. In this manner, different sorts of metacognitive skills are realized. Both, processes and skills hold feedback relationships with each other, and also with the basic components. This dynamic schema constitutes the metacognition as a whole LS that is, a *unity*!

Metacognition is a kind of mechanistic system whose organization is specifiable in terms of relationships between metacognitive processes generated by the interactions of their components. As a consequence, the product of the metacognition operation as a system is *always metacognition*!

In this sense, metacognition holds an *AO* due to it being defined as a *unity* by a network of productions of basic components organized as a structure determined system. The metacognitive components recursively participate in the same network of productions of components that produced them (i.e., new metacognitive

Fig. 3.4 A graphical
evolution of the
metacognition according
to the CMMA

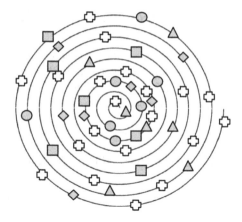

knowledge and metacognitive activities are produced) and realize the network of productions as a unity in the space in which the components exist.

Based on the stated arguments, metacognition is an *AS*. Therefore, it is a system closed in its states dynamics. In the sense that it is alive only while all its structural changes are structural changes that conserve their autopoiesis.

Metacognition, as a kind of cognitive activity, operates and evolves in a spiral way, as is pictured in Fig. 3.4. During a given cycle, different outcomes are produced (e.g., new or transformed metacognitive activities, knowledge, processes, and skills, and new or altered FOC activities and abilities) as well as feedback and recursive flows are triggered.

For instance, *monitoring* is one of the triggered skills (e.g., it is pictured through gray circles in Fig. 3.4). In addition, other functionalities are concurrently or sequentially fulfilled (e.g., they are sketched as the cross symbol in Fig. 3.4). One of them corresponds to the *regulation*, which is drawn as a gray square in Fig. 3.4. As a result of the activation of several functionalities, *awareness* (e.g., it is shaped as a diamond) is developed to get consciousness of the FOC activities being fulfilled to deploy the FOC ability of learning. In order to support the execution of those functionalities *knowledge*, illustrated by a triangle in Fig. 3.4, is retrieved. In reciprocity new knowledge is added, and current knowledge is updated and deleted.

3.5 Analysis of Metacognition Models

Once a series of classic, descriptive, and dynamic models of metacognition, as well as the propose CMMA, have been described, a comparative analysis of the traits they represent is outlined in this section. The aim is to provide another conceptual element for contributing to develop a holistic baseline of the metacognition field.

3.5.1 A Comparative Profile of Metacognition Models

In order to facilitate the comparison of the fifteen models stated in Sect. 3.3, a profile is tailored to characterize their main traits. Such a profile is presented in Tables 3.1, 3.2 and 3.3; where the left column identifies the attributes and the remaining unveil the characteristics of their respective model.

3.5.2 An Analysis of the Metacognition Models

Based on the description given for the fifteen models in Sect. 3.3 and in Tables 3.1, 3.2 and 3.3; an analysis of their nature is stated in this subsection. The purpose is to summarize the attributes used to model metacognition. In addition, a comparison of such pattern of traits versus the CMMA is also achieved. The aim is to continue the provision of conceptual items to shape a theoretical baseline for the metacognition arena.

With relation to the sample of classic metacognition models, psychology and neurology are the disciplines taken into account. The object to study is metacognition and some facets. Most of the models only identify and depict facets; whilst a few also explain how the components interact. The relationship between components is diverse without a prevalent type. Also the main components are heterogeneus; some are metacognitive facets, others are conceptual components or brain areas. The subcomponents are also diverse (e.g., variables, facets, and conceptual items).

Table 3.1 A comparative profile of classic metacognitive models

Trait	Flavel [152]	Brown [152]	Nelson and Narens [154]	Norman and Shallice [155]	Shimamura [156]
Baseline	Psychology	Psychology	Neurology	Neurology	Neurology
Target	Monitoring	Metacognition	Metacognition	Attention–perception	Metacognition
Nature	Descriptive	Descriptive	Descriptive–dynamic	Descriptive–dynamic	Descriptive
Relationship	Relational	Hierarchical	• Bottom-up	• Bottom-up	• Spatial
Components	• Knowledge	• Knowledge	• Meta-level	• Executive system	• Prefrontal cortex
	• Experiences	• Regulation	• Object-level	• Instance–level	• Posterior cortex
	• Goals-tasks				
	• Strategies				
Sub-components	Variables:	• Person	• Monitor	• Schemas	• Monitoring
	• Person	• Task	• Object model		• Control
	• Task		• Control		
	• Strategies				

Table 3.2 A comparative profile of descriptive metacognitive models

Traits	Kuhn [157]	Alexander and Schwanenflugel [158]	Tobias and Everson [159]	Schraw et al. [84]	Zohar [85]	Efklides [4]
Baseline	Psychology	Psychology	Psychology	Psychology	Psychology	Psychology
Target	Metacognition	Metacognition	Monitoring	Metacomprehension	Metastrategic knowledge	Metacognition
Nature	Descriptive	Descriptive	Descriptive–dynamic	Descriptive	Descriptive	Descriptive–dynamic
Relationship	Hierarchical	Hierarchical	Bottom-up	Hierarchical	Hierarchical	Hierarchical
Components	Metacognitive: • Knowing • Metastrategic • Epistemological	• Declarative knowledge • Cognitive monitoring • Regulation of strategies	• Monitoring • Evaluation • Selection • Make plans • Control	• Metacognition • Metamemory	Knowledge: • Persons • Tasks • Strategies	• Monitoring • Control
Sub-components				Knowledge: • Declarative • Procedural • Conditional Skills: • Information management • Debugging • Planning • Monitoring • Evaluation		Monitoring: • Knowledge • Experiences Control: • Skills

Table 3.3 A comparative profile of procedural metacognitive models

Traits	Veenman [160]	Zelazo [136, 162]	Flavell [136, 163]	Efklides [164]
Baseline	Psychology	Psychology	Psychology	Psychology
Target	Metacognition	Conscious–awareness	Awareness –uncertainty	Metacognition, affection, SRL
Nature	Dynamic	Evolutionary	Evolutionary	Dynamic
Relationship	• Bottom-up • Top-down	Sequential	Sequential	• Bottom-up • Top-down
Components	• Meta-level • Obeject-level	• Conscious • Awareness	• Awareness • Uncertainty	• Person • Task x person
Sub-components	• Task performance • Monitoring • Control • Program of self-instructions			Person: • Cognitive ability, self-concept • Metacognitive knowledge–skills • Perceptions of control, attitudes • Emotions, and motivation Task x person: • Metacognitive experiences • Online affective states
Sequence or stages	1. Task performance 2. Monitoring 3. Control 4. Program of self-instructions	1. At birth: minimal consciousness 2. 1st year: recursive consciousness 3. 2nd year: self-consciousness 4. Proceeding years: conscious awareness	1. At birth: non experience of uncertainty 2. Young children: unaware of subjective uncertain experiences 3. Children: lightly aware of uncertain experiences 4. Later: children are aware of uncertain experiences	1. Task activates person and task x person 2. Reciprocal relationships between person and task x person 3. Reciprocal relationships between person facets 4. Later: reciprocal relationships between task x person facets • Cognition/meta-cognition–affect • Metacognition–affect/SR of affect–effort

In resume, the classic models offer different viewpoints to depict metacognition and consider its components. Thus, they lack an evident pattern to characterize them.

Concerning the descriptive metacognition models, psychology is the only discipline; whilst besides the metacognition several facets are the object to model. Most of the models are only descriptive and a pair includes dynamic aspects. The typical relationship between components is hierarchical. Diverse kinds of metacognitive knowledge are prevalent, as well as monitoring and control. Also, as subcomponents, they appear as the most considered. As a summary, the sample of descriptive models offers an essential pattern of traits that suggest a common study target, as well as a viewpoint to characterize metacognition.

In another vein, procedural models of metacognition are characterized as founded on psychology, considering metacognition and awareness as their target. The models are expressed as dynamic and evolutionary; while bottom-up and sequential are the classic relationships between components. Awareness and abstract concepts are the typical components; whereas metacognitive facets are the typical subcomponents. Two views are considered to explain how metacognition is achieved: one corresponds to conceptual interaction of metacognitive components; whilst the other to the natural maturing of the individual whose awareness is increasing whilst the person grows up. As a synopsis, procedural models are psychological representations of metacognition, and particularly awareness. They focus on conceptual workflows or chronological maturity of the individual.

Finally, the CMMA is contrasted against the three types of models to highlight its contribution to the metacognition field. The CMMA offers a different perspective: It takes into account the biological viewpoint to characterize metacognition. It depicts a cyclical workflow, where components of a closed network interact through direct, feedback, and recursive relationships. Moreover, conceptual relationships are established between metacognitive components, processes, and strategies, as well as FOC activities. Thus, instead of specifying particular metacognitive facets, the CMMA includes all the possible manifestations of metacognition. The sequence of stages is characterized as a closed and permanent AS.

3.6 Conclusions

This work is an attempt to deal with the fuzziness of the metacognition concept and the lack of a theoretical and well sounded baseline for the metacognition field. It provides some reasons that reveal the complexity of the concept through a revision of research in the metacognition field and models oriented to describe metacognition. Furthermore, a conceptual model, named CMMA, to characterize the metacognitive activity has been introduced. Moreover, an analysis of the surveyed models is made and a comparison between them and the CMMA is fulfilled to highlight the contribution of the proposed model. Thus, this work concludes with a summary of its achievements, a series of observations about the metacognition field through a brief SWOT analysis, and the identification of future work to ground the CMMA.

An Account of the Achievements. The starting point tackled in the work concerns the fuzziness of the metacognition concept; particularly, its real meaning, nature, traits, performance, application, scope as well as its interrelationship with other constructs are just a sample of issues. Such a claim has been illustrated by the exposition of related works that highlight the background, appearance of the term, development, profile, and two polemic study targets of the metacognition.

As a second step towards an integral foundation, a survey of works oriented to shape three kinds of metacognition models has been outlined. The idea is to show the traits of the metacognition, its components, structure, and way of interaction. The sample includes well-known classic, descriptive, and procedural models in order to provide diverse perspectives of modeling and study.

The third step has contributed by means of introducing a new model, named CMMA, to tailor a holistic metacognition ground. It offers a conceptual shape of the metacognition activity from the neurological and biological perspectives. The proposed model conceives metacognition as an AS, whose components are able to produce new ones as well as interact with FOC activity. The metacognition activity is shaped as a closed network of components that interact to fulfill metacognitive processes and develop metacognitive skills oriented to monitor and regulate ordinary cognitive activities such as learning.

A fourth step to provide conceptual elements for the metacognition basis corresponds to an analysis of the traits that characterize the sample of surveyed works. In addition, a comparison of the models attributes against the ones of the CMMA has been made to reveal the contributions of the proposed model.

A SWOT Analysis of the Metacognition Arena. With the purpose of describing the status of the metacognition field, some strengths, weaknesses, opportunities, and threats are presented as part of the conclusions. As for the *strengths*, metacognition is based on well-sounded disciplines: psychology and neurology. Mature research in progress is being carried out. Scientists of diverse fields have been working on metacognition and contributing to extend the scope. New ways to improve knowledge acquisition in formal and long-life settings are being explored. Applications of the metacognition are varied, not only for learning and education purposes, but also for health sciences, business, marketing, and social networks. Results of empirical studies are frequently published to share findings. Stimulation and measurement of metacognition is supported by computer-based tools.

With relation to the *weakness*, the first two are the complexity of the mind and the challenge that represents to organize its components and performance in diverse tiers. Another is the difficulty faced by human beings to think about their thoughts and the mental overload produced when people are aware of their ordinary cognition prior, during, and after it happens. Other issues include: the lack of an interdisciplinary study to provide the theoretical and well-sounded basis of metacognition; the diversity of viewpoints applied to carry out research; the plurality of metacognitive jargons used for the practitioners to define their research; the methods, instruments, and criteria for developing research are quite heterogeneous, so it is difficult to transfer their application in other settings. Another weakness is the complex interactions between cognitive, metacognitive, motivational, and emotional processes that bias individual's behavior and cognitive achievements.

Concerning the *opportunities*, metacognition field has the chance to receive the formal acknowledgment of educational programs interested in improving the teaching-learning processes by including metacognitive practices in their curricula. Pedagogues, educators, and teachers interested in enhancing their labor are the potential users to be trained in metacognitive practices. The demand for upgrading the quality and practices of education represents an opportunity to include metacognition as a normal guide for all. Learners are the main target for study, stimulation, and assessment of their outcomes through metacognitive behavior that are useful in formal and informal settings. The application of experiences gained in diverse fields of study provides a valuable source of knowledge. The call for teaching metacognitive skills is considered one of the main implications for instruction.

Regarding the *threats* some of them are: the unsatisfied need for theoretical clarity and well-known accepted definitions and descriptions of the components of metacognition. Others consist in the widespread proliferation of terms, constructs, methods, and processes that are used in literature. The incompetence of current metacognition baseline, theories, frameworks, models, and methods to tackle issues such as level of granularity, collaboration, descriptiveness, social networking, ubiquitous learning, comprehensiveness, and dynamic processes. Although researchers have been engaged in a considerable amount of research, to date there is still no work that examines this body of research. Theoreticians in the field of metacognition endorse different theoretical perspectives. Some of these perspectives reflect the prevalent "fuzziness" in the field. Others may be internally clear, but they have not been well understood how the various perspectives relate to each other.

Future Work to Support the CMMA. The proposed model pursues to characterize and explain how the metacognitive activity is accomplished. However, more theoretical work is needed to enhance the neurological and biological basis, as well as the psychological constructs. In addition, a formal characterization is also required to provide a mathematical representation of the structure, performance, and outcomes.

Moreover, experimental trials are claimed to provide empirical evidence to support the concepts, statements, and suppositions stated by the CMMA. The consideration of other disciplines such as system engineering is needed to develop a systemic and holistic model of the metacognition. Finally, the implementation of a computer-based system to provide metacognitive stimuli based on the CMMA is required.

Acknowledgments The author gives testimony of the strength given by his Father, Brother Jesus and Helper, as part of the research projects of World Outreach Light to the Nations Ministries (WOLNM). Moreover, author recognizes the help provided by Mr. Lawrence Whitehill to edit the chapter. This work holds a partial support from grants: CONACYT-SNI-36453, IPN-COFAA-SIBE-DEBEC/647-391/2013, IPN/SIP/DI/DOPI/EDI/848/14, IPN-SIP-20141249, CONACYT 289763, and IPN-SIP-PIFI-20141249.

References

1. Klein, E.D., Evans, T.A., Schultz, N.B., Beran, M.J.: Learning how to "make a deal": human and monkey performance when repeatedly faced with the Monty Hall Dilemma. J. Comp. Psychol. **127**, 103–108 (2013)
2. Metcalfe, J., Shimamura, A.P.: Metacognition: Knowing About Knowing. MIT Press, Cambridge (1994)
3. Papaleontiou-Louca, E.: The concept and instruction of metacognition. Teach. Dev. **7**(1), 9–30 (2003)
4. Efklides, A.: Metacognition and affect: what can metacognitive experiences tell us about the learning process? Educ. Res. Rev. **1**, 3–14 (2006)
5. Veenman, M.V.J.: Metacognition in science education: definitions, constituents, and their intricate relation with cognition. In: Zohar, A., Dori, Y.J. (eds.) Metacognition in Science Education: Trends in Current Research, Contemporary Trends and Issues in Science Education, vol. 40, pp. 21–36. Springer, Heidelberg (2012)
6. Zohar, A., Dori, Y.J.: Introduction. In: Zohar, A., Dori, Y.J. (eds.) Metacognition in Science Education: Trends in Current Research, Contemporary Trends and Issues in Science Education, vol. 40, pp. 1–19. Springer, Heidelberg (2012)
7. Whitebread, D., Almeqdad, Q., Bryce, D., Demetriou, D., Grau, V., Sangster, C.: Metacognition in young children: current methodological and theoretical developments. In: Efklides, A., Misailidi, P. (eds.) Trends and Prospects in Metacognition Research, pp. 233–258. Springer Science+Business Media, New York (2010)
8. Efklides, A., Misailidi, P.: Introduction: The present and the future in metacognition. In: Efklides, A., Misailidi, P. (eds.) Trends and Prospects in Metacognition Research, pp. 1–18. Springer Science+Business Media, New York (2010)
9. Azevedo, R., Aleven, V.: Metacognition and learning technologies: an overview of current interdisciplinary research. In: Azevedo, R., Aleven, V. (eds.) International Handbook of Metacognition and Learning Technologies. Springer International Handbooks of Education, vol. 26, pp. 1–16. Springer Science+Business Media, New York (2013)
10. Beran, M.J., Brandl, J., Perner, J., Proust, J.: On the nature, evolution, development, and epistemology of metacognition: Introductory thoughts. In: Beran, M.J., Brandl, J., Perner, J., Proust, J.: Foundations of Metacognition, pp. 1–18. Oxford University Press, Oxford (2012)
11. James, W.: Principles of psychology, vol. 1. Holt, New York (1890)
12. Hart, J.T.: Memory and the feeling-of-knowing experience. J. Educ. Psychol. **56**, 208–216 (1965)
13. Flavell, J.H.: The Developmental Psychology of Jean Piaget. D. Van Nostrand, New York (1963)
14. Tulving, E., Madigan, S.A.: Memory and verbal learning. Annu. Rev. Psychol. **21**, 437–484 (1970)
15. Flavell, J.H.: First discussant's comments: what is memory development the development of? Hum. Dev. **14**, 272–278 (1971)
16. Fischhoff, B.: Hindsight = foresight: The effect of outcome knowledge on judgment under uncertainty. J Exp Psychol Hum Percept Perform. **1**, 288–299 (1975)
17. Brown, A.L.: Knowing when, where, and how to remember a problem of metacognition. Technical Report No. 47, University of Illiniois at Urbana-Champaign (1977)
18. Flavell, J.H., Wellman, H.M.: Metamemory. In: Kail Jr, R.V., Hagen, J.W. (eds.) Perspectives on the Development of Memory and Cognition, pp. 3–33. Erlbaum, Hillsdale (1977)
19. Lachman, J.L.: Metamemory through the adult life span. Dev. Psychol. **15**, 543–551 (1979)
20. Flavell, J.H.: Metacognitive aspects of problem solving. In: Resnick, L.B. (ed.) The Nature of Intelligence, pp. 231–236. Erlbaum, Hillsdale (1976)
21. Flavell, J.H.: Metacognition and cognitive monitoring: a new area of cognitive-developmental inquiry. Am. Psychol. **34**, 906–911 (1979)

22. Markman, E.M.: Comprehension monitoring. In: Dickson, W.P. (ed.) Children's Oral Communication Skills, pp. 61–84. Academic Press, New York (1981)
23. Kluwe, R.H.: Cognitive knowledge and executive control: metacognition. In: Griffin, D.R. (ed.) Animal Mind-Human Mind, pp. 201–224. Springer, New York (1982)
24. Brown, A.L., Bransford, J.D., Ferrara, R.A., Campione, J.C.: Learning, remembering, and understanding. In: Flavell, J.H., Markman, E.M. (eds) Handbook of Child Psychology, vol. 3, pp. 77–166. Cognitive Development, 4th edn. Wiley, New York (1983)
25. Forrest-Pressley, D.L., MacKinnon, G.E., Waller, T.G. (eds.): Metacognition, Cognition, and Human Performance, vol. 2. Academic Press, Orlando (1985)
26. Duell, O.K.: Metacognitive skills. In: Phye, G.D., Andre, T. (eds.): Cognitive Classroom Learning: Understanding, Thinking, and Problem Solving, pp. 205–242. Academic Press, Orlando (1986)
27. Garner, R.: Metacognition and Reading Comprehension. Ablex, Norwood (1987)
28. Pressley, M., Borkowski, J.J., Schneider, W.: Cognitive strategies: good strategy users coordinate metacognition and knowledge. In: Vasta, R., Whitehurst, G. (eds.) Annals of Child Development, vol. 5, pp. 89–129. JAI Press, Greenwich (1987)
29. Weinert, F.E., Kluwe, R.H. (eds.): Metacognition, Motivation, and Understanding. Erlbaum, Hillsdale (1987)
30. Brown, A.L., Palincsar, A.S.: Guided cooperative learning and individual knowledge acquisition. In: Resnick, L.B. (ed.) Knowing, Learning, and Instruction: Essays in Honor of Robert Glaser, pp. 393–451. Erlbaum, Hillsdale (1989)
31. Paris, S.G., Winograd, P.: How metacognition can promote academic learning and instruction. In: Jones, B.F., Idol, L. (eds.) Dimensions of Thinking and Cognitive Instruction, pp. 15–51. Erlbaum, Hillsdale (1990)
32. Borkowski, J.G., Carr, M., Rellinger, E., Pressley, M.: Self-regulated cognition: interdependence of metacognition, attributions, and self-esteem. In: Jones, B.F., Idol, L. (eds.) Dimensions of Thinking and Cognitive Instruction, pp. 53–92. Erlbaum, Hillsdale (1990)
33. Nelson, T.O., Narens, L.: Metamemory: a theoretical framework and new findings. In: Bower, G. (ed.) The Psychology of Learning and Motivation, vol. 26, pp. 125–141. Academic Press, New York (1990)
34. Nelson, T.O. (ed.): Metacognition: Core Readings. Allyn and Bacon, Boston (1992)
35. Schwartz, B.L.: Sources of information in metamemory: judgments of learning and feelings of knowing. Psychon. Bull. Rev. 1(3), 357–375 (1994)
36. Alexander, J.M., Carr, M., Schwanenflugel, P.J.: Development of metacognition in gifted children: directions for future research. Dev. Rev. 15, 1–37 (1995)
37. Jarman, R.F., Vavrik, J., Walton, P.D.: Metacognitive and frontal lobe processes: at the interface of cognitive psychology and neuropsychology. Genet. Soc. Gen. Psychol. Monogr. 121, 153–210 (1995)
38. Schraw, G., Moshman, D.: Metacognitive theories. Educ. Psychol. Rev. 7, 351–371 (1995)
39. Reder, L.M. (ed.): Implicit Memory and Metacognition. Erlbaum, Mahwah (1996)
40. Shimamura, A.P.: The role of the prefrontal cortex in controlling and monitoring memory processes. In: Reder, L.M. (ed.) Implicit Memory and Metacognition, pp. 259–274. Erlbaum, Mahwah (1996)
41. Chambres, P., Izaute, M., Marescaux, P.J. (eds.): Metacognition: Process, Function and Use. Kluwer Academic Publishers, Norwell (2002)
42. Levin, D.T. (ed.): Thinking and Seeing Visual Metacognition in Adults and Children. MIT Press, Cambridge (2004)
43. Terrace, H.S., Metcalfe, J. (eds.): The Missing Link in Cognition. Origins of Self-Reflective Consciousness. Oxford University Press, New York (2005)
44. Dunlosky, J., Bjork, R.A. (eds.): Handbook of Metamemory and Memory. Psychology Press, New York (2008)
45. Dunlosky, J., Metcalfe, J.: Metacognition. SAGE, California (2009)
46. Khine, M.S., Saleh, I.M. (eds.): New Science of Learning Cognition, Computers and Collaboration in Education. Springer Science+Business Media, New York (2010)

47. Efklides, A., Misailidi, P.: (eds.): Trends and Prospects in Metacognition Research. Springer Science+Business Media, New York (2010)
48. Cox, M.T., Raja, A., Horvitz, E. (eds.): Metareasoning. Thinking about Thinking. MIT Press, Cambridge (2011)
49. Zohar, A., Dori, Y.J. (eds.): Metacognition in Science Education: Trends in Current Research, Contemporary Trends and Issues in Science Education, vol. 40. Springer, Heidelberg (2012)
50. Beran, M.J., Brandl, J., Perner, J., Proust, J.: Foundations of Metacognition. Oxford University Press, Oxford (2012)
51. Azevedo, R., Aleven, V. (eds.): International Handbook of Metacognition and Learning Technologies. Springer International Handbooks of Education, vol. 26, Springer Science+Business Media, New York (2013)
52. Cognitive Development. An Elsevier journal. TR-JCR® IF (TR-JCR, IF.: Thompson Reuters, Journal Citations Report®, Impact Factor.). 1.446 http://www.journals.elsevier.com/cognitive-development
53. Journal of Cognitive Neuroscience. An IEEE journal. TR-JCR® IF. 4.493 http://ieeexplore.ieee.org/xpl/aboutJournal.jsp?punumber=6720218
54. Trends in Cognitive Sciences. An Elsevier journal. TR-JCR® IF. 16.008 http://www.journals.elsevier.com/trends-in-cognitive-sciences/
55. Journal of Cognition and Development. A Taylor & Francis Journal. TR-JCR® IF. 1.096 http://www.tandfonline.com/toc/hjcd20/current#.U86X3ZUU_IU
56. Metacognition and Learning. A Springer Journal. TR-JCR® IF. 1.516. http://link.springer.com/journal/11409
57. Journal of Applied Research in Memory and Cognition. An Elsevier Journal. SCImago Journal Rank (SJR) IF. 0.871 http://www.journals.elsevier.com/journal-of-applied-research-in-memory-and-cognition/
58. The 2nd International Conference on Metacognition. http://metacog2014.lapsco.fr/
59. International Association for Metacognition. http://www.personal.kent.edu/%7Ejdunlosk/metacog/
60. Kuhn, D., Dean Jr, D.: Metacognition: a ridge between cognitive psychology and educational practice. Theor. Pract. 43(4), 268–273 (2004)
61. Flavell, J.H., Miller, P.H., Miller, S.A.: Cognitive Development, 4th edn. Prentice Hall, Upper Saddle River (2002)
62. Hacker, D.J., Dunlosky, J., Graesser, A.C. (eds.): Handbook of Metacognition in Education (Educational Psychology). Routledge, New York (2009)
63. Fernandez-Duque, D., Baird, J.A., Posner, M.I.: Executive attention and metacognitive regulation. Conscious. Cogn. 9, 288–307 (2000)
64. de Jong, T., van Gog, T., Jenks, K., van Manlove, S., Hell, J., Jolles, J., van Merrienboer, J., van Leeuwen, T., Boschloo, A.: Explorations in Learning and the Brain: On the Potential of Cognitive Neuroscience for Educational Science. Springer Science+Business Media, New York (2014)
65. Shea, N., Boldt, A., Bang, F., Yeung, N., Reyes, C., Frith, C.D.: Supra-personal cognitive control and metacognition. Trends Cogn. Sci. 18(4), 186–193 (2014)
66. Crowder, J.A., Carbone, J.N., Friess, S.N.: Artificial consciousness. In: Crowder, J.A., Carbone, J.N., Friess, S.N. (eds.) Artificial Cognition Architectures, pp. 79–108. Springer Science+Business Media, New York (2014)
67. Eggen, P., Schellenberg, S.: Human memory and the new science of learning. In: Khine, M.S., Saleh, I.M.: New Science of Learning Cognition, Computers and Collaboration in Education, pp. 79–107. Springer Science+Business Media, New York (2013)
68. Bacon, E.: Further insight into cognitive and metacognitive processes of the tip-of-the-tongue state with an amnesic drug as cognitive tool. In: Efklides, A., Misailidi, P. (eds.) Trends and Prospects in Metacognition Research, pp. 81–104. Springer Science+Business Media, New York (2010)

69. Efklides, A., Touroutoglou, A.: Prospective memory failure and the metacognitive experience of blank in the mind. In: Efklides, A., Misailidi, P. (eds.) Trends and Prospects in Metacognition Research, pp. 105–126. Springer Science+Business Media, New York (2010)

70. Izaute, M., Elisabeth Bacon, E.: Metamemory in schizophrenia: monitoring or control deficit? In: Efklides, A., Misailidi, P. (eds.) Trends and Prospects in Metacognition Research, pp. 127–147. Springer Science+Business Media, New York (2010)

71. Briñol, P., DeMarree, G. (eds.): Social Metacognition (Frontiers of Social Psychology). Psychology Press, New York (2011)

72. Frith, C.D.: The role of metacognition in human social interactions. Philos. Trans. R. Soc. Lond. B Biol. Sci. **367**(1599), 2213–2223 (2012)

73. Dinsmore, D.L., Alexander, P.A., Loughlin, S.M.: Focusing the conceptual lens on metacognition, self-regulation, and self-regulated learning. Educ. Psychol. Rev. **20**, 391–409 (2008)

74. Tatar, N., Akpınar, E., Feyzioglu, E.Y.: The effect of computer-assisted learning integrated with metacognitive prompts on students' affective skills. J. Sci. Educ. Technol. **22**, 764–779 (2013)

75. Carneiro, R., Lefrere, P., Steffens, K., Underwood, J.: Self-regulated Learning in Technology Enhanced Learning Environments A European Perspective. Sense Publishers, Rotterdam (2011)

76. Schwartz, N.H., Scott, B.M., Holzberger, D.: Metacognition: A closed-loop model of biased competition–evidence from neuroscience, cognition, and instructional research. In: Azevedo, R., Aleven, V. (eds.) International Handbook of Metacognition and Learning Technologies. Springer International Handbooks of Education, vol. 26, pp. 79–94. Springer Science+Business Media, New York (2013)

77. Ford, C.L., Yore, L.D.: Toward convergence of critical thinking, metacognition, and reflection: Illustrations from natural and social sciences, teacher education, and classroom practice. In: Zohar, A., Dori, Y.J. (eds.) Metacognition in Science Education: Trends in Current Research, Contemporary Trends and Issues in Science Education, vol. 40, pp. 251–271. Springer, Heidelberg (2012)

78. Flavell, J.H.: Theory-of-mind development: retrospect and prospect. Merrill-Palmer Q. **50**(3), 274–290 (2004)

79. Misailidi, P.: Children's metacognition and theory of mind: bridging the gap. In: Efklides, A., Misailidi, P. (eds.) Trends and Prospects in Metacognition Research, pp. 279–291. Springer Science+Business Media, New York (2010)

80. Egré, P., Bonnay, D.: Metacognitive perspectives on unawareness and uncertainty. In: Beran, M.J., Brandl, J., Perner, J., Proust, J. (eds.) Foundations of Metacognition, pp. 322–342. Oxford University Press, Oxford (2012)

81. Scott, B.M., Schwartz, N.H.: Navigational spatial displays: the role of metacognition as cognitive load. Learn. Instr. **1**(17), 89–105 (2007)

82. Maier, J., Richter, T.: Fostering multiple text comprehension: how metacognitive strategies and motivation moderate the text-belief consistency effect. Metacognition Learn. **9**(1), 51–74 (2014)

83. Buratti, S., Allwood, C.M., Kleitman, S.: First- and second-order metacognitive judgments of semantic memory reports: the influence of personality traits and cognitive styles. Metacognition Learn. **8**, 79–102 (2013)

84. Schraw, G., Olafson, L., Weibel, M., Sewing, D.: Metacognitive knowledge and field-based science learning in an outdoor environmental education program. In: Zohar, A., Dori, Y.J. (eds.) Metacognition in Science Education: Trends in Current Research, Contemporary Trends and Issues in Science Education, vol. 40, pp. 57–77. Springer, Heidelberg (2012)

85. Zohar, A.: Explicit teaching of metastrategic knowledge: definitions, students' learning, and teachers' professional development. In: Zohar, A., Dori, Y.J. (eds.) Metacognition in Science Education: Trends in Current Research, Contemporary Trends and Issues in Science Education, vol. 40, pp. 196–223. Springer, Heidelberg (2012)

86. D'Mello, S.K., Strain, A.C., Olney, A., Graesser, A.C.: Affect, meta-affect, and affect regulation during complex learning. In: Azevedo, R., Aleven, V. (eds.) International Handbook of Metacognition and Learning Technologies. Springer International Handbooks of Education, vol. 26, pp. 668–682. Springer Science+Business Media, New York (2013)

87. Burleson, W.: Affective learning companions and the adoption of metacognitive strategies. In: Azevedo, R., Aleven, V. (eds.) International Handbook of Metacognition and Learning Technologies. Springer International Handbooks of Education, vol. 26, pp. 645–657. Springer Science+Business Media, New York (2013)

88. Schwartz, B.L., Efklides, A.: Metamemory and memory efficiency: implications for student learning. J. Appl. Res. Mem. Cogn. 1, 145–151 (2012)

89. Vierkant, T.: What metarepresentation is for. In: Beran, M.J., Brandl, J., Perner, J., Proust, J. (eds.) Foundations of Metacognition, pp. 279–280. Oxford University Press, Oxford (2012)

90. Verpoorten, D., Westera, W., Specht, M.: Using reflection triggers while learning in an online course. Br. J. Educ. Technol. 43(6), 1030–1040 (2012)

91. Lopera, A.B., Hallahana, D.P.: Meta-attention: the development of awareness of the attentional process. J. Gen. Psychol. 106(1), 27–33 (1982)

92. Kim, B., Park, H., Baek, Y.: Not just fun, but serious strategies: using meta-cognitive strategies in game-based learning. Comput. Educ. 52, 800–810 (2009)

93. van Gog, T., Jarodzka, H.: Eye tracking as a tool to study and enhance cognitive and metacognitive processes in computer-based learning environments. In: Azevedo, R., Aleven, V. (eds.) International Handbook of Metacognition and Learning Technologies. Springer International Handbooks of Education, vol. 26, pp. 143–156. Springer Science+Business Media, New York (2013)

94. Allwood, C.M.: The realism in children's metacognitive judgments of their episodic memory performance. In: Efklides, A., Misailidi, P. (eds.) Trends and Prospects in Metacognition Research, pp. 148–169. Springer Science+Business Media, New York (2010)

95. Bannert, M., Mengelkamp, C.: Scaffolding hypermedia learning through metacognitive prompts. In: Azevedo, R., Aleven, V. (eds.) International Handbook of Metacognition and Learning Technologies. Springer International Handbooks of Education, vol. 26, pp. 171–186. Springer Science+Business Media, New York (2013)

96. Dabbagh, N., Kitsantas, A.: Using learning management systems as metacognitive tools to support self-regulation in higher education contexts. In: Azevedo, R., Aleven, V. (eds.) International Handbook of Metacognition and Learning Technologies. Springer International Handbooks of Education, vol. 26, pp. 197–211. Springer Science+Business Media, New York (2013)

97. Narciss, S., Koerndle, H., Proske, A.: Challenges of investigating metacognitive tool use and effects in (rich) web-based learning environments. In: Azevedo, R., Aleven, V. (eds.) International Handbook of Metacognition and Learning Technologies. Springer International Handbooks of Education, vol. 26, pp. 243–260. Springer Science+Business Media, New York (2013)

98. van Joolingen, W., de Jong, T.: Model-based diagnosis for regulative support in inquiry learning. In: Azevedo, R., Aleven, V. (eds.) International Handbook of Metacognition and Learning Technologies. Springer International Handbooks of Education, vol. 26, pp. 589–600. Springer Science+Business Media, New York (2013)

99. Carr, A., Luckin, R. Yuill, N., Avramides, K.: How mastery and performance goals influence learners' metacognitive help-seeking behaviours when using Ecolab II. In: Azevedo, R., Aleven, V. (eds.) International Handbook of Metacognition and Learning Technologies. Springer International Handbooks of Education, vol. 26, pp. 658–668. Springer Science+Business Media, New York (2013)

100. Norman, E., Price, M.C., Duff, S.C.: Fringe consciousness: a useful framework for clarifying the nature of experience-metacognitive feelings. In: Efklides, A., Misailidi, P. (eds.) Trends and Prospects in Metacognition Research, pp. 63–80. Springer Science+Business Media, New York (2010)

101. Litman, D., Forbes-Riley, K.: Towards improving (meta) cognition by adapting to student uncertainty in tutorial dialogue. In: Azevedo, R., Aleven, V. (eds.) International Handbook of Metacognition and Learning Technologies. Springer International Handbooks of Education, vol. 26, pp. 385–396. Springer Science+Business Media, New York (2013)
102. Roll, I., Holmes, N.G., Day, J., Bonn, D.: Evaluating metacognitive scaffolding in guided invention activities. Instr. Sci. **40**, 691–710 (2012)
103. Roll, I., Aleven, V., McLaren, B.M., Koedinger, K.R.: Improving students' help-seeking skills using metacognitive feedback in an intelligent tutoring system. Learn. Instr. **21**, 267–280 (2011)
104. Thillmann, H., Gößling, J., Marschner, J., Wirth, J., Leutner, D.: Metacognitive knowledge about and metacognitive regulation of strategy use in self-regulated scientific discovery learning: new methods of assessment in computer- based learning environments. In: Azevedo, R., Aleven, V. (eds.) International Handbook of Metacognition and Learning Technologies. Springer International Handbooks of Education, vol. 26, pp. 575–588. Springer Science+Business Media, New York (2013)
105. Sperling, R.A., Howard, B.C., Miller, L.A.: Measures of children's knowledge and regulation of cognition. Contemp. Educ. Psychol. **27**(1), 51–79 (2002)
106. Clarebout, G., Elen, J., Collazo, N.A., Lust, G., Jiang, L.: Metacognition and the use of tools. In: Azevedo, R., Aleven, V. (eds.) International Handbook of Metacognition and Learning Technologies. Springer International Handbooks of Education, vol. 26, pp. 187–195. Springer Science+Business Media, New York (2013)
107. Veenman, M.V.J.: Assessing metacognitive skills in computerized learning environments. In: Azevedo, R., Aleven, V. (eds.) International Handbook of Metacognition and Learning Technologies. Springer International Handbooks of Education, vol. 26, pp. 157–168. Springer Science+Business Media, New York (2013)
108. Carneiro, R., Steffens, K.: Research on self-regulated learning in technology enhanced learning environments: two examples from Europe. In: Azevedo, R., Aleven, V. (eds.) International Handbook of Metacognition and Learning Technologies. Springer International Handbooks of Education, vol. 26, pp. 601–614. Springer Science+Business Media, New York (2013)
109. Finley, J.R., Tullis, J.G., Benjamin, A.S.: Metacognitive control of learning and remembering. In: Khine, M.S., Saleh, I.M. (eds.) New Science of Learning Cognition, Computers and Collaboration in Education, pp. 108–131. Springer Science+Business Media, New York (2013)
110. Touroutoglou, A., Efklides, A.: Cognitive interruption as an object of metacognitive monitoring: Feeling of difficulty and surprise. In: Efklides, A., Misailidi, P. (eds.) Trends and Prospects in Metacognition Research, pp. 171–208. Springer Science+Business Media, New York (2010)
111. Kinnunen, R., Vauras, M.: Tracking on-line metacognition: monitoring and regulating comprehension in reading. In: Efklides, A., Misailidi, P. (eds.) Trends and Prospects in Metacognition Research, pp. 209–229. Springer Science+Business Media, New York (2010)
112. van den Boom, G., Paas, F., van Merriënboer, J.J.G., van Gog, T.: Reflection prompts and tutor feedback in a web-based learning environment: effects on students self-regulated learning competence. Comput. Hum. Behav. **20**, 551–567 (2004)
113. Koriat, A.: The feeling of knowing: some metatheoretical implications for consciousness and control. Conscious. Cogn. **9**, 149–171 (2000)
114. Brady, M., Seli, H., Rosenthal, J.: "Clickers" and metacognition: a quasi-experimental comparative study about metacognitive self-regulation and use of electronic feedback devices. Comput. Educ. **65**, 56–63 (2013)
115. Pellas, N.: The influence of computer self-efficacy, metacognitive self-regulation and self-esteem on student engagement in online learning programs: evidence from the virtual world of second life. Comput. Hum. Behav. **35**, 157–170 (2014)

116. Sun, Chih-Yuan: J., Rueda, R.: Situational interest, computer self-efficacy and self-regulation: their impact on student engagement in distance education. Br. J. Educ. Technol. **43**(2), 191–204 (2012)

117. Chiu, J.L., Linn, M.C.: The role of self-monitoring in learning chemistry with dynamic visualizations. In: Zohar, A., Dori, Y.J. (eds.) Metacognition in Science Education: Trends in Current Research, Contemporary Trends and Issues in Science Education, vol. 40, pp. 133–163. Springer, Heidelberg (2012)

118. Kleitman, S., Moscrop, T.: Self-confidence and academic achievements in primary-school children: their relationships and links to parental bonds, intelligence, age, and gender. In: Efklides, A., Misailidi, P. (eds.) Trends and Prospects in Metacognition Research, pp. 293–326. Springer Science+Business Media, New York (2010)

119. Renkl, A., Berthold, K., Grosse, C.S., Schwonke, R.: Making better use of multiple representations: how fostering metacognition can help. In: Azevedo, R., Aleven, V. (eds.) International Handbook of Metacognition and Learning Technologies. Springer International Handbooks of Education, vol. 26, pp. 397–408. Springer Science+Business Media, New York (2013)

120. Dokic, J.: Seeds of self-knowledge: noetic feelings and metacognition. In: Beran, M.J., Brandl, J., Perner, J., Proust, J. (eds.) Foundations of Metacognition, pp. 302–321. Oxford University Press, Oxford (2012)

121. Roebers, C.M., Cimeli, P., Röthlisberger, M., Neuenschwander, R.: Executive functioning, metacognition, and self-perceived competence in elementary school children: an explorative study on their interrelations and their role for school achievement. Metacognition and Learning. **7**, 151–173 (2012)

122. Koriat, A.: The relationships between monitoring, regulation and performance. Learn. Instr. **22**, 296–298 (2012)

123. Rivers, W.P.: Autonomy at all costs: an ethnography of metacognitive self-assessment and self-management among experienced language learners. Mod. Lang. J. **85**(2), 279–290 (2001)

124. Rookhached, C., Suksringarm, P., Singseewo, A.: The development model of 7 steps learning cycle using multiple intelligences and metacognitive techniques in the subject of life and environment, scientific education group, secondary school, year 2. Soc. Sci. **5**(6), 477–480 (2010)

125. Vollmeyer, R., Rheinberg, F.: The role of motivation in knowledge acquisition. In: Azevedo, R., Aleven, V. (eds.) International Handbook of Metacognition and Learning Technologies. Springer International Handbooks of Education, vol. 26, pp. 697–707. Springer Science+Business Media, New York (2013)

126. Norris, S.P., Phillips, L.M.: Reading science: how a naive view of reading hinders so much else. In: Zohar, A., Dori, Y.J. (eds.) Metacognition in Science Education: Trends in Current Research, Contemporary Trends and Issues in Science Education, vol. 40, pp. 37–56. Springer, Heidelberg (2012)

127. Grotzer, T., Mittlefehldt, S.: The role of metacognition in students' understanding and transfer of explanatory structures in science. In: Zohar, A., Dori, Y.J. (eds.) Metacognition in Science Education: Trends in Current Research, Contemporary Trends and Issues in Science Education, vol. 40, pp. 79–99. Springer, Heidelberg (2012)

128. Herscovitz, O, Kaberman, Z., Saar, L., Dori, Y.J.: The relationship between metacognition and the ability to pose questions in chemical education. In: Zohar, A., Dori, Y.J. (eds.), Metacognition in Science Education: Trends in Current Research, Contemporary Trends and Issues in Science Education, vol. 40, pp. 165–195. Springer, Heidelberg (2012)

129. Dettori, G., Lupi, V.: Self-observation and shared reflection to improve pronunciation in L2. In: Azevedo, R., Aleven, V. (eds.) International Handbook of Metacognition and Learning Technologies. Springer International Handbooks of Education, vol. 26, pp. 615–625. Springer Science+Business Media, New York (2013)

130. Vanderswalmen, R., Vrijders, J., Desoete, A.: Metacognition and spelling performance in college students. In: Efklides, A., Misailidi, P. (eds.) Trends and Prospects in Metacognition Research, pp. 367–394. Springer Science+Business Media, New York (2010)

131. Colombo, B., Iannello, P., Antonietti, A.: Metacognitive knowledge of decision-making: an explorative study. In: Efklides, A., Misailidi, P. (eds.) Trends and Prospects in Metacognition Research, pp. 445–472. Springer Science+Business Media, New York (2010)
132. Hollingworth, R.W., McLoughlin, C.: Developing science students' metacognitive problem solving skills online. Aust. J. Educ. Technol. **17**(1), 50–63 (2001)
133. Roll, I., Aleven, V., McLaren, B.M., Koedinger, K.R.: Designing for metacognition—applying cognitive tutor principles to the tutoring of help seeking. Metacognition Learn. **2**(2–3), 125–140 (2007)
134. Lajoie, S.P., Jingyan, L.: Supporting collaboration with technology: does shared cognition lead to co-regulation in medicine? Metacognition Learn. **7**, 45–62 (2012)
135. Franks, B.A., Therriault, D.J., Buhr, M.I., Chiang, E.S., Gonzalez, C.M., Kwon, H.K., Schelble, J.L., Wang, X.: Looking back: reasoning and metacognition with narrative texts. Metacognition Learn. **8**, 145–171 (2013)
136. Lyons, K.E., Ghetti, S.: Metacognitive development in early childhood: new questions about old assumptions. In: Efklides, A., Misailidi, P. (eds.) Trends and Prospects in Metacognition Research, pp. 259–278. Springer Science+Business Media, New York (2010)
137. Kolić-Vehovec, S., Bajšanski, I., Zubković, B.R.: Metacognition and reading comprehension: age and gender differences. In: Efklides, A., Misailidi, P. (eds.) Trends and Prospects in Metacognition Research, pp. 327–344. Springer Science+Business Media, New York (2010)
138. Csíkos, C., Steklács, J.: Metacognition-based reading intervention programs among fourth-grade Hungarian students. In: Efklides, A., Misailidi, P. (eds.) Trends and Prospects in Metacognition Research, pp. 345–366. Springer Science+Business Media, New York (2010)
139. Sodian, B., Thoermer, C., Kristen, S., Perst, H.: Metacognition in infants and young children. In: Beran, M.J., Brandl, J., Perner, J., Proust, J. (eds.) Foundations of Metacognition, pp. 119–133. Oxford University Press, Oxford (2012)
140. Esken, F.: Early forms of metacognition in human children. In: Beran, M.J., Brandl, J., Perner, J., Proust, J. (eds.) Foundations of Metacognition, pp. 134–145. Oxford University Press, Oxford (2012)
141. Kloo, D., Rohwer, M.: The development of earlier and later forms of metacognitive abilities: Reflections on agency and ignorance. In: Beran, M.J., Brandl, J., Perner, J., Proust, J. (eds.) Foundations of Metacognition, pp. 167–180. Oxford University Press, Oxford (2012)
142. Bryce, D., Whitebread, D.: The development of metacognitive skills: evidence from observational analysis of young children's behavior during problem-solving. Metacognition Learn. **7**, 197–217 (2012)
143. Krebs, S.S., Roebers, C.M.: The impact of retrieval processes, age, general achievement level, and test scoring scheme for children's metacognitive monitoring and controlling. Metacognition Learn. **7**, 75–90 (2012)
144. Barfurth, M.A., Ritchie, K.C., Irving, J.A., Shore, B.M.: A metacognitive portrait of gifted learners. In: Shavinina, L.V. (ed.) International Handbook on Giftedness, pp. 397–417. Springer Science+Business Media, New York (2009)
145. Morgan, C.L.: An introduction to comparative psychology. Walter Scott, London (1906)
146. Beran, M.J., Couchman, J.J., Coutinho, M.V.C., Boomer, J., Smith, J.D.: Metacognition in nonhumans: methodological and theoretical issues in uncertainty monitoring. In: Efklides, A., Misailidi, P. (eds.) Trends and Prospects in Metacognition Research, pp. 21–35. Springer Science+Business Media, New York (2010)
147. Couchman, J.J., Beran, M.J., Couthino, M.V.C., Boomer, J., Smith, J.D.: Evidence for animal metaminds. In: Beran, M.J., Brandl, J., Perner, J., Proust, J. (eds.) Foundations of Metacognition, pp. 21–35. Oxford University Press, Oxford (2012)
148. Crystal, J.D.: Validating animal models of metacognition. In: Beran, M.J., Brandl, J., Perner, J., Proust, J. (eds.) Foundations of Metacognition, pp. 36–49. Oxford University Press, Oxford (2012)
149. Fujita, K., Nakamura, N., Iwasaki, S., Watanabe, S.: Are birds metacognitive? In: Beran, M.J., Brandl, J., Perner, J., Proust, J. (eds.) Foundations of Metacognition, pp. 50–61. Oxford University Press, Oxford (2012)

150. Call, J.: Seeking information in non-human animals. In: Beran, M.J., Brandl, J., Perner, J., Proust, J. (eds.) Foundations of Metacognition, pp. 62–75. Oxford University Press, Oxford (2012)

151. Carruthers, P., Ritchie, B.: The emergence of metacognition: affect and uncertainty in animals. In: Beran, M.J., Brandl, J., Perner, J., Proust, J. (eds.) Foundations of Metacognition, pp. 76–93. Oxford University Press, Oxford (2012)

152. Flavell, J.H.: Speculation about the nature and development of metacognition. In: Winert, F., Kluwe, R. (eds.) Metacognition, Motivation, and Understanding, pp. 21–29. Lawrence Erlbaum, Hillsdale (1987)

153. Brown, A.: Metacognition, executive control, self-regulation and other more mysterious mechanisms. In: Weinert, F.E., Kluwe, R.H. (eds.) Metacognition, Motivation, and Understanding, pp. 65–116. Lawrence Erlbaum, Hillsdale (1987)

154. Nelson, T.O., Narens, L.: Why investigate metacognition? In: Metcalfe, J., Shimamura, A.P. (eds.) Metacognition, pp. 1–25. MIT Press, Massachussetts (1994)

155. Norman, D.A., Shallice, T.: Attention to action willed and automatic control of behavior. In: Davidson, R.J., Schwartz, G.E., Shapiro, D. (eds.) Consciousness and Self Regulation: Advances in Research and Theory, vol. 4, pp. 1–18. Plenum, New York (1986)

156. Shimamura, A.P.: A neurocognitive approach to metacognitive monitoring and control. In: Dunlosky, J., Bjork, R. (eds.) Handbook of Memory and Metamemory: Essays in Honor of Thomas O. Nelson, pp. 373–390. Psychology Press, New York (2008)

157. Kuhn, D.: Metacognitive development. Curr. Dir. Psychol. Sci. **9**, 178–181 (2000)

158. Alexander, J.M., Schwanenflugel, P.J.: Development of metacognitive concepts about thinking in gifted and non-gifted children: Recent research. Learn. Individ. Differ. **4**, 305–326 (1996)

159. Tobias, S., Everson, H.T.: Knowing what you know and what you don't: further research on metacognitive knowledge monitoring. Technical report. College Board Research Report 2002-3, College Entrance Examination Board, New York (2002)

160. Veenman, M.V.J.: Learning to self-monitor and self-regulate. In: Mayer, R., Alexander, P. (eds.) Handbook of Research on Learning and Instruction, pp. 197–218. Routledge, New York (2011)

161. Anderson, J.R., Schunn, C.D.: Implications of the ACT-R learning theory: no magic bullets. In: Glaser, R. (ed.) Advances in Instructional Psychology, vol. 5, pp. 1–33. Erlbaum, Mahwah (2000)

162. Zelazo, P.D.: The development of conscious control in childhood. Trends Cogn. Sci. **8**, 12–17 (2004)

163. Flavell, J.H.: Varieties of uncertainty monitoring. Brain Behav. Sci. **26**, 344 (2003)

164. Efklides, A.: Interactions of metacognition with motivation and affect in self-regulated learning: the MASRL model. Educ. Psychol. **46**(1), 6–25 (2011)

165. Maturana, H.: Metadesign. Instituto de Terapia Cognitiva INTECO, Chile. http://www.inteco.cl/articulos/006/texto_ing.htm

166. Varela, F., Maturana, H., Uribe, R.: Autopoiesis: the organization of living systems, its characterization and a model. BioSystems **5**, 187–196 (1974)

167. Barghgraeve, P.: Mechanistic explanations and structure determined systems Maturana and the human sciences. In: Van de Vijver, G. (ed.) New Perspectives on Cybernetics: Self-Organization, Autonomy and Connectionism, vol. 220, pp. 207–217. Kluwer Academic Publishers, Dordrecht (1992)

168. Maturana, H., Varela, F.G.: De máquinas y seres vivos. Editorial Universitaria, Santiago de Chile (1973)

169. Maturana, H., Varela, F.G.: El Árbol del Conocimiento. Editorial Lumen, Buenos Aires (1984)

170. Bishop, C.M.: Neural Networks for Pattern Recognition. Oxford University Press, New York (1995)

171. Lai, E.R.: Metacognition: a literature review research report. Technical Report. Pearson (2011)

Part II
Framework

Chapter 4
Metacognitive Skill Development and Applied Systems Science: A Framework of Metacognitive Skills, Self-regulatory Functions and Real-World Applications

Michael J. Hogan, Christopher P. Dwyer, Owen M. Harney, Chris Noone and Ronan J. Conway

Abstract Metacognition, or thinking about thinking [1], refers to the application and regulation of cognitive processes. According to Boekaerts and Simons [2], Brown [3] and Ku and Ho [4], individuals think metacognitively in two ways: first, individuals must be aware of their own cognitive processes (e.g., through self-monitoring or self-regulation); second, individuals must be able to apply available cognitive processes for purposes of learning or devising solutions to problems [e.g., using critical thinking or reflective judgment [5]. Though a topic of research interest for almost 40 years, debate continues as to how best to conceptualise metacognition and cultivate metacognitive ability. However, due to what can be considered an exponential increase in the creation of new information every year [6, 7], higher-order, metacognitive skills are needed more than ever in order to aid individuals and groups in becoming more adaptable, flexible and better able to cope in the context of a rapidly evolving information society. In order to help guide the facilitation of metacognitive understanding in educational and applied settings, this chapter draws upon recent research in the learning sciences to propose a new framework of metacognition. Specifically, we outline a model of metacognition that integrates processes associated with self-other representations,

M.J. Hogan (✉) · C.P. Dwyer · O.M. Harney · C. Noone · R.J. Conway
School of Psychology, National University of Ireland, Galway County, Galway,
Republic of Ireland
e-mail: michael.hogan@nuigalway.ie

C.P. Dwyer
e-mail: cdwyer@nuigalway.ie

O.M. Harney
e-mail: o.harney2@nuigalway.ie

C. Noone
e-mail: noonec@gmail.com

R.J. Conway
e-mail: rojconway@gmail.com

© Springer International Publishing Switzerland 2015
A. Peña-Ayala (ed.), *Metacognition: Fundaments, Applications, and Trends*,
Intelligent Systems Reference Library 76, DOI 10.1007/978-3-319-11062-2_4

executive function, emotion regulation, mindfulness, motivation, thinking dispositions, critical thinking, reflective judgment and collaborative systems thinking skills, which can be fostered in the context of individual and team-based tool use. Two cases studies are presented, which provide evidence for the value of both argument mapping and collective intelligence tools in facilitating the development of higher-order critical, reflective and collaborative metacognitive thinking skills. The chapter concludes with a number of recommendations for future research on applied systems science and metacognitive skills development.

Keywords Critical thinking · Reflective judgment · Mindfulness · Executive functioning · Self-regulation · Argument mapping · Collective intelligence

Abbreviations

AM   Argument mapping
CACS   Conversational argument coding scheme
CSCL   Computer-supported collaborative learning
CT   Critical thinking
IM   Interactive management
RJ   Reflective judgment
RJM   Reflective judgment model

4.1 Introduction

One of the defining features of human evolution is the emergent capacity of human beings to think about thinking. Cultural evolution is itself a metacognitive process, as each new generation thinks about the thinking of previous generations—the contents of thinking, the process of thinking, the products of thinking—and modifies the culture of thinking in multifarious ways. However, the ontogenesis of metacognitive skills is a gradual and uncertain process that is contingent upon the quality of education and training children and adults receive. Like other artefacts of culture, systems of education and educational practices are products of our cultural evolution that are open to metacognitive reflection. Nevertheless, for those of us who have received a science education, or who are immersed in acts of inquiry as part of our work, we may take for granted the curious nature of metacognitive acts and not seek to modify educational practices. We may assume that metacognitive action is a cultural given—a commonplace and everyday activity.

Congruent with Jean Piaget's theory of cognitive development, we may assume that the development of critical, reflective and collaborative metacognitive thinking skills in adolescence and adulthood is an inevitable product of an orderly, cumulative, directional sequence of change—a movement from reflexes, simple representations and intuitive logic toward increasingly logical and adaptive representations of reality. Not only is this view fundamentally false, culturally bounded, and idealistic, it distracts us, with its wonderful simplicity, from the complexity of metacognitive skill development and the educational challenges associated with cultivating these skills. The continuous development and cultural evolution of metacognitive skills is always a hard earned path—the skill and determination of each generation is needed to foster the skill and determination of the next generation. Understanding and influencing metacognitive skill development is essential for further cultural evolution, and ultimately, our survival, adaptation and flourishing as a species.

Due to what can be considered an exponential increase in the creation of new information every year [6, 7], higher-order, metacognitive skills are needed more than ever in order to aid individuals and groups in becoming more adaptable, flexible and better able to cope in the context of a rapidly evolving information society. Thus, it is widely recognized that the challenge of education implies more than a focus on developing domain-specific knowledge—also needed is metacognitive knowledge and skill, and approaches to enquiry that allow children and adults to think about their thinking and approach problems, solutions, experiments, explanations, and simulated actions with a mindful, reflective, collaborative sensibility that facilitates adaptive action at individual and group levels.

In this chapter, we outline a model of metacognition that integrates processes associated with self-other representations, executive function, emotion regulation, mindfulness, motivation, thinking dispositions, critical thinking, reflective judgment and collaborative systems thinking skills. Building upon research highlighting the importance of self-regulation skills as basic building blocks of higher-order thinking skills, we also highlight the value of mindfulness as an intra-personal and inter-personal asset that facilitates enhanced executive functioning and emotion regulation skill in both individual and group learning contexts.

Our framework proposes an integration of organismic and pragmatic models of skill development. Consistent with organismic models [8], we assume that skills can unfold at greater levels of integration and complexity; however, skilled action is a function of training, support and the collaborative dynamics of social groups. Consistent with pragmatism and functional contextualism, we also propose a pragmatic systems science framework [9] that highlights the value of specific tools and processes to support the training and collaborative dynamics of metacognitive skill development [10]. Ultimately, we argue that metacognitive skills can be seen as part of a collection of potentially synergistic talents that can be fostered in the context of individual and team-based tool use. We present two cases studies that provide evidence for the value of both argument mapping tools and collective intelligence tools in facilitating the development of higher-order critical, reflective, and collaborative metacognitive thinking skills. We close the chapter with a number of recommendations for future research on applied systems science and metacognitive skills development.

4.2 Metacognition: Foundational Definitions and Perspectives

Though the term metacognition is relatively novel, first described by Flavell [1] as "knowledge concerning one's own cognitive processes and products or anything related to them; and the active monitoring, consequent regulation and orchestration of these processes", the construct has been the focus of attention in recent years. Though a large body of research and theory on metacognition exists [e.g., 11, 12, 13, 14, 15, 16, 17, 18], the majority of scholars envisage a two-component framework. For example, according to Boekaerts and Simons [2], Brown [3] and Ku and Ho [4], individuals think metacognitively in two ways: first, individuals must be aware of their own cognitive processes (e.g., through self-monitoring or self-regulation); second, individuals must be able to apply available cognitive processes appropriately for purposes of learning or devising solutions to problems (e.g., using critical thinking or reflective judgment [5, 19, 20]).

According to Schraw and Dennison [18], the two ways in which individuals think metacognitively can be described as the regulation of cognition and knowledge about cognition, respectively. Though Schraw and Dennison [18] suggested the existence of eight sub-components of these two types of metacognition (e.g., planning, information management strategies, comprehension monitoring, debugging strategies, evaluation, declarative knowledge, procedural knowledge and conditional knowledge), the results of factor analyses revealed little evidence to support the eight component model; however, the two-factor model proposed was confirmed.

From a developmental and educational perspective, Kuhn [13, 21] defines metacognition by reference to three types of knowing, which differ in terms of their declarative, procedural, and epistemological focus. First, *metacognitive knowing* is a type of declarative knowledge—the knowledge a person may possess in relation to cognition. For example, Kuhn describes *Theory of Mind* as a form of metacognitive knowing, specifically, knowledge that mental states exist—with personal metacognitive knowing referring to knowledge about one's own mental states, and impersonal metacognitive knowing referring to knowledge about others' mental states. Theory of Mind is therefore seen as a fundamental building block of higher-order critical, reflective and collaborative metacognitive thinking, as having knowledge of mental states is a prerequisite for the ability to think about mental states and thus think about thinking.

The second form of metacognition, *metastrategic knowing*, involves procedural knowledge—a person's knowledge about cognitive processes and of their impact on performance. Metastrategic knowledge comes in two forms, according to Kuhn, meta-task knowledge about task goals and metastrategic knowledge about the strategies one has available to address these goals. Metastrategic knowledge is seen as a fundamental driver of cognitive development in Kuhn's scheme, as thinking skills cannot develop in the absence of awareness and control over goals and strategies. Finally, the third form of metacognition, *epistemological knowing*, refers to an individual's understanding of what knowledge and knowing are in general, and how one comes to know. As argued below, the development of epistemological knowing is fundamental to high-level critical, reflective, and collaborative thinking skills.

Rather than constituting a single transition from one way of being to another, development, according to Kuhn [21], entails a shifting distribution in the frequencies with which more or less adequate strategies are applied, with inferior strategies increasingly inhibited as more superior strategies are acquired. This is a view consistent with the tenets of Dynamic Skill Theory [8, 22], which characterizes skill structures that vary in terms of how key principles, abstractions, representations, and actions are coordinated in context, with the context itself having a significant influence on the level of coordination achieved. Transfer and generalization of skill from one context to another also depends in part on whether or not key transferable principles and abstractions that may serve to metacognitively regulate action are available as part of coordinated action sequences across contexts [22]. According to Kuhn, this developmental 'shifting' in strategy and skill involves meta-level actions that dictate which strategies are selected for use and applied on a given occasion. Increasing meta-level awareness and control has been identified as an important target for interventions focused on promoting cognitive development [13]; and thus, we argue below that tools that facilitate meta-level awareness and control (e.g., awareness of argument structures and control over critical and reflective thinking processes) may provide a tool-mediated context in which skills can be transferred from one thinking context to another. Furthermore, dispositions, principles and practices associated with self-regulation of individual or collaborative action may support the transfer and generalization of key processes (e.g., mindful observation and non-reactivity) that facilitate awareness and control across different thinking contexts.

4.3 The Synergy of Teams, Tools and Talents in Metacognition

There are many ways to conceptualise meta-level awareness and control; and an applied systems science perspective highlights the active and applied nature of these acts in context. Consistent with Skill Theory [22], there are a variety of principles and abstractions that may be coordinated with action in context and that may be critical for understanding meta-level awareness and control. These include *epistemological principles and abstractions* associated with critical thinking and reflective judgment, and *dispositional principles and abstractions* associated with key motivations and values that frame critical, reflective, and collaborative action in context. Furthermore, at a lower level in the skill hierarchy, there are a number of fundamental self-regulation skills that influence the development of meta-level awareness and control, including self-other representation, executive control, and emotion regulation.

The emergence of higher-order skills or talents in an educational context often occurs in a social context and in the context of tool use. Social and tool affordances serve to accelerate the development of skills or talents, including lower-order and higher-order metacognitive skills or talents. But, effective tool design needs to be coupled with effective instructional design and management of teams in order for individual talents to be cultivated and coordinated in a group learning setting [10, 23].

While the learning sciences have provided a great deal of insight into the cultivation of key talents at the individual level, including the cultivation of critical thinking skills using argument mapping training programs [24], more work is needed to understand the optimal conditions needed to coordinate talents in a group learning level ([25]; see below). Moreover, while tools and teams provide a supportive context for the cultivation of skills or talents (Fig. 4.1), a broader understanding of developmental trajectories of child and adult development is needed to optimize the fit between individuals, teams, and the tools they are using to facilitate skill development. Notably, tools and teams provide a supportive context that facilitates intentional self-regulation and metacognitive control of key learning processes. Researchers have highlighted both the conscious and non-conscious aspects of intentional self-regulation that can be conceptualised in a *dual process* model of decision-making [e.g., 26, 27]. The common feature of these dual process models is that decision-making is subject to two systems: a reflexive, *intuitive* route of decision-making and a reasoned, *reflective* route of decision-making [26, 28, 29, 30].

Intuitive judgment has been described as automatic cognitive processing, which generally lacks effort, intention, awareness or voluntary control—usually experienced as perceptions or feelings [26, 28]. This mode of decision-making has low consistency and is moderately accurate [30]. On the other hand, reflective judgment has been described as slow, conscious and consistent [31] and associated

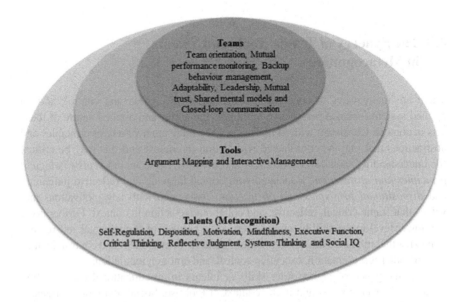

Fig. 4.1 The synergy of teams, tools and talents in metacognition

with high levels of cognitive control and conscious awareness and slow data processing [31, 32] and is generally quite accurate [30]. Essentially, reflective judgment allocates attention to the effortful mental activities that demand it ([e.g. complex computation; 26]). Notably, the types of thinking necessary to make reflective and intuitive judgments are often considered as working in parallel, and are therefore capable of reaching different decisions [31, 32].

Given that intuitive judgment operates automatically and cannot be voluntarily 'turned off', associated errors and unsupported biases are difficult to prevent. Even when errors appear obvious, in hindsight, they can only be prevented through the careful, self-regulated monitoring and control afforded by reflective judgment. Thus, a key developmental challenge involves the cultivation of self-regulation skills that support metacognitive awareness and control and the slow, effortful, and intentional processes that are often needed for high-level reflective judgment, as well as critical and collaborative thinking.

In the next section, we outline some of the key developmental milestones relevant for understanding the emergence of higher-order, metacognitive thinking skills. We focus in particular on aspects of self-regulation, including self-other representation and executive control skills. We also describe how *mindfulness* (i.e., observation and non-judgmental awareness of ongoing thought processes) may support aspects of self- and emotion-regulation and thus support higher-order thinking skills. We propose that mindfulness is a skill or talent that can operate at both an individual and group level to facilitate adaptive action in context. Following on from this, we examine how specific team and tool affordances can serve to accelerate the development of lower-order and higher-order metacognitive skills or talents.

4.4 Milestones in the Development of Metacognition: An Integrated Framework

A key underlying feature of metacognitive skill is the development of self-regulation. During childhood, changes in self-regulation primarily encompass changes in controlling emotions, attention, and behaviour, including self-monitoring and response inhibition [33]. Two inter-related aspects of self-regulation facilitate the development of metacognitive abilities underlying skills such as reflective judgment: the development of self-representations and the development of executive functioning. A central aspect of self-regulation is the development of self-representation [12] and the differentiation of self from others [34]. Mental representations are necessary for self-regulation as they facilitate goal selection, monitoring of progress, and the development of problem-solving strategies.

In order to think metacognitively, individuals must be aware of their own cognitive processes and have a basic concept of the self as an active agent [2]. Throughout infancy and childhood, individuals develop increasingly higher-order representational systems, which by middle-late childhood, integrates self-evaluations and self-representations [35]. This functions as a general self-monitoring and self-evaluation agent that influences goal pursuit and increased meta-level awareness and control [12]. The abilities to self-monitor and self-correct are central to self-regulation as individuals must first obtain awareness of their current state in relation to a desired goal in order to identify potential conflicts and discrepancies, which may precede subsequent goal-correcting behaviour [22]. This is central to reflective judgment as it facilitates an individual's understanding of the nature, limits and certainty of knowing. By middle adolescence, simple abstractions are organised into higher order abstract mapping in relation to both self and others ([8]; e.g., "I am intelligent because I am smart and creative; and Dad is ignorant because he relies on prior learning and lacks creativity"). Furthermore, some aspects of the developing self-concept may include epistemological principles and abstractions associated with critical and reflective thinking and problem solving, or dispositional principles and abstractions associated with key motivations and values that frame critical, reflective, and collaborative action in context (e.g., "Personality shapes learning in context: I'm an ambivert who is extroverted when with friends, introverted when with a group of strangers initially, but I transition from introverted to extroverted in collaborative learning environments over time and this change dynamically influences my learning in context").

The executive functioning is essential for the metacognitive processes involved in self-regulatory functioning, strategic planning and application of higher-order thinking. Although many models of executive functioning have been developed [36–41], there is a consensus that executive functioning involves three basic processes: updating, inhibition and shifting. Updating refers to the active revision and maintenance of working memory representations. Shifting refers to switching between tasks or mental sets; and inhibition refers to the active, deliberate suppression of thoughts or responses and the maintenance of attention on goal-relevant information [39]. These processes have been shown to interact differentially to control higher-order problem-solving [42], and decision-making [43]. Del Missier et al. [43] found that, in a task where the application of decision strategies became increasingly complex, inhibition was the most crucial process observed, as attention must be directed away from no longer relevant information; while in a separate task, where subjects were asked to assess the probability of risky events, shifting was identified as a process observed, as attention must be switched between varying judgment contexts.

Executive functioning skills are particularly important in situations or tasks which are novel, where automatic, intuitive or learned automatic responses are not adaptive or where a conflict between goals and the situation has been detected; and therefore, additional control is required [44]. Hofmann et al. [44] have outlined how integral executive functioning is to self-regulation. They argue that executive functions allow us to organise our thoughts in a goal-directed manner and are essential for success in education, work and everyday living. The updating and maintenance of working memory is vital for the accurate active representation of goals and goal-related information in situations where self-regulatory routines have not been automatized [44]. Furthermore, greater working memory capacity protects goal representations from thought intrusions and decreases mind-wandering. This is often referred to as goal shielding [45]. The importance of inhibition is clear here, with greater inhibitory control linked to more successful self-regulation in behaviours ranging from eating behaviour to sexual fidelity [46–48].

Though certain aspects of executive control are slow to develop throughout childhood [49], such as inhibiting learned responses, consistent with Fischer's Skill theory, children aged 9 years and older become increasingly effective at monitoring (e.g., "I'm good at this") and regulating their actions (e.g., "I need to slow down here"), thereby moving toward increasingly complex representational and abstract control skills that build upon less complex skill structures [8, 49, 50]. A consequence of these aspects of skill development is improved goal setting skills in children. For example, although children younger than 4-years old struggle to plan and organize actions in advance, simple planning skills, such as planning moves in the Tower of Hanoi task, are exhibited by older children [51]. Further development of executive function skills emerges as young children utilize increasingly complex strategies for planning and organisation. For example, when asked to copy a drawing of a complex figure (i.e., the Rey-Osterrieth Complex Figure Test), children younger than 7-years old may use simple strategies that are inefficient, inflexible, or haphazard, whereas children aged 7–11 years of age use strategies that are more organized, efficient, and flexible [52]. Improved organization and decision-making skills continues into adolescence [52] and correlates with ongoing biological changes within the adolescent brain, including continued myelination of nerve fibres, increased connectivity between frontal brain regions and other brain regions, and significant and localized synaptic pruning, especially in frontal areas, all of which are crucial for executive functioning [e.g., 53] and the coordination of affect and cognition [54].

Overall, the concurrent development of self-representations and executive functioning are fundamental for the emergence of meta-level awareness and control and the development of higher-order thinking skills. However, high-order reflective judgment, critical thinking and collaborative thinking do not automatically emerge, but rather, develop alongside effortful self-regulation. Such effortful self-regulation of thought and behaviour can be developed through mindfulness which involves deployment of the executive functions to ensure attention is fully focused on current experience in a non-judgmental and non-reactive manner. As such, mindfulness acts as a form of intentional self-regulation that can facilitate executive control, reflective judgment, critical thinking and self-regulation in collaborative learning contexts.

4.4.1 Mindfulness and Metacognition

In considering the development of metacognition, mindfulness is a functionally relevant construct [55, 56]. Much like critical and reflective thinking skills and dispositions, there is no guarantee that mindfulness will emerge as a result of development. Though conceptualisations of mindfulness vary, all highlight the role of mindfulness in enhanced self-regulation of thought, behaviour and emotional and physiological reactivity and all contrast mindful information processing with automatic, habitual or heuristic information processing, often referred to as mindlessness [57, 58].

Specifically, mindfulness is often operationalised as a two-component process. The first component involves the deployment of attention to both internal and external experience in the present moment [55]. This component has been associated with greater sensitivity to perceptual cues and facilitates conflict monitoring [59–61] and executive control skills, particularly switching between and updating the contents of working memory [55, 62]. The second component of mindfulness is characterised by non-judgmental acceptance of emotions and thoughts. This involves executive control to inhibit elaboration and/or suppression of affective cues and allows for the early engagement of emotion regulation before intense emotional reactivity to the attended thoughts, feelings and sensations can occur [61]. Notably, these skills of attention and non-judgment follow distinct developmental trajectories which may vary across individuals [63]. This operationalisation of mindfulness implies both monitoring and control skills which are inherently metacognitive [55]. Often, studies investigating the mechanisms and outcomes of mindfulness either involve brief meditation inductions or more long-term meditation interventions. However, it is also possible to take a dispositional approach where measures of trait mindfulness are associated with specific outcome variables [64].

Evidence for improved self-regulation of behaviour as a result of mindfulness comes from studies on procrastination [65], smoking cessation [66], persistence [67], and alcohol intake [68], while evidence for improved self-regulation of thoughts and emotions comes from the extensive literature on the clinical benefits of mindfulness (see review by Hofmann et al. [69]). Both components of mindfulness involve executive control [55, 60]. The sustained attention to current experience developed through mindfulness practice requires the ability to switch attention between stimuli in current experience and back to current experience when the mind wanders, updating the contents of working memory and inhibiting elaborative processing [56]. Notably, research on mindfulness and inhibition has found better performance on the Stroop task in experienced meditators [70], following a 6-week mindfulness training [71], after a brief mindfulness induction [72], and in association with dispositional mindfulness [73]. Studies focusing on the ability to shift between mental sets have shown improvements following mindfulness interventions [74]. Studies also show the benefits of mindfulness for working memory capacity as well as the updating of working memory [74, 75].

Evidence also suggests that it is through the enhancement of executive functioning that mindfulness facilitates successful self-regulation and that these effects may extend to higher-order thinking skills, including problem-solving, decision-making and both divergent and convergent thinking, and social cognitive skills [64, 76]. Mindfulness training is also beneficial for higher-order thinking skills, and has been found to be related to performance improvements on insight problem-solving tasks ($[r = 0.30, p < 0.01$; [77]), moral reasoning and ethical decision-making ($r = 0.43, p < 0.01$, [78]; $p < 0.05, \eta^2 = 0.20$ [79]) and creative thinking (i.e., creative thinking flexibility: $p < 0.001, \eta^2 = 0.37$; creative thinking fluency: $p < 0.001$, $\eta^2 = 0.33$; and creative thinking originality: $p < 0.001$, $\eta^2 = 0.32$ [80]). Each of these studies highlighted the non-automatic, reflective processing of information associated with mindfulness practice as the key to improved thinking outcomes. Hart et al. [58] suggested Kahneman's description of dual-process theory as a framework for explaining the cognitive benefits of mindfulness interventions. They suggested that mindfulness prompts the self-regulation of attention which activates system 2 or reflective thinking. Default-interventionist dual process theories of higher-order cognition offer a more refined view [81], arguing that intuitive processing is always engaged and generates automatic responses by default. Further reflective processing allows for higher-order thinking skills to be used but requires the engagement of executive functions which may or may not intervene depending on the individual and context. So, enhancement of executive functions may be the primary mechanism through which mindfulness facilitates higher-order thinking skills.

Metacognitive monitoring of others thinking and representations of others' thinking strategies are key in interpersonal relationships and especially in social situations where higher-order thinking skills are applied interactively such as in group problem-solving [82]. Research into mindful leadership, mindfulness in the workplace, mindfulness in relationships and social mindfulness suggests mindfulness could enhance self-regulation and metacognitive monitoring in social interactions [83].

In addition to factors such as enhanced positive affect, decreased stress and increased acceptance, constructs of a metacognitive nature such as perspective-taking, theory of mind and empathy are key in explaining how mindfulness can enhance social interaction [76, 84]. In delineating social mindfulness, Van Doesum et al. [84] explain how perspective-taking and theory of mind involve the mental representation of others' beliefs, needs, intentions, desire and knowledge and that people who are socially mindful maintain these representations accurately, consider them and act in a manner which maximises other people's autonomy in cases of interdependence.

Studies on the use of mindfulness in close relationships have also shown that trait mindfulness is positively related to relationship satisfaction and quality of communication [85] as well as successful identification of and communication regarding emotional states [86]. Research on mindfulness in the workplace and mindful leadership is in its infancy but initial empirical work suggests that where leaders have higher levels of mindfulness, employees have higher levels of psychological need satisfaction, job satisfaction and successful job performance [83]. It was argued that the broadened attention cultivated by mindfulness attuned leaders to employee needs which allowed them to support their employees more. This suggests a degree of skill in perspective-taking and empathy. Thus, mindfulness may extend from the metacognitive monitoring of one's own thinking to the metacognitive monitoring of representations of other's thinking and social, collaborative dynamics.

It is for this reason that mindfulness is central to our framework on meta-cognition and applied systems science in an educational context and we will return to this issue below. However, before returning to this issue, we describe the meta-cognitive nature of higher-order critical and reflective thinking skills and their central role in collaborative systems thinking, and we present two case studies describing how we have used tools to foster these high-levels skills.

4.4.2 Higher-Order Thinking Skills: Critical Thinking and Reflective Judgment

Critical thinking (CT) is a metacognitive process, consisting of a number of sub-skills: analysis, evaluation and inference; that, through purposeful, reflective judgment, increase the chances of producing a logical conclusion to an argument or solution to a problem [19, 87, 88, 89, 90]. Analysis is used in the context of argumentation to identify: the role statements play in an argument (e.g., central claim, reason, objection or rebuttal); the inferential relationships among statements (e.g., between or among the central claim, reasons, objections and/or rebuttals); the source(s) of information presented in an argument; and the balance, or imbalance, of contrasting information presented [87, 88, 90]. Evaluation is a CT skill that is used in the assessment of the strengths and/or weaknesses of information resulting from: the credibility of information, with respect to its sources(s); the relevance of information to another presented in the same context; the logical strength of inferential relationships among propositions within an argument; and the potential for bias, for example, purposefully pitting weak arguments that lack credibility and relevance, against strong stronger arguments [19, 87, 91].

The final CT skill, *inference,* refers to the "gathering" of credible, relevant and logical evidence based on the previous analysis and evaluation of available evidence, for the purposes of "drawing a reasonable conclusion" [87, p. 9]. This may imply accepting a conclusion pointed to by an author in light of the evidence they present, or "conjecturing an alternative", equally logical, conclusion or argument based on the available evidence [87]. Notably, though CT may be best cultivated, initially, in an educational environment, the importance of CT utilisation is not restricted to such settings [4, 19, 91, 92]. Rather, CT is required for use in many every day, real-world situations, for example, when problem-solving, hypothesis testing, analysing arguments, and assessing risks and probabilities [19, 89, 91, 92, 93].

Though one may possess the cognitive skills necessary to conduct CT, the willingness to conduct these skills ultimately dictates how well they are performed [92, 94, 95, 96]. Along with the ability to perform CT skills, "a critical thinker must also have a strong intention to recognise the importance of good thinking and have the initiative to seek better judgment" [97, p. 71]. This willingness to self-regulate can be described in terms of executive function, key dispositions towards thinking, the motivation to think and learn, and the perceived need to use specific cognitive processes when solving problems [19].

Possessing this willingness, or disposition towards thinking, refers to the extent to which an individual is inclined to perform a given thinking skill [98]. Specific dispositions have also been described, including dispositions toward clarity, systematisation, truth-seeking, open-mindedness, scepticism, reflection, perseverance and confidence in reason [92, 99, 100, 101]. Notably, a large body of research has demonstrated significant correlations between CT dispositions and CT ability [91, 102, 103, 104, 105, 106].

The ability to metacognitively think about thinking [1, 20] and the ability to apply critical thinking skills to a particular problem also implies a reflective sensibility and the capacity for reflective judgment [107]. Reflective judgment (RJ) is a fundamental aspect of metacognition that is used in the context of critical thinking (see Fig. 4.2) to judge and make decisions in a reflective manner [5, 14]. Like critical thinking, RJ is an important skill for students to acquire and practice, because it may facilitate their ongoing acquisition and application of knowledge both inside and outside of school and university [108, 109].

According to King and Kitchener [107], reflective judgment is an individuals' understanding of the nature, limits, and certainty of knowing and how this can affect how they defend their judgments and reasoning in context. Moreover, reflective judgment involves the ability of an individual to acknowledge that their views might be falsified by additional evidence obtained at a later time [107]. The ability to acknowledge levels of certainty and uncertainty when engaging in critical thinking is important because sometimes the information a person is presented with (along with that person's pre-existing knowledge) provides only a limited source of information from which to draw a conclusion. This is often the case when a person is presented with an ill-structured problem [110], that is, a problem that cannot be solved with absolute certainty [111]. In the context of uncertainty, a combination of critical thinking skills (i.e., analysis, evaluation and inference) and reflective judgment is often necessary in situations where one seeks to arrive at a reasonable conclusion or decide upon a reasonable course of action [111–114].

Research suggests that child and adult development may see a progressive development of RJ ability toward greater levels of complexity and skill. Kitchener and King [114] created the Reflective Judgment Model (RJM) in order to characterise the development of people's RJ ability. The RJM describes changes in the thinker's recognition of limited knowledge (i.e., uncertainty) and how these changes influence other thinking skills, such as analysis, evaluation and inference. Specifically, the RJM is a seven stage model that is broken down into three periods of development (i.e., pre-reflective thought, quasi-reflective thought and reflective thought; see [115]). Progress on the RJM (from one stage to another; and from one period to another) is a type of evolution of RJ, in which each progression marks the increasing complexity of the thinking required to justify a belief. The more developed one's RJ, the better able one is to present "a more complex and effective form of justification, providing more inclusive and better integrated assumptions for evaluating a view" [107, p. 13].

Within the RJM, a series of developmental changes occur in the way people come to understand the process of knowing and reasoning. More importantly, research supports a developmental trajectory of RJ along the lines described by King and Kitchener [107]. However, RJ development is not a simple function of age or time, but more so a function of the amount of interaction, or active engagement an individual has in working on ill-structured problems [5, 23, 115], such that the development of higher levels of reasoning and RJ ability can emerge, see Fig. 4.2.

Notably, a large body of research suggests that RJ and CT are inter-related [5, 107, 110, 115] and RJ may be a component of CT [110, 116, 117]. Furthermore, positive dispositions towards thinking are positively correlated with CT-related abilities [91, 102, 103, 104, 105, 106]; and both CT abilities and possessing the disposition to apply CT skills together determine a person's actual thinking performance [4, 92, 94, 95, 96, 118]. It is widely accepted that CT and RJ skills require special training and support if they are to emerge during development [5, 91]. Below we describe how we have used argument mapping tools to enhance higher-order thinking skills in students. We also describe how one collaborative thinking tool and process, Interactive Management, can be used to cultivate higher-order, metacognitive thinking skills and effective team dynamics at the group level.

Case Study 1—Argument Mapping for Critical Thinking. Argument mapping (AM) is a method of visually representing a text-based argument via 'boxes-and-arrows', wherein the boxes are used to highlight propositions and the arrows are used to highlight the inferential relationships that link the propositions together [14, 19, 89, 90, 119]. Specifically, an arrow between two propositions is used to indicate that one is evidence for or against another. Similarly, colour can be used in AM to distinguish evidence for a claim from evidence against a claim (i.e., green represents a support

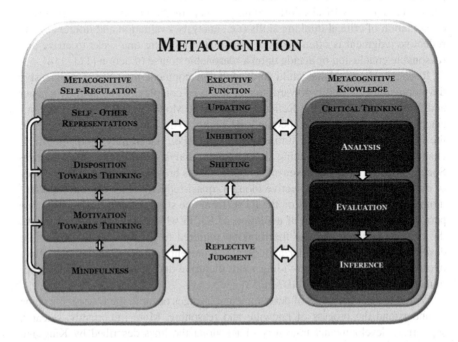

Fig. 4.2 An integrated model of metacognition

and red represents an objection). AM is designed in such a way that if one proposition is evidence for another, the two will be appropriately juxtaposed [120]; and the link explained via a relational cue, such as *because, but and however* (see Fig. 4.3).

These AM features have been hypothesized to facilitate metacognitive acts of critical thinking, both by making the structure of the argument open to deliberation and assessment; and by revealing strengths and weaknesses in the credibility, relevance, and logical soundness of arguments in the argument structure. Recent research by Dwyer et al. [89, 90] indicates that argument mapping may have beneficial effects on CT ability. Two separate studies examined the effects of a six-week argument mapping-infused CT training intervention on CT performance. Specifically, research by Dwyer et al. [89] examined the effects of a six-week Argument Mapping-infused CT training course, compared with a traditional (Hierarchical Outlining) CT training course and a no-CT training control condition on CT ability, as assessed by the California Critical Thinking Skills Test [121]. Results revealed that though the CT course taught through AM did not facilitate overall CT performance over and above the traditional HO CT course or the control condition (i.e., no-CT course), participants in the AM training group performed significantly better on inductive reasoning than those in the control group ($p = 0.038$, $d = 0.67$). However, given that the HO group also scored significantly higher than the control group on inductive reasoning ($p = 0.037$, $d = 0.55$), findings suggested that teaching CT through AM was no better than teaching CT through HO. Furthermore, when analysed together, the CT course attendees (i.e., both AM and HO groups combined) outperformed the control condition on analysis, evaluation and inductive reasoning ($p < 0.05$ for all three), suggesting that exposure to CT training infused with some hierarchical organisation strategy may have beneficial effects on CT performance.

Overall, these findings indicate that though the use of colour and spatial organisation of propositions within AMs may provide beneficial visual cues to both

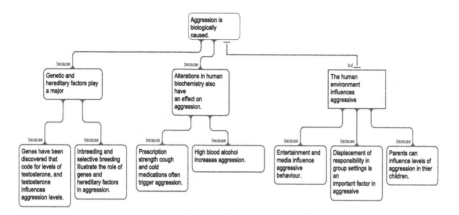

Fig. 4.3 Example of an argument map [89] using Rationale™ [119]

the reader and constructor of AMs, the hierarchical organisation of propositions within AMs and HOs may be the critical feature associated with the beneficial effects observed for CT learning outcomes. Further research by Dwyer et al. [122] examined the effects of an AM-infused CT e-learning course on CT performance, as assessed by the Halpern CT Assessment [93], by comparing the effects of an AM-infused CT e-learning course with a no-CT course control group on measures of CT ability. The relationship between student engagement in the AM-infused CT course and CT performance changes over time was also evaluated. Results revealed that those who participated in the AM-infused CT group outperformed those in the control group on overall CT ($p = 0.018$, $d = 0.60$) and argument analysis ($p = 0.025$, $d = 0.54$).

Results also revealed that performance on overall CT and all CT sub-scales (i.e., hypothesis testing, argument analysis, verbal reasoning, assessing likelihood and uncertainty, and problem-solving; $p < 0.05$ for all three) of those in the AM-infused CT group were enhanced from pre-to-post-testing. Results further revealed a time x engagement interaction effect for the CT sub-skill of problem-solving, with those in the high-engagement group showing a greater gain in problem-solving ability than those in the low-engagement group ($p = 0.004$, partial $\eta^2 = 0.19$). The positive effect of engagement on problem-solving performance in the AM group is promising, as it is broadly consistent with the suggestion that more intensive levels of engagement may be crucial for the development of higher-level thinking skills.

Recently, we have begun to examine collaborative critical and system thinking and we have developed a new tool that integrates argument mapping software with a system thinking tool, Interactive Management. Below we consider the issue of collaborative metacognition and we describe a second series of studies that highlight the value of our collaborative systems thinking tools and key processes associated with successful team dynamics.

4.4.3 Collaborative Metacognition

Collaborative learning is becoming increasingly prevalent within education. In educational settings, where foundational skills in collaborative learning may be developed for application in work settings, various forms of collaborative learning methodologies, centred on peer interaction and pooling of resources have been implemented and evaluated [123]. Many of these methodologies, especially those which facilitate collaborative problem-solving, are metacognitive in nature. Metacognitive collaboration refers to the process of team or group members thinking about, and reflecting on, how their team processes information, works on problems, and feels about the collaborative process [124, 125]. Effective metacognitive collaboration requires the coordination of a number of factors; effective facilitation, feedback and instruction regarding the collaborative process and goals; cultivation of enhanced team functioning in the collaborative context, including the promotion of cooperative, exploratory discourse; and the use of tools and methodologies which facilitate group coherence, the management of complexity and group problem-solving.

Dynamics of metacognitive team functioning. Metacognition in teams can promote effective teamwork [126]. McIntyre and Salas [127] identified four essential principles of effective teamwork which reflect metacognitive principles: Mutual performance monitoring, feedback, closed-loop communication and back up behaviours. Mutual performance monitoring refers to the ability of group members to monitor the work of their peers while simultaneously carrying out their own work [128]. This aspect of team functioning is closely related to the second aspect of effective teamwork, which involves the process of feedback. Notably, research has shown that although individuals are often unaware of their performance deficiencies [e.g., 129], peer feedback can result in individuals becoming more aware of their performance.

The third aspect of effective teamwork, closed-loop communication, has been suggested as an effective method for reducing inefficient information sharing, and ensuring that communications are adequately delivered, received and understood. Closed-loop communication involves (a) the sender initiating a message, (b) the receiver receiving the message, interpreting it, and acknowledging its receipt, and (c) the sender following up to insure the intended message was received [127]. Research conducted by Siegel and Federman [130] found that teams who received training in this type of communication outperformed teams who did not receive such training. Finally, back-up behaviour refers to the provision of resources and effort by team members upon recognition of inadequate workload distribution in their team [131]. Mutual performance monitoring allows team-members to recognise if and when the workload of others in their team has exceeded their capacity [128]. On recognition of such overload, team-members can provide three means of back-up behaviour: provision of performance-related feedback; assistance in completing the task, or complete the task for the team member [129]. Various research programs have highlighted the importance of these metacognitive processes in successful team performance. While these processes occur naturally in some teams, this is not always the case, and conscious effort must be made to promote them [126]. So an important role of the metacognitive facilitator, involves the cultivation and coordination of these processes.

Feedback and facilitation in metacognitive teams. Fruitful metacognitive collaboration does not occur spontaneously; it is not enough to provide learners with a method or tool. Much like in the development of critical thinking skills, it is necessary to guide learners through the process of metacognitive collaboration by means of effective feedback and facilitation. Feedback is considered to be a vital instructional tool in education. The beneficial effects of receiving feedback in a learning context are widely reported, including positive effects on learning, performance, and satisfaction [132]. Feedback can be administered for a number of purposes. For example, task-level feedback aims to provide declarative knowledge or information about the correctness of a learner's response. Such feedback provides the learner with the correct answer; however, it does not provide the learner with prompts or strategies to guide future efforts. As such, simple, basic, forms of feedback may fail to help learner's to monitor, be aware of, or adjust, their learning strategies according to how effective their current learning strategy is [e.g., 133, 134]. Lee et al. [135] suggest that one potential solution here is to provide meta-cognitive feedback about cognitive strategies.

Metacognitive feedback is designed to make the learner conscious of the learning strategies being used and their degree of success. Such feedback reminds the learner to consider and evaluate the suitability of strategies employed. As noted by Kuhn [21], this type of meta-strategic awareness is a critical driver of cognitive development. Process-level feedback and self-regulation feedback are two types of metacognitive feedback. Through the use of such feedback strategies, learners can be guided through skills associated with critical thinking, including: analysis, evaluation, inference, and reflective judgment. Process-level feedback provides prompts or strategies that learners can use for error search, information search, or review and revision of work done. Examples include prompting students to consider other questions that could be asked in relation to a problematic situation or asking what other strategies could be used to complete a task or solve a problem. Feedback at the self-regulation level provides conditional knowledge, generally in the form of probing questions. The probes or prompts are designed to guide the learner on when, where are why they should select a certain strategy. It seeks to help the learner to identify the demands of the task at hand, and prompt them towards appropriate strategies. Examples may include questions asking students to compare the current problem to previous problems they have worked on or reflective questions such as "What am I doing here?", or "What is the purpose of what I am doing here?"

Tool use in team contexts. Effective collaboration is rarely achieved in the absence of guidance [136]. Just as instruction and feedback from a facilitator can enhance the efficacy of collaboration, various forms of support or instruction tools have been developed to promote metacognitive collaboration [137]. One such methodology is Interactive Management (IM). IM is a computer facilitated thought and action mapping technique that enhances group creativity, group problem solving, group design, and collective action in the context of complexity. There are a series of steps in the IM process. First, a group of key stakeholders with an interest in resolving a problematic situation come together in a situation room and are asked to generate a set of 'raw' ideas (commonly 50–200) about what might potentially have a bearing on the problem they all agree exists. Next, under the guidance of a trained facilitator, group discussion and voting helps the group to clarify the sub-set of ideas that bear upon the most critical problem issues.

Throughout the course of this discussion, the facilitator cultivates a sense of critical engagement and reflective judgment in relation to the ideas suggested. Once a set of critical problem issues has been agreed upon, the idea structuring phase begins. Using IM software, each of the critical issues is compared systematically in pairs and the same question is asked of each in turn: "Does A influence B?" Again, the facilitator uses process-level and self-regulatory feedback and instruction to guide the group through this process, while also encouraging the use of strategies including: mutual performance monitoring, backup behaviours, and closed-loop communication. The facilitator encourages learners to reflect upon, and critically evaluate each relational statement presented by the IM, and cultivates a sense of cooperative and exploratory collaboration and collective mindfulness and awareness of ongoing thought and action in the room. Unless there is majority consensus within the group that one issue impacts upon another, the relation does not appear in the final analysis.

After all the critical issues have been compared in this way, IM software generates a problem structure (or problematique) showing how the issues are interrelated. The problematique can be viewed and printed for discussion. The problematique becomes the launch pad for planning solutions to problems within the problem field. The logical structure of problems is visible in the problematique and when generating solutions, action plans are aimed at resolving problems in a logical and orderly manner. When the group is happy that they have modelled both the problem field and the best possible set of solutions, the IM session closes and each member leaves with a detailed action plan, a specific set of goals to work on, and the roadmap and logic describing how all the various plans and goals of each member will work together to resolve the original problem.

As it is currently used, IM is a deeply engaging and cooperative process. However, when it is further merged with cooperative argument mapping (AM) work, the cooperative enquiry process is transformed into a process that explicitly links the metacognitive processes of critical thinking with ongoing metacognitive collaboration. More specifically, students who are mapping out a problematic situation are called upon to source and evaluate scientific evidence to support their beliefs as to the nature of discrete paths of influence in a problematique. Also, for problematic situations that draw upon multiple sciences of description, it is evident that students working in multidisciplinary teams will be exposed to arguments from multiple scientific domains and will have to learn to analyse and evaluate these arguments in cooperation with others. While it might be assumed that only those with specialized knowledge of domain-specific science content will be able to analyse domain-specific arguments and evidence, we believe that the knowledge and perspective of students from multiple scientific backgrounds can help to enhance the creativity and the overall quality of evaluation and inference work [23].

By way of example of an application of IM, presented below is the outcome of a session conducted in a *Thinking, Modelling and Writing in Psychology* module in NUI Galway, in response to the trigger question, *What are the most important skills and dispositions of good critical thinkers?*. Students used the IM software to structure the interdependencies among the highest ranked skills and dispositions (see Fig. 4.4). The problematique is to be read from left to right, with paths in the model interpreted as 'significantly enhances'. Once the problematique had been designed, students used cooperative AM to explore and evaluate the paths of influence in the IM problematique.

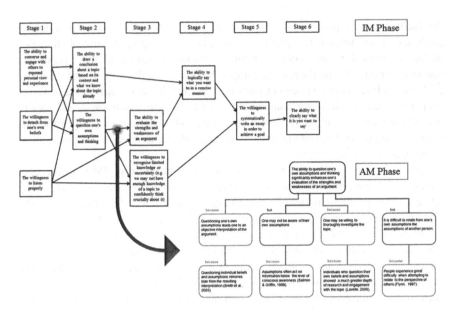

Fig. 4.4 Sample enhancement structure of skills and dispositions required for critical thinking (*top panel*; paths in the model are to be interpreted as 'significantly enhances'); (*bottom panel*) linked argument map exploring how the ability to question one's assumptions and thinking enhances evaluation skills

Case Study 2—Interactive Management and Collaborative Enquiry. We have conducted a series of studies to evaluate key aspects of our collaborative systems thinking tool and learning process. In the first study, Harney et al. [25] investigated the effect of open versus closed IM voting and dispositional trust on perceived consensus, objective consensus and perceived efficacy of IM technology. Two groups of 15 undergraduate students came together to structure the interdependencies between positive and negative aspects of social media. Participants high and low on dispositional trust were identified and were randomly assigned to either an open or closed voting condition. Those in the closed voting group were not permitted to discuss the problem relations, but consensus votes were recorded by the group design facilitator.

This scenario simulated an online voting system where participants converged upon a decision without direct contact or open dialogue in advance of voting. The open group was allowed to discuss the relations before voting. Notably, the results of this study indicated that participants with higher dispositional trust, and those in more open working groups, reported higher levels of perceived consensus ($p = 0.007$, partial $\eta^2 = 0.25$; $p = 0.011$, partial $\eta^2 = 0.22$, respectively) and higher levels of perceived efficacy of the IM technology ($p = 0.047$, partial $\eta^2 = 0.14$; $p = 0.02$, partial $\eta^2 = 0.19$, respectively).

These results, combined with results from studies showing increased learning gains in more open and interactive groups [e.g., 138], suggest that open discussion and dialogue are critical factors in the success of collaboration using tools such as IM. These results have implications for the design of computer-supported collaborative learning (CSCL) tools and how they are used. Peterson [139] highlighted the effect that a facilitator's directiveness has on the results of collaborative working sessions. Peterson found that process-directive leaders, who aimed to foster the decision-making process by encouraging discussion and by remaining open regarding their position, facilitated more positive group processes and outcomes than outcome-directive leaders who did not encourage discussion in the group and who focused solely on reaching a decision.

In the context of CSCL, it is imperative that the facilitator of any group session provide ample opportunities for open dialogue and discussion in the group. In the absence of open dialogue and discussion, metacognitive collaboration will not emerge. Harney et al. [140] investigated the effects of generic versus metacognitive feedback on levels of perceived and objective consensus and argumentation of subjects high and low in dispositional trust, in the context of an IM. Four groups of undergraduate psychology students ($N = 75$) came together to discuss the negative consequences of online social media usage. After screening for trust scores, participants high and low on dispositional trust were randomly assigned to either a generic or metacognitive feedback condition. In each feedback condition, an independent facilitator was given a specific set of prompts or instructions which could be used as part of the feedback process.

A second facilitator was present to oversee the feedback process, and assist with the input of ideas into the IM. In both conditions, participants were asked to silently generate ideas which they felt had a significant impact on the problem at hand. This is referred to as the *Idea Generation* phase of IM. Once this phase was complete, and each participant had their own list of ideas to offer, the facilitator went around the room, to each participant asking them to present their idea to the rest of the group. If necessary, they explained their idea clearly and succinctly to the rest of the group.

The facilitator would then open the discussion up to the group, by asking "Does anyone else have an opinion on this idea?" While these guidelines were also followed by the facilitator in the metacognitive feedback group, there was also the addition of further process-level and self-regulatory feedback prompts. Through the use of such feedback, the facilitator aimed to enhance the critical thinking of the group in relation to the suggested ideas. Specifically, the facilitator used these prompts to cultivate a higher level of analysis and evaluation as the group generated a set of problem issues. The facilitator could, where necessary, request further clarification, suggest that some ideas offered may be similar in nature and require further examination, suggest merging of ideas or breaking down of ideas which appear to have multiple components. The facilitator could also suggest considering the relevance of the idea offered in the problem context, and suggest considering the generalizability of the idea offered. Next, in the *Idea Structuring* phase, as each relation "Does issue A significantly influence issue B?" was presented on the screen, the facilitator would open the discussion to the room, and ask if anyone has a yes or no preference and this stage.

As participants indicated their preference, the facilitator would ask for some insight into their reasoning for this preference, and then request other opinions from the group. The facilitator would then request a show of hands from the group, and a vote would be taken. Again, these guidelines were also followed by the facilitator in the metacognitive feedback condition, but with the addition of some further prompts and instructions, designed to enhance the critical thinking and reflective judgment of the group as they discuss, and decide upon, the relational statements. In the metacognitive feedback condition, the facilitator could, where necessary, ask for contrary opinions, ask for further analysis and evaluation, suggest considering the relevance of arguments and suggest considering the generalizability of the opinions offered.

Levels of perceived consensus, objective consensus and perceived efficacy of the collaborative learning methodology were measured before and after the IM session. Results indicated that those in the metacognitive feedback condition, and those with higher levels of dispositional trust, reported higher levels of perceived consensus ($p < 0.001$, partial $\eta^2 = 0.18$; $p = 0.035$, partial $\eta^2 = 0.61$, respectively) in response to the group design problem. Furthermore, those in the metacognitive feedback condition also reported significantly higher levels of perceived efficacy of the ISM process ($p = 0.009$, partial $\eta^2 = 0.093$) compared to those in the generic feedback condition. Finally, analysis of the dialogue from the IM sessions revealed that those in the information feedback condition exhibited higher levels of sophistication in their arguments, as revealed by the Conversational Argument Coding Scheme (CACS) [141].

Notably, of the 16 CACS possible argument codes, which comprise the five argument categories, 13 were observed in the metacognition feedback condition at least once; eight were observed in the generic feedback condition at least once; and three were not observed in any condition. Significant differences were reported across conditions for three argument codes, showing higher levels of incidence in the information feedback condition in each case, specifically, for amplifications ($p = 0.002$, $V = 0.123$), challenges ($p = 0.006$, $V = 0.118$) and propositions ($p = 0.012$, $V = 0.108$). In each of the remaining codes, higher incidence was also reported in the metacognitive feedback condition than in the generic feedback condition; however, these differences were not significant.

These results show that the metacognitive feedback group were engaging at a higher-level with the information at hand, and made more effective moves towards reaching a level of understanding and consensus within the group. For example, while elaborations—statements that support other statements by providing evidence, reasons or other support—were high in both groups, amplifications—statements that explain or expound upon other statements to establish the relevance of an argument through inference, were observed more frequently in the metacognitive feedback group. In this way, those in the metacognitive feedback group were moving beyond accumulation of evidence and support, and establishing how this reasoning relates to the problem at hand, that is, how relevant it is. Similarly, while levels of objections—statements that deny the truth or accuracy of an arguable—were almost identical across the two groups, challenges—statements that offer problems or questions that must be solved if agreement is to be secured on an arguable—occurred significantly more often in the metacognitive feedback condition.

This again shows that those in the information feedback condition engaging more critically with the information at hand, and thereby engaging in more productive argumentation. Finally, of the 16 types of argument codes with comprise the CACS, 13 were observed at least once in the metacognitive feedback condition, whereas only 8 were observed at least once in the generic feedback condition. This highlights the fact that those in the metacognitive feedback condition were engaging in more complex and varied argumentation, relative to the generic feedback group.

Our third study examined the effects of feedback on the emergence of cooperative conversational dynamics in groups high and low in dispositional trust. Specifically, we used Mercer's Categories of Classroom Talk [142] and Multiple Episode Protocol Analysis [143] in the context of a 2 (Generic Feedback versus Peer Feedback) × 2 (High Trust versus Low Trust) design to examine the individual and combined effects of feedback and trust on the emergence of exploratory, cumulative, and disputational talk within teams, and the replicability of these effects across multiple applied systems science project domains.

Generic feedback was operationalised in the same manner as in the previous study. Peer feedback, however, represented an extension of the metacognitive feedback strategy used in the previous study. In this case, the facilitator first administered metacognitive feedback to the group, thereby familiarising them with the metacognitive prompts and strategies. The facilitator then handed over the tasks of feedback and performance monitoring to the group, thereby empowering them to engage in peer-centred metacognitive feedback. Preliminary results indicate that process and self-regulatory feedback, provided by the facilitator, and subsequently by peers themselves, resulted in enhanced cooperative conversational dynamics, increased argumentation duration, intensity, and complexity, and significantly more episodes of exploratory talk than in the generic feedback condition (Feeney et al. [144], The effects of peer feedback on cooperative conversational dynamics in collaborative learning, Unpublished manuscript).

4.5 Conclusions

Human evolution has resulted in the emergence of our capacity to think about thinking. However, the development of metacognitive skills and talents is contingent upon the availability of quality educational supports. Promoting the development of individual and collaborative metacognitive skills and talents during child and adult development is essential for survival, adaptation and flourishing. Higher-order, metacognitive skills are needed now more than ever in order to aid individuals and groups in becoming more adaptable in the context of a rapidly evolving information society. Promoting successful team dynamics and effective tool use can help children and adults to think about their thinking and approach problems with a mindful, reflective, collaborative sensibility that facilitates adaptive action at individual and group levels.

In this chapter, we highlight the importance of basic building blocks of self-regulation, including self-other representation, executive functions, and goal setting in supporting both lower-order and higher-order metacognitive regulation of ongoing action and problem solving. We also highlighted the important role of mindfulness and thinking dispositions, both for executive functioning and for higher order critical thinking and reflective judgment. We presented evidence to suggest that teaching critical thinking skills using argument mapping tools can facilitate the development of critical thinking skills, particularly when students engage with and use the tool regularly.

We further highlighted the importance of metacognitive feedback in the context of collaborative tool use and we provided evidence to suggest that metacognitive feedback promotes trust and consensus along with increased complexity of collaborative argumentation and increased tendency to move beyond disputational and cumulative styles of talk to exploratory modes of talk. Central to our view on metacognition is the idea that mindfulness can emerge as a development asset that supports intra-personal and inter-personal meta-strategic awareness and control in both individual and group learning contexts. Given that mindfulness along with critical thinking, reflective judgment and collaborative enquiry may require explicit training to develop, we propose an integration of organismic and pragmatic models.

Consistent with neo-Piagetian organismic models [8], we assume that skills can unfold at greater levels of integration and complexity; however, skilled action is a function of training, support and the collaborative dynamics of social groups. Consistent with pragmatism and a pragmatic systems science framework [9] we have argued that specific tools and group-mediated processes can be used to support metacognitive skill development [10, 23]. However, we also believe that specific higher-order principles and practices, including those associated with mindfulness, can transfer from one context to another in the support of skilled action. Ongoing research in our lab is examining the relationship between mindfulness dispositions, executive functioning and higher-order critical thinking processes.

Furthermore, consistent with the idea that mindfulness can be trained, future work in our lab will examine if training in mindfulness enhances executive functioning and critical thinking abilities and if there are key underlying brain processes associated with the transferable benefits of mindfulness from one problem solving context to another. Ultimately, we believe that metacognitive skills display complex interdependencies in the context of lower-order and higher-order skill structures. In this way, metacognitive skills can be viewed as part of a collection of potentially synergistic talents that can be fostered in the context of individual and team-based tool use and problem solving.

Future research adopting an organismic and applied systems science perspective may seek to examine the efficacy of other environmental affordances and supports designed to enhance individual and collaborative metacognition, including training programmes focused on emotional intelligence and social intelligence. For example, while mindfulness training may enhance critical and collaborative thinking via improvements in executive functioning and emotion regulation, emotional intelligence and social intelligence programmes may be effective in transmitting key principles and practices that support perspective taking, empathy, listening skills, team orientation, and other processes that also enhance individual and collaborative metacognition.

One principle of organismic and systems models of development is relevant here, that is, the principle of *equifinality*. Simply stated, the principle of equifinality implies that there are many different ways in which the same developmental outcomes can be arrived at. As such, there may be many different ways to develop metacognitive skills and talents that support adaptive functioning and effective problem solving in context. We have focused here on the developmental affordances associated with a focus on teams and tools. However, within and outside of this broad focus there are many ways to develop metacognitive skills other than those addressed in this chapter.

As noted in the introduction, we believe that cultural evolution is itself a metacognitive process. It is the task of each new generation to think about the thinking of previous generations and modify the culture of thinking in ways that foster adaptive development. Like other artefacts of culture, educational practices are products of our cultural evolution that are open to metacognitive reflection. This offers us hope but also highlights the need for mindful skill and creativity, and determination such that we can continue to improve the education system for the benefit of the current and the next generation.

References

1. Flavell, J.: Metacognitive aspects of problem solving. In: Resnick, L. (ed.) The Nature of Intelligence, pp. 231–236. Lawrence Erlbaum Associates, Hillsdale (1976)
2. Boekaerts, M., Simons, P.R.J.: Leren en Instructie: Psychologie van de Leerling en het leerproces [Learning and Instruction: Psychology of the Learner and His Learning Process]. Lawrence Erlbaum Associates, Hillsdale (1993)
3. Brown, A.: Metacognition, executive control, self-regulation, and other more mysterious mechanisms. In: Weinert, F., Kluwe, R. (eds.) Metacognition, Motivation, and Understanding, pp. 65–116. Erlbaum, Hillsdale (1987)
4. Ku, K.Y.L., Ho, I.T.: Dispositional factors predicting chinese students' critical thinking performance. Personal. Individ. Differ. **48**(1), 54–58 (2010)
5. Dawson, T.L.: Metacognition and learning in adulthood. Prepared in response to tasking from ODNI/CHCO/IC Leadership Development Office, Developmental Testing Service, LLC Northhampton, MA (2008)
6. Darling-Hammond, L.: How can we teach for meaningful learning? In: Darling-Hammond, L., Barron, B., Pearson, P.D., Schoenfeld, A.H., Stage, E.K., Zimmerman, T.D., Cervetti, G.N., Tilson, J.L. (eds.) Powerful Learning: What We Know About Teaching for Understanding, pp. 1–10. Jossey-Bass, San Francisco (2008)
7. Jukes, I., McCain, T.: Minds in Play: Computer Game Design as a Context Of Children's Learning. Erlbaum, New Jersey (2002)
8. Fischer, K.W.: A theory of cognitive development: the control and construction of hierarchies of skills. Psychol. Rev. **87**(6), 431–477 (1980)

9. Warfield, J.N.: An Introduction to Systems Science. World Scientific, Singapore (2006)
10. Hogan, M.J., Harney, O., Broome B.: Integrating argument mapping with systems thinking tools—advancing applied systems science. In: Okada, A., Buckingham Shum, S., Sherborne, T. (eds.) Knowledge Cartography: Software Tools and Mapping Techniques. Springer, Heidelberg (in press)
11. Anderson, L.W., Krathwohl, D.R., Airasian, P.W., Cruikshank, K.A., Mayer, R.E., Pintrich, P.R., Wittrock, M.C.: A Taxonomy for Learning, Teaching, and Assessing: A Revision of Bloom's Taxonomy of Educational Objectives, abridged edition. Longman, White Plains (2001)
12. Demetriou, A.: Organisation and development of self-understanding and self-regulation: toward a general theory. In: Boekaerts, M., Pintrich, P.R., Zeidner, M. (eds.) Handbook of Self-regulation, pp. 209–251. Academic Press, London (2000)
13. Kuhn, D.: A Developmental model of critical thinking. Educ. Res. **28**(2), 16–25 (1999)
14. Hofer, B.K.: Epistemological understanding as a metacognitive process: Thinking aloud during online searching. Educ. Psychol. **39**(1), 43–55 (2004)
15. Magno, C.: The role of metacognitive skills in developing critical thinking. Metacogn. Learn. **5**(2), 137–156 (2010)
16. Pintrich, P.R.: The role of goal orientation in self-regulated learning. In: Boekaerts, M., Pintrich, P.R., Zeidner, M. (eds.) Handbook of Self-regulation, pp. 451–502. Academic Press, San Diego (2000)
17. Presseisen, B.Z.: Thinking skills: meanings and models revisited. In: Costa, A.L. (ed.) Developing Minds: A Resource Book for Teaching Thinking, vol. 1, pp. 47–53. ASCD Publications, Alexandria (2001)
18. Schraw, G., Dennison, R.S.: Assessing metacognitive awareness. Contemp. Educ. Psychol. **19**(4), 460–475 (1994)
19. Dwyer, C.P., Hogan, M.J., Stewart, I.: An integrated critical thinking framework for the 21st Century. Think. Skills Creativity **12**, 43–52 (2014)
20. Ku, K.Y.L., Ho, I.T.: Metacognitive strategies that enhance critical thinking. Metacogn. Learn. **5**(3), 251–267 (2010)
21. Kuhn, D.: Metacognitive development. Curr. Dir. Psychol. Sci. **9**(5), 178–181 (2000)
22. Fischer, K.W., Bidell, T.R.: Dynamic development of action, thought, and emotion. In: Damon W., Lerner, R.M. (eds.) Handbook of Child Psychology: Theoretical Models of Human Development, 6th edn., pp. 313–399. Wiley, New York (2006)
23. Hogan, M.J., Harney O.: A proposal for systems science education. In: Wegerif, R., Kaufman, J., Li, L. (eds.) The Routledge Handbook of Research on Teaching Thinking (in press)
24. Dwyer, C.P., Hogan, M.J., Stewart, I.: An examination of the effects of argument mapping on students' memory and comprehension performance. Think. Skills Creativity. **8**, 11–24 (2013)
25. Harney, O., Hogan, M.J., Broome, B.: Collaborative learning: the effects of trust and open and closed dynamics on consensus and efficacy. Soc. Psychol. Educ. **15**(4), 517–532 (2012)
26. Kahneman, D.: Thinking Fast and Slow. Penguin, Great Britain (2011)
27. Reyna, V.F., Farley, F.: Risk and rationality in adolescent decision making: implications for theory, practice, and public policy. Psychol. Sci. Public Interest **7**(1), 1–44 (2006)
28. Lieberman, M.D.: Reflexive and reflective judgment processes: a social cognitive neuroscience approach. In: Forgas, J.P., Villiams, K.D., Von-Hippel, W. (eds.) Social Judgments: Implicit and Explicit Processes, Vol. 5, pp. 44–67. Cambridge University Press, Cambridge (2003)
29. Stanovich, K.E., West, R.F.: Individual differences in reasoning: implications for the rationality debate? Behav. Brain Sci. **23**(5), 645–665 (2000)
30. Hammond, K.R.: Upon reflection. Think. Reason. **2**(2–3), 239–248 (1996)
31. Hamm, R.M.: Clinical intuition and clinical analysis: expertise and the cognitive continuum. Professional judgment: A Reader in Clinical Decision Making, pp. 78–105 (1988)
32. Cader, R., Campbell, S., Watson, D.: Cognitive continuum theory in nursing decision-making. J. Adv. Nurs. **49**(4), 397–405 (2005)

33. Gestsdottir, S., Lerner, R.M.: Positive development in adolescence: the development and role of intentional self-regulation. Hum. Dev. **51**(3), 202–224 (2008)
34. Labouvie-Vief, G.: Self-with-other representations and the organization of the self. J. Res. Pers. **39**(1), 185–205 (2005)
35. Harter, S.: Causes, correlates, and the functional role of global self-worth: a life-span perspective. In: Sternberg, R., Kolligian, J. (eds.) Competence Considered, pp. 67–97. Yale University Press, New Haven (1990)
36. Baddeley, A.D., Hitch, G.: Working memory. In: Bower, G.H. (ed.) The Psychology of Learning and Motivation: Advances in Research and Theory, vol. 8, pp. 47–90. Academic Press, New York (1974)
37. Cowan, N.: What are the differences between long-term, short-term, and working memory? Prog. Brain Res. **169**, 323–338 (2008)
38. Duff, K., Schoenberg, M.R., Scott, J.G., Adams, R.L.: The relationship between executive functioning and verbal and visual learning and memory. Arch. Clin. Neuropsychol. **20**(1), 111–122 (2005)
39. Miyake, A., Friedman, N.P., Emerson, M.J., Witzki, A.H., Howerter, A., Wager, T.D.: The unity and diversity of executive functions and their contributions to complex "frontal lobe" tasks: a latent variable analysis. Cogn. Psychol. **41**(1), 49–100 (2000)
40. Norman, D., Shallice, T.: Attention to action: Willed and automatic control of behaviour. In: Gazzaniga, M. (ed.) Cognitive Neuroscience: A Reader. Blackwell, Oxford (2000)
41. Norman, D.A., Shallice, T.: Attention to Action. Springer, US (1986)
42. Burton, C.L., Strauss, E., Hultsch, D.F., Hunter, M.A.: Cognitive functioning and everyday problem solving in older adults. Clin. Neuropsychol. **20**(3), 432–452 (2006)
43. Del Missier, F., Mäntylä, T., Bruine de Bruin, W.: Executive functions in decision making: an individual differences approach. Think. Reason. **16**(2), 69–97 (2010)
44. Hofmann, W., Schmeichel, B.J., Baddeley, A.D.: Executive functions and self–regulation. Trends Cogn. Sci. **16**(3), 174–180 (2012)
45. Brewin, C.R., Smart, L.: Working memory capacity and suppression of intrusive thoughts. J. Behav. Ther. Exp. Psych. **36**(1), 61–68 (2005)
46. Hofmann, W., Gschwendner, T., Friese, M., Wiers, R.W., Schmitt, M.: Working memory capacity and self-regulatory behavior: toward an individual differences perspective on behavior determination by automatic versus controlled processes. J. Pers. Soc. Psychol. **95**(4), 962–977 (2008)
47. Nederkoorn, C., Houben, K., Hofmann, W., Roefs, A., Jansen, A.: Control yourself or just eat what you like? Weight gain over a year is predicted by an interactive effect of response inhibition and implicit preference for snack foods. Health Psychol. Off. J. Div. Health Psychol. **29**(4), 389–393 (2010)
48. Pronk, T.M., Karremans, J.C., Wigboldus, D.H.J.: How can you resist? Executive control helps romantically involved individuals to stay faithful. J. Pers. Soc. Psychol. **100**(5), 827–837 (2011)
49. Anderson, P., Anderson, V., Northam, E., Taylor, H.: Standardization of the contingency naming test (CNT) for school-aged children: a measure of reactive flexibility. Clin. Neuropsychol. Assess. **1**, 247–273 (2000)
50. Anderson, P., Anderson, V., Lajoie, G.: The tower of london test: validation and standardization for pediatric populations. Clin. Neuropsychol. **10**(1), 54–65 (1996)
51. Welsh, M.C., Pennington, B.F., Groisser, D.B.: A normative developmental study of executive function: a window on prefrontal function in children. Develop. Neuropsychol. **7**(2), 131–149 (1991)
52. Anderson, P., Anderson, V., Garth, J.: Assessment and development of organizational ability: the rey complex figure organizational strategy score (RCF-OSS). Clin. Neuropsychol. **15**(1), 81–94 (2001)
53. Sowell, E.R., Trauner, D.A., Gamst, A., Jernigan, T.L.: Development of cortical and subcortical brain structures in childhood and adolescence: a structural mri study. Dev. Med. Child Neurol. **44**(1), 4–16 (2002)

54. Keating, D.P.: Cognitive and brain development. In: Lerner, R., Steinberg, L. (eds.) Handbook of Adolescent Psychology, vol. 2, pp. 45–84. Wiley, Hoboken (2004)

55. Bishop, S.R., Lau, M., Shapiro, S., Carlson, L., Anderson, N.D., Carmody, J., Segal, Z.V., Abbey, S., Velting, D, Devins, G.: Mindfulness: a proposed operational definition. Clin. Psychol. Sci. Practice. **11**(3), 230–241 (2004)

56. Holas, P., Jankowski, T.: A cognitive perspective on mindfulness. Int. J. Psychol. **48**(3), 232–243 (2012)

57. Brown, K.W., Ryan, R.M., Creswell, J.D.: Mindfulness: theoretical foundations and evidence for its salutary effects. Psychol. Inquiry Int. J. Adv. Psychol.Theor. **18**(4), 211–237 (2007)

58. Hart, R., Ivtzan, I., Hart, D.: Mind the gap in mindfulness research: a comparative account of the leading schools of thought. Rev. Gen. Psychol. **17**(4), 453–466 (2013)

59. Anicha, C.L., Ode, S., Moeller, S.K., Robinson, M.D.: Toward a cognitive view of trait mindfulness: distinct cognitive skills predict its observing and nonreactivity facets. J. Pers. **80**(2), 255–285 (2011)

60. Teper, R., Inzlicht, M.: Meditation, mindfulness and executive control: the importance of emotional acceptance and brain-based performance monitoring. Soc. Cogn. Affect. Neurosci. **8**(1), 85–92 (2012)

61. Teper, R., Segal, Z.V., Inzlicht, M.: Inside the mindful mind: how mindfulness enhances emotion regulation through improvements in executive control. Curr. Dir. Psychol. Sci. **22**(6), 449–454 (2013)

62. Moore, A., Malinowski, P.: Meditation, mindfulness and cognitive flexibility. Conscious. Cogn. **18**(1), 176–186 (2009)

63. Lilja, J.L., Lundh, L.G., Josefsson, T., Falkenström, F.: Observing as an essential facet of mindfulness: a Comparison of FFMQ patterns in meditating and non-meditating individuals. Mindfulness **4**(3), 203–212 (2013)

64. Chiesa, A., Calati, R., Serretti, A.: Does mindfulness training improve cognitive abilities? A systematic review of neuropsychological findings. Clin. Psychol. Rev. **31**(3), 449–464 (2011)

65. Sirois, F.M., Tosti, N.: Lost in the moment? An investigation of procrastination, mindfulness, and well-being. J. Rational-Emot. Cogn. Behav. Ther. **30**(4), 237–248 (2012)

66. Libby, D.J., Worhunsky, P.D., Pilver, C.E., Brewer, J.A.: Meditation-induced changes in high-frequency heart rate variability predict smoking outcomes. Front. Human Neurosci. **6**, 54 (2012)

67. Evans, D.R., Baer, R.A., Segerstrom, S.C.: The effects of mindfulness and self-consciousness on persistence. Personal. Individ. Differ. **47**(4), 379–382 (2009)

68. Ostafin, B.D., Bauer, C., Myxter, P.: Mindfulness decouples the relation between automatic alcohol motivation and heavy drinking. J. Soc. Clin. Psychol. **31**(7), 729–745 (2012)

69. Hofmann, S.G., Sawyer, A.T., Witt, A.A., Oh, D.: The effect of mindfulness-based therapy on anxiety and depression: a meta-analytic review. J. Consult. Clin. Psychol. **78**(2), 169–183 (2010)

70. Chan, D., Woollacott, M.: Effects of level of meditation experience on attentional focus: is the efficiency of executive or orientation networks improved? J. Altern. Complement. Med. **13**(6), 651–657 (2007)

71. Allen, M., Dietz, M., Blair, K.S., van Beek, M., Rees, G., Vestergaard-Poulsen, P., Roepstorff, A.: Cognitive-affective neural plasticity following active-controlled mindfulness intervention. J. Neurosci. **32**(44), 15601–15610 (2012)

72. Wenk-Sormaz, H.: Meditation can reduce habitual responding. Altern. Ther. Health Med. **11**(2), 42–58 (2005)

73. Oberle, E., Schonert-Reichl, K.A., Lawlor, M.S., Thomson, K.C.: Mindfulness and inhibitory control in early adolescence. J. Early Adoles. **32**(4), 565–588 (2011)

74. Chambers, R., Lo, B.C.Y., Allen, N.B.: The impact of intensive mindfulness training on attentional control, cognitive style, and affect. Cogn. Ther. Res. **32**(3), 303–322 (2008)

75. Zeidan, F., Johnson, S.K., Diamond, B.J., David, Z., Goolkasian, P.: Mindfulness meditation improves cognition: evidence of brief mental training. Conscious. Cogn. **19**(2), 597–605 (2010)
76. Glomb, T.M., Duffy, M.K., Bono, J.E., Yang, T.: Mindfulness at work. Res. Personnel Human Res. Manage. **30**, 115–157 (2011)
77. Ostafin, B.D., Kassman, K.T.: Stepping out of history: mindfulness improves insight problem solving. Conscious. Cogn. **21**(2), 1031–1036 (2012)
78. Ruedy, N.E., Schweitzer, M.E.: In the moment: the effect of mindfulness on ethical decision making. J. Bus. Ethics **95**(1), 73–87 (2011)
79. Shapiro, S.L., Jazaieri, H., Goldin, P.R.: Mindfulness-based stress reduction effects on moral reasoning and decision making. J. Pos. Psychol. **7**(6), 504–515 (2012)
80. Colzato, L.S., Ozturk, A., Hommel, B.: Meditate to create: the impact of focused-attention and open-monitoring training on convergent and divergent thinking. Front. Psychol. **3**, 116–131 (2012)
81. Evans, J.S.B.T., Stanovich, K.E.: Dual-process theories of higher cognition: advancing the debate. Perspect. Psychol. Sci. **8**(3), 223–241 (2013)
82. Jost, J.T., Kruglanski, A.W., Nelson, T.O.: Social metacognition: an expansionist review. Personal. Soc. Psychol. Rev. **2**(2), 137–154 (1998)
83. Reb, J., Narayanan, J., Chaturvedi, S.: Leading mindfully: two studies on the influence of supervisor trait mindfulness on employee well-being and performance. Mindfulness **5**(1), 36–45 (2012)
84. Van Doesum, N.J., Van Lange, D.W., van lange, P.M.: social mindfulness: skill and will to Navigate the Social World. Journal of Personality and Social Psychology. **105**(1), 86–103 (2013)
85. Barnes, S., Brown, K.W., Krusemark, E., Campbell, W.K., Rogge, R.D.: The role of mindfulness in romantic relationship satisfaction and responses to relationship stress. J. Marital Family Ther. **33**(4), 482–500 (2007)
86. Wachs, K., Cordova, J.V.: Mindful relating: exploring mindfulness and emotion repertoires in intimate relationships. J. Marital Family Ther. **33**(4), 464–481 (2007)
87. Facione, P.A.: The Delphi Report: Committee on Pre-college Philosophy. California Academic Press, Millbrae (1990)
88. Reeves, M.F.: An application of bloom's taxonomy to the teaching of business ethics. J. Bus. Ethics **9**(7), 609–616 (1990)
89. Dwyer, C.P., Hogan, M.J., Stewart, I.: The Promotion of critical thinking skills through argument mapping. In: Horvart, C.P., Forte, J.M. (eds.) Critical Thinking, pp. 97–122. Nova Science Publishers, New York (2011)
90. Dwyer, C.P., Hogan, M.J., Stewart, I.: The Evaluation of argument mapping-infused critical thinking instruction as a method of enhancing reflective judgment performance (in preparation)
91. Dwyer, C.P., Hogan, M.J., Stewart, I.: The evaluation of argument mapping as a learning tool: Comparing the effects of map reading versus text reading on comprehension and recall of arguments. Think. Skills Creativity **5**(1), 16–22 (2010)
92. Halpern, D.F.: Teaching Critical Thinking for Transfer Across Domains: Disposition, Skills, Structure Training, and Metacognitive Monitoring. Am. Psychol. **53**(4), 449 (1998)
93. Halpern, D.F.: Manual: The Halpern Critical Thinking Assessment. Schuhfried, Vienna (2010)
94. Halpern, D.F.: Is intelligence critical thinking? Why we need a new definition of intelligence. In: Kyllonen, P.C., Roberts, R.D., Stankov, L. (eds.) Extending Intelligence: Enhancement and New Constructs, pp. 293–310. Taylor & Francis Group, New York (2008)
95. Facione, P.A., Facione, N.C., Blohm, S.W., Giancarlo, C.A.: The California Critical Thinking Skills Test: CCTST. Form A, Form B, and Form 2000. Test Manual, 2002 Updated Edition. Insight Assessment, Millbrae (2002)
96. Ennis, R.H.: Is critical thinking culturally biased? Teach. Philos. **21**(1), 15–33 (1998)
97. Ku, K.Y.L.: Assessing Students' critical thinking performance: urging for measurements using multi-response format. Think. Skills Creativity **4**(1), 70–76 (2009)

98. Valenzuela, J., Nieto, A.M., Saiz, C.: Critical thinking motivational scale: a contribution to the study of relationship between critical thinking and motivation. J. Res. Educ. Psychol. 9(2), 823–848 (2011)
99. Ennis, R.H.: The nature of critical thinking: an outline of critical thinking dispositions and abilities. University of Illinois, http://faculty.education.illinois.edu/rhennis/documents/TheN atureofCriticalThinking_51711_000.pdf (2011)
100. Facione, P.A., Facione, N., Giancarlo, C.A.F.: The California Critical Thinking Disposition Inventory: CCTDI Test Manual, California Academic Press (1996)
101. Paul, R., Elder, L.: A Guide for Educators to Critical Thinking Competency Standards: Standards, Principles, Performance Indicators, and Outcomes with a Critical Thinking Master Rubric, Vol. 8. Foundation Critical Thinking (2005)
102. Colucciello, M.L.: Critical thinking skills and dispositions of baccalaureate nursing students—a conceptual model for evaluation. J. Prof. Nurs. 13(4), 236–245 (1997)
103. Facione, P.A., Facione, N.C., Giancarlo, C.A.: The Disposition toward critical thinking: its character, measurement, and relationship to critical thinking skill. Inf. Logic 20(1), 61–84 (2000)
104. Facione, N.C., Facione, P.A., Sanchez, C.A.: Critical thinking disposition as a measure of competent clinical judgment: the development of the california critical thinking disposition inventory. J. Nurs. Educ. 33(8), 345–350 (1994)
105. Profeto-McGrath, J.: The relationship of critical thinking skills and critical thinking dispositions of baccalaureate nursing students. J. Adv. Nurs. 43(6), 569–577 (2003)
106. Rimiene, V.: Assessing and developing students' critical thinking. Psychol. Learn. Teach. 2(1), 17–22 (2002)
107. King, P.M., Kitchener, K.S.: Developing Reflective Judgment: Understanding and Promoting Intellectual Growth and Critical Thinking in Adolescents and Adults. Jossey Bass, San Francisco (1994)
108. Huffaker, D.A., Calvert, S.L.: The new science of learning: active learning, metacognition and transfer of knowledge in e-learning applications. J. Educ. Comput. Res. 29(3), 325–334 (2003)
109. U.S. National Research Council.: Technology and Assessment: Thinking Ahead. Board on Testing and Assessment Center for Education, Washington, DC (2002)
110. King, P.M., Wood, P.K., Mines, R.A.: Critical thinking among college and graduate students. Rev. High. Educ. 13(2), 167–186 (1990)
111. Wood, P.K.: Context and Development of Reflective Thinking: A Secondary Analysisof the Structure of Individual Differences. Unpublished manuscript. University of Missouriat Columbia, Missouri (1993)
112. Dewey, J.: How We Think: A Restatement of the Relation of Reflective Thinking to the Educative Process. Heath & Co., Lexington (1933)
113. King, P.M., Kitchener, K.S.: Reflective judgment: theory and research on the development of epistemic assumptions through adulthood. Educ. Psychol. 39(1), 5–18 (2004)
114. Kitchener, K.S., King, P.M.: Reflective judgment: Concepts of justifications and their relation to age and gender. J. Appl. Develop. Psychol. 2, 89–116 (1981)
115. Brabeck, M.M.: The Relationship between critical thinking skills and development of reflective judgment among adolescent and adult women. In: 89th Annual Convention of the American Psychological Association, Los Angeles (1981)
116. Huffman, K., Vernoy, M., Williams, B., Vernoy, J.: Psychology in Action. Wiley, New York (1991)
117. Baril, P.B., Cunningham, B.M., Fordham, D.R., Gardner, R.L., Wolcott, S.K.: Critical thinking in the public accounting profession: aptitudes and attitudes. J. Account. Educ. 16(3–4), 381–406 (1998)
118. Perkins, D.N., Ritchhart, R.: When is good thinking? In: Dai, D.Y., Sternberg, R.J. (eds.) Motivation, Emotion, and Cognition: Integrative Perspectives on Intellectual Functioning and Development, pp. 351–384. Erlbaum, Mawah (2004)

119. van Gelder, T.J.: Enhancing Deliberation through computer supported argument mapping. In: Kirschner, P.A., Buckingham-Shum, S., Carr, C. (eds.) Visualizing Argumentation: Software Tools for Collaborative and Educational Sense-Making, pp. 97–115. Springer, London (2003)

120. van Gelder, T.J.: How to improve critical thinking using educational technology. In: Kennedy, G., Keppell, M., McNaught, C., Petrovic, T. (eds.) Meeting at the Crossroads: Proceedings of the 18th Annual Conference of the Australian Society for Computers in Learning in Tertiary Education, pp. 539–548. Biomedical Multimedia Unit, University of Melbourne, Melbourne (2001)

121. Facione, P.A.: The California Critical Thinking Skills Test (CCTST): Forms A and B. The CCTST Test Manual. California Academic Press, Millbrae (1990)

122. Dwyer, C.P., Hogan, M.J., Stewart, I.: An evaluation of argument mapping as a method of enhancing critical thinking performance in e-learning environments. Metacogn. Learn. 7(3), 219–244 (2012)

123. Dillenbourg, P.: What do you mean by collaborative learning? In: Dillenbourg, P. (ed.) Collaborative Learning: Cognitive and Computational Approaches, pp. 1–19. Elsevier, Oxford (1999)

124. Hinsz, V.B., Tindale, R.S., Vollrath, D.A.: The Emerging conceptualization of groups as information processors. Psychol. Bull. 121(1), 43 (1997)

125. Hinsz, V.B.: Metacognition and mental models in groups: an illustration with metamemory of group recognition memory. In: Salas, E., Fiore, S.M. (eds.) Team Cognition: Understanding the Factors that Drive Process and Performance, pp. 33–58. American Psychological Association, Washington (2004)

126. Thompson, L., Cohen, T.R.: Metacognition in teams and organizations. In: Briñol, P., DeMarree, K.G. (eds.) Social Metacognition, pp. 283–302. Psychology Press, New York (2012)

127. McIntyre, R.M., Salas, E.: Team performance in complex environments: what we have learned so far. In: Guzzo, R., Salas, E. (eds.) Team Effectiveness and Decision Making in Organizations, pp. 333–380. Josey-Bass, San Francisco (1995)

128. Salas, E., Sims, D.E., Burke, C.S.: Is there a "big five" in teamwork? Small Group Res. 36(5), 555–599 (2005)

129. Doten, G.W., Cockrell, J.T., Sadacca, R.: The Use of Teams in Image Interpretation: Information Exchange, Confidence, and Resolving Disagreements (Tech. Rep. 1151). U.S. Army Personnel Research Office, Washington, DC (1966)

130. Siegel, A.I., Federman, P.J.: communications content training as an ingredient in effective team performance. Ergonomics 16(4), 403–416 (1973)

131. Porter, C.O.L.H., Hollenbeck, J.R., Ilgen, D.R., Ellis, A.P.J., West, B.J., Moon, H.: Backup behaviors in teams: the role of personality and legitimacy of need. J. Appl. Psychol. 88(3), 391–403 (2003)

132. Scheuer, O., Loll, F., Pinkwart, N., McLaren, B.M.: Computer-supported argumentation: a review of the state of the art. Int. J. Comput. Support. Collabor. Learn. 5(1), 43–102 (2010)

133. Azevedo, R., Cromley, J.G.: Does training of self-regulated learning facilitate students' learning with hypermedia? J. Educ. Psychol. 96(3), 523–535 (2004)

134. Kramarski, B., Mevarech, Z.R.: Enhancing mathematical reasoning in the classroom: effects of cooperative learning and metacognitive training. Am. Educ. Res. J. 40(1), 281–310 (2003)

135. Lee, H.W., Lim, K.Y., Grabowski, B.L.: Improving self-regulation, learning strategy use, and achievement with metacognitive feedback. Educ. Tech. Res. Dev. 58(6), 629–648 (2010)

136. Kreijns, K., Kirschner, P.A., Jochems, W.: The sociability of CSCL environments. Educ. Technol. Soc. 5(1), 8–22 (2002)

137. Kwon, K., Hong, R.Y., Laffey, J.M.: The educational impact of metacognitive group coordination in computer-supported collaborative learning. Comput. Hum. Behav. 29(4), 1271–1281 (2013)

138. Ada, W.W.: Computer-supported collaborative learning and higher-order thinking skills: a case study of textile studies. Interdisc. J. E-Learning Learn. Objects **5**, 145–167 (2009)
139. Peterson, R.S.: A directive leadership style in group decision-making can be both virtue and vice: evidence from elite and experimental groups. J. Pers. Soc. Psychol. **72**(5), 1107 (1997)
140. Harney, O., Hogan, M.J., Broome, B.: Investigating the effects of feedback on argumentation style, consensus and perceived efficacy in collaborative learning. In Gómez Chova, L.A., López Martínez, Candel Torres, I. (eds.). INTED2014 Proceedings, pp. 1638–1647. IATED Academy, Valencia (2014)
141. Seibold, D.R., Meyers, R.A.: Group argument: a structuration perspective and research program. Small Group Res. **38**, 312–336 (2007)
142. Mercer, N.: Sociocultural discourse analysis. J. Appl. Linguistics **1**(2), 137–168 (2004)
143. Erkens, G.: Multiple episode protocol analysis (MEPA). Version 4.10. http://edugate.fss.uu.nl/mepa/ (2014)
144. Feeney, P., Harney, O., Hogan, M.J.: The effects of peer feedback on cooperative conversationaldynamics in collaborative learning. Unpublished manuscript. (2014)

Chapter 5
Metacognition, Self-regulation and Assessment in Problem-Solving Processes at University

María Consuelo Sáiz-Manzanares and Eduardo Montero-García

Abstract We firstly present an analysis of the most recent research in the field of metacognition and self-regulation linked to teaching-learning processes. Our approach involves the assessment of these processes through the use of rubrics for students to conduct self-assessments (at the start-during-and at the end) of their own learning processes and for teachers to evaluate these processes. We also refer to the processes of calibration and feedback as essential components in the assessment and learning processes. In second place, we examine possible methods of metacognitive intervention in self-regulation. Finally, we present an example application of these methods to the solution of a Thermodynamics problem from a degree course in Engineering Electronics and Automation. In this last section, we present the instruments for the measurement of the evaluation: questionnaires on the prior knowledge of students, and rubrics for the assessment of the conceptual and the procedural content of the solution as well as a short summary of the results of three research projects.

Keywords Metacognition · Self-regulation · Higher education · Rubrics · Calibration · Feedback · Training in solving problem

Abbreviations

A   Area
C   Speed
h Enthalpy

M.C. Sáiz-Manzanares (✉)
Department of Educational Sciences, Faculty of Humanities and Education,
University of Burgos, C/Villadiego S/N, 09001 Burgos, Spain
e-mail: mcsmanzanares@ubu.es

E. Montero-García
Department of Electromechanical Engineering, Higher Polytechnic School,
University of Burgos, Avenida de Cantabria, S/N, 09006 Burgos, Spain
e-mail: emontero@ubu.es

© Springer International Publishing Switzerland 2015
A. Peña-Ayala (ed.), *Metacognition: Fundaments, Applications, and Trends*,
Intelligent Systems Reference Library 76, DOI 10.1007/978-3-319-11062-2_5

K	Kelvin
kg	Kilogram
kJ	Kilojoule
kPa	Kilopascal
kW	Kilowatt
m	Meter
m^2	Square meter
Pa	Pascal
Q	Heat
SI	International system of units
SRL	Self-regulated learning
T	Temperature
W	Power
W_{vc}	Power that the compressor consumes
WWW&H	Which, when, why and how

5.1 Introduction

Metacognition is a term defined by Flavell [1] that refers to the conscious reflection of individuals on their own cognition processes. It implies awareness of what we do and how we do it. From this perspective, metacognition has a relation with individual knowledge of their own mental processes, on how these processes function in the mind. Knowledge of their own cognition gives individuals greater fluidity in their mental thought processes. However, knowledge of those non-visible thought processes is a complex and difficult task. We cannot observe those processes directly and we have to make assumptions about them through different mental observations. The factors that constitute the metacognitive world according to Flavell [2] are:

- *Cognitive ends or goals* related to the tasks that we have to solve.
- *Metacognitive experiences* that refer to information held by individuals on the use of their own cognitive skills in similar tasks.
- *Cognitive actions* are the strategies that an individual employs in the solution of a problem.
- *Metacognitive knowledge* is knowledge concerning the three variables: person, task, and strategies.

In his approach leaves the door open to learning, as he connects metacognitive capability with the possibility of instruction and therefore of improvement. Another relevant name in this field is Ann Brown [3], an author who relates metacognition to self-regulation. Her approach leads to self-analysis of problem-solving heuristics related to planning processes, feedback and assessment in learning processes [4].

Recent studies [5, 6] acknowledge metacognition as a very important predictor of learning. Their findings suggest that metacognitive skills explained 40 % of the variance in learning results. Veenman et al. [8] distinguished between the following integral elements within the term metacognition:

- Knowledge of cognition: includes *declarative knowledge* on the cognitive system, *procedural knowledge* that refers to the completion of cognitive strategies and *conditional knowledge* that refers to the utility of those strategies.
- Regulation cognition: refers to the metacognitive skills needed for control over which strategy to use in the solution of tasks and includes planning, monitoring and assessment strategies.
- Metacognitive knowledge: based on the self-awareness that individuals have of their own mental states [9]. This knowledge refers to *declarative knowledge* and includes the interrelation between the variables person, task and strategies, in the same terms as those used by Veenman and Elshout [10]. As a concept, it has links to early experiences in task solving and involves the memory in the solution of similar tasks. Metacognitive knowledge can be correct or incorrect, as students can underestimate or overestimate their own competencies in the completion of a task. This concept has to do with individual subjective perception of task complexity and is therefore a poor predictor of learning results. Likewise, metacognitive knowledge interacts with other aspects of metacognition such as metacognitive experiences and metacognitive skills.
- Conditional knowledge: a part of *declarative knowledge* that refers to when and why we can use certain metacognitive strategies in problem solving. This type of knowledge will not guarantee effective completion.
- Metastrategic knowledge: includes conditional knowledge and procedural knowledge.
- Metacognitive experiences: related to the experience of an individual in the solution of similar tasks. This concept is interrelated with personal and task-related aspects. It includes metacognitive feelings, metacognitive judgments and task-specific knowledge. They all calibrate the effort and the time taken to solve a task. Both metacognitive experiences and metacognitive knowledge arise from mental monitoring processes.
- Metacognitive skills: refer to *procedural knowledge* acquired from observation. These skills guide and control self-learning and problem solving. It is important to follow the use of these strategies at the start, during and at the end of the problem-solving process. These skills refer to purposeful control of individuals over their own cognition.

  - *At the start of the task*, the individual analyzes the task and activates any prior knowledge, sets goals, and applies problem-solving plans and strategies for regulation of the cognitive processes.
  - *During the development of the task*, the individual follows the established plan in a systematic manner, by monitoring and checking the process. These activities guide and control the completion of the task.
  - *At the end of the development of the task*, the individual evaluates and considers the problem-solving process and the final results. Thus, learning from the procedure takes place that will help find solutions to similar tasks.

- Metacognitive activities: these vary in accordance with the knowledge domain. Metacognitive activities activate prior knowledge and orient goal setting. They likewise have implications for the processes of generalization and transference of learning, facilitate the construction of new learning and generate new meta-cognitive knowledge.

5.2 Metacognitive Skills and Self-regulation

The self-regulation of learning has links to individual self-perception of competence in a given subject or a task that the individual has to solve. The metacognitive experiences that the individual will apply to solve the task will mediate that individual's self-perception [11].

Metacognition appears to have a direct influence on the self-regulation of learning processes. Metacognitive experiences influence both declarative and procedural knowledge. Individuals have to review their personal experience in the solution of similar tasks and their prior knowledge in that subject, in order to solve a task. In Fig. 5.1, we present a schematic diagram showing the relations between all of the concepts discussed above.

Self-regulated learning is an active and constructive process in which we can distinguish different levels of control. Knowledge on regulation of self-learning is necessary, for self-regulation to be effective. Students have to know how they are learning;

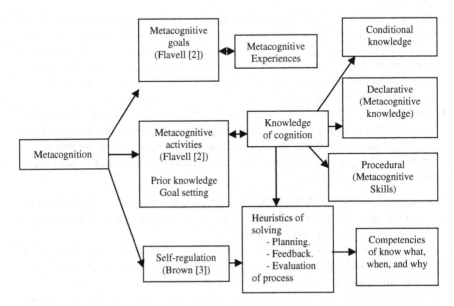

Fig. 5.1 Relations between the concept of metacognition and the learning process adapted by Sáiz et al. [11]

also, they have to use self-regulation strategies in an effective way in the learning processes. This type of knowledge is especially relevant in university areas [12].

According to Wäschle et al. [13], we can distinguish between the following components in a self-regulation process:

- *Goal-setting*: helps learners to decide which learning strategies are the most adequate to resolve a problem in an effective way [14, 15].
- *Self-set learning goals*: guide the learning process and promote reflection on how and why to choose goal success.
- *Learning strategies*: metacognitive strategies of acquisition, codification, recovery and support for information processing.
- *Type of tasks*: deep learning is different from superficial learning and the strategies involved in the solution of one or another.
- *Self-reflection on goal achievement*: helps students to regulate their learning.

Along these lines, Zimmerman and Moylan [16] proposed a cyclic-phase self-regulatory feedback loop model. These authors refer to three interconnected phases.

- *Forethought phase*: composed of two categories: *task analysis processes* (implying a breakdown of the learning tasks and a personal construction of the strategies on the basis of prior knowledge). This analysis includes setting goals, refers to specific outcomes and produces a specific feedback loop that requires self-evaluation. Likewise, it requires *strategic planning* (referring to the selection and the construction of task and context-specific learning methods and *sources of self-motivation beliefs* that include self-efficacy perceptions, outcome expectations, intrinsic interest and learning orientation).
- *Performance phase*: this phase includes *self-control* methods (referring to task strategies that include self-instruction, imagery, time management, environmental structuring, help-seeking methods, interest incentives and self-consequences. The use of self-regulation strategies increases the quality of student learning in the different domains. The strategies have to adapt to forms of learning of the students and specific learning domain) and *self-observation* (these processes play a role in student self-control over the effort they invest in learning. In these processes, there are two factors that are metacognitive-monitoring that improve self-recording, which in turn analyzes the type of responses and any possible error).
- *Self-Reflection phase*: this phase consists of *self-judgments* (which include self-evaluation, comparisons of the development and evaluation of learning on the basis of standard criteria. Such criteria may be set by the teacher and by fellow students in the completion of the tasks) and *self-reactions* (which refer to self-satisfaction which is to say both the cognitive and affective reactions of the self-judgments which can be adaptive and defensive. Motivation is a powerful predictor of successful learning on the basis of metacognitive learning strategies).

In addition to the above-mentioned factors, we should consider that the generalization of learning and learning transference processes are not automatic, even when learning may have been successful. The learning responses depend on the knowledge that is available in the learning domain on the cognitive and metacognitive spaces of the students. Likewise, we have found the following difficulties in these studies:

- The reliability of the protocols that we use to study awareness in self-analysis of the processes and of the learning results.
- The complexity of the analysis of the self-regulation processes as it includes metacognitive, cognitive, motivational, contextual and behavioral aspects.
- Teaching staff should direct instruction in self-regulation towards providing guidance and support for the self-learning processes, which not only implies instruction, but also self-analysis of the instruction processes at the start, during and at the end of the teaching-learning process.

5.3 Instruments for the Evaluation of Aspects Relating to Processes of Self-regulation

As we have seen, SRL refers to the processes of monitoring and control of cognitive and metacognitive processes throughout a learning situation. SRL includes aspects of planning, goal setting, implementing strategies and self-monitoring [17–19].

There are different theoretical models that seek to explain the relation between self-regulation processes and meta-cognitive processes. However, the common denominator in all of them is the analysis of the following variables:

- *Knowledge*: includes facts, concepts and schemes developed in learning tasks.
- *Metaknowledge*: refers to self-knowledge in self-regulation. This knowledge of cognition includes a greater variety of general strategies in each knowledge domain which are strategies for goal-setting, planning, implementing strategies, monitoring, and evaluating self-learning.
- *Motivation*: refers to the beliefs and the strategies of self-efficacy that allow perseverance and constancy in learning based on error analysis.

Van der Stel and Veenman [5] distinguished three models that can explain the relation between metacognitive skills and the development of learning. One would be the cognitive model in which the metacognitive skills would form part of the cognitive skills and these would not contribute in a special way to explaining the learning process. The second model advocates independence between the metacognitive skills and the learning processes and the third is a mixed model in which a correlation exists between both. The authors highlighted the difficulty that exists in the evaluation of metacognitive processes. Likewise, they pointed out that they found the mixed model more often among secondary school students and university students.

Longitudinal studies have shown moderate correlations between cognitive and metacognitive skills and learning, the highest of which were correlations between metacognitive skills and learning with a degree of independence of cognitive skills [6]. According to Schraw et al. [20], there are different forms of measuring self-regulation processes in learning tasks.

Reference is made both to *on-line* (e.g., think aloud protocols and calibration judgments) and to *off-line* (e.g., questionnaires and interviews) methods. It is important to collect information on self-regulation processes from different measurement methods, in order to have enough data to allow a triangulation analysis of the results.

The *off-line* methods refer to evaluations through questionnaires and interviews that the learner completes before and after the completion of a learning task or a problem-solving process. The questions refer to the frequency of use of the procedural strategies. The *on-line* methods refer to the evaluation during the completion of the tasks, which use self-observation techniques, think aloud protocols or logline records on the computer. The advantage of the *off-line* methods is that we can administer them to broad samples of individuals as opposed to the *on-line* methods, administration of which is individual.

The second problem is validity. Evaluation with *off-line* methods involves information that the individual retains in the memory and is not always crystal clear. Evaluation with the *on-line* methods is by means of individual introspection of thought processes during the completion of the task and implies a suitable verbal level to express those thoughts out aloud. Both methods differ in their external validity.

This validity is important for the prediction of the learning competencies [20]. Traditionally, educators have used *off-line* methods for the evaluation of conceptual competences in Higher Education. Research has indicated that university students use different types of learning strategies according to the course they are following; in other words, in accordance with the type of content that they have to process [21]. This is why self-perception of knowledge is a variable that is subject to individual learning experience.

This aspect can on occasions lead to error or the distortion of reality by the learner, as self-perception is conditioned by motivational variables related to the history of earlier learning [22–24]. The learning processes may be clearly dependent on the learning context and the type of task [25]. Likewise, the results of the learning evaluation appear to vary in accordance with the method that is used. The evaluation of conceptual knowledge (conceptual competencies) is not the same as the evaluation of problem-solving skills (procedural competencies). A summary of the above may be seen in Tables 5.1 and 5.2.

5.3.1 Student Self-assessment of the Self-learning Process and Evaluation by the Teacher of the Student Learning Process

Self-evaluation is a tool available to teachers to evaluate the teaching-learning process of their students and to guide their instruction, an aspect that we will examine later on.

With regard to the first criteria, referred to earlier on, Zimmerman and Moylan [16] described the relation between self-regulation and self-assessment in three phases: Forethought phase, Performance phase and Self-reflection phase.

The difference between self-assessment and self-evaluation is that the first is a more qualitative evaluation that does not necessarily imply the use of scores. The second refers to a method of reflection on previously defined assessment criteria.

Table 5.1 Relation between the different aspects of metacognition and the different methods of evaluation

Type of metacognition		Implications	Evaluation procedures
Declarative knowledge	Metacognitive knowledge	May be correct or incorrect, and depends on the learning history, the self-knowledge of the individual, as well as motivational factors. It is subjective and cannot guarantee acceptable completion of the task (Veenman [7])	*Off-line* methods
	Cognitive knowledge		
	Metacognitive experiences [1, 9]	These arise from observation and depend on motivation	*Off-line* methods
	Self-knowledge	Resistant to change, especially when the learner attributes error to external causes	*Off-line* methods
	Conditional knowledge	Knowledge of which metacognitive strategies to apply for each purpose. It cannot guarantee acceptable processes for completion and it represents a bridge between meta-cognitive knowledge and procedural knowledge in which how to use the strategies (conscious) and the skills (conscious and automatic) would be used	*Off-line* methods
Procedural knowledge	Self-Regulation of cognition	• Orientation	*On-line* methods
		• Planning	
	Metacognitive skills	• Evaluation	*Off-line* methods
		• Elaboration	

The first analyzes the learning process (start, development and end) and the basis of the second is fundamentally the final product. Planning of an acceptable self-assessment is necessary before the development of the learning process and should continue throughout the development of the task.

5.3.2 Instruments to Measure Self-assessment

In this section, we discuss the rubrics and the calibration and feedback procedures.

Rubrics. Rubrics are a suitable tool for the analysis of self-assessment. However, if they are to be effective, they should comply with the following points [26]:

• A list of criteria specified in the form of achievable goals in the problem-solving tasks. A good instrument for the design of those evaluation criteria is the taxonomy of Bloom, as it cognitively defines the difficulty of each problem-solving step [27].

Table 5.2 Evaluation indicators of the metacognitive skills in the *on-line* evaluation methods

Evaluation methods of metacognitive skills	Evaluation criteria	Evaluation indicators
Quantitative *on-line* method (analyzes the frequency of appearance of the indicators of metacognitive activities)	Orientation	Reads the problem or the task during the problem-solving activity
		Activates prior task-related knowledge
		Prepares a representative scheme of the problem
		Activates prior knowledge of the goals
		Construes the essential elements of the problem and what it requires in order to predict a response
	Planning	Expresses the problem-solving strategies while analyzing the problem
		Identifies each of the steps to solve the problem one by one
	Evaluation	Observes and detects errors
		Compares the sentences of the problem with the responses that are given
		Notes any errors
	Elaboration	Completes a summary and a plan of the conclusions
Qualitative *on-line* method (analyze the depth of the metacognitive activities on a scale of 0-none-to 4 maximum)	Orientation	Instruments of observation: methods of thinking out loud and *log-line* registers on the computer

- A graduation scale with different levels of task completion measured at a qualitative level. Students can compare their work, placing it next to one of the different evaluation standards defined in accordance with their own perceptions of how it was completed.
- A design that allows teachers and students to study both the process and the product of the learning results. The teacher has to use the rubrics before the student starts the learning activity, as it will help to centre both learning goals and the problem-solving task [28]. This self-assessment procedure has an important effect on the processes of self-regulation, as it allows reflection on self-learning and facilitates its improvement [29]. The rubrics assist the development of planning and self-assessment strategies, are of assistance for reflection on solving tasks, and likewise allow self-correction of possible errors [30].
- A combination of rubrics and metacognitive activities such as modeling, self-instruction and self-monitoring.

In conclusion, we can recommend rubrics as excellent tools with which to orient the learning goals and for effective planning at each stage of the solution. These will help with the processes of instruction, to clarify how to do a task and to specify the evaluation criteria at different qualitative levels of achievement [31].

Calibration and Feedback. We can define calibration as the relation that exists between the perceptions that students have towards their own learning and its real results [32]. Calibration plays an important role in the acquisition of skills by the students that allows them to follow through successful self-regulation processes in self-learning. Calibration is not a unidimensional concept, as it contains others: development, strategies, confidence in what we know, problem solving, learning tasks that imply a challenge. Effective calibration should take account of the next points [33]:

- Measures on knowledge.
- Correction of responses.
- Effectiveness of strategies.
- Achievement of specific learning goals.

However, objective measurement of all these aspects is difficult. On the other hand, we can define feedback as the influence with greatest learning power [34], although it has a series of limitations:

- We find it in laboratory contexts more so than in ecological learning contexts.
- It relates to the perception that individuals have towards their own learning competency.
- Effective feedback has to give information on the following questions:
  - Where am I going?
  - How will I go there?
  - Where to next?

We can distinguish between four levels of feedback:

- Task performance.
- A process of understanding.
- Regulatory or metacognitive process.
- Self-regulation.

The third and the fourth points generate the highest levels of feedback. Those levels include [35]:

- *Product-oriented components*: offering information on a specific domain, on a specific task to complete.
- *Orientation towards the process*: providing responses to such questions as when, how and why certain strategies are appropriate. Those components combine the demands and the responses to the three feedback questions and focus on self-regulation processes.

Likewise, those orientation processes include two important sub-processes:

- *Knowledge of mistakes*: informing students of their weaknesses.
- *Knowledge on how to continue*: giving students task-solving strategies.

According to Narciss, [36] feedback with an orientation function provides reinforcements for the learner, facilitates the solution of the tasks and increases levels of motivation towards learning. The professor should therefore provide instruction that is directed at giving information on the learning process of the students, in relation to the acquisition of competencies and the development of their learning process in relation to those competencies.

The problem centers on the construction of instruments that permit the evaluation of feedback processes during the instruction [37, 38]. The professor should therefore take great care in designing the instruction by taking note of the following points [39].

- Clearly explain the learning expectations to students before beginning the instruction.
- Give precise criteria to students on what successful learning entails.
- Gain familiarity with the beliefs of the students with regard to their learning goals.
- Guarantee that the teaching strives to reduce the distance between what students know and understand and what they believe they know and understand with regard to the object of knowledge.
- Guarantee feedback on the steps directed at the reduction of that distance.

However, teachers should have a realistic idea of the following points, so that calibration and feedback are sufficiently well developed:

- How their students learn.
- How to help their students to construct their learning.
- How to help their students to comprehend better, to learn better.
- And, how to measure the calibration and prior knowledge of their students.

We can affirm that the feedback has a powerful influence on the development of successful learning responses. Hattie and Gan [40] reviewed 12 studies on the meta-analysis of feedback based on 196 studies and 6,972 values of the effect. They found a high effect value (explained variance $d = 0.79$) and found that not all the types of feedback are equal. The most successful feedback took the following into account:

- The relation between individual learning characteristics of the learners and the feedback strategies that they employ.
- Cognitive and metacognitive strategies and the motivations of the learner.
- The frequency and the duration time of the feedback.

Feedback can reduce the distance between what the student knows from prior learning in a specific domain in the here and how and the goals of that learning. We can distinguish between three levels of feedback:

- *Task-level feedback*: gives information on a correct response. This level is important for new learners as it implies the relation between the type of task and its learning.
- *Process-level feedback*: this process relates to their self-efficacy or the self-confidence that individuals have in their own learning.
- *Feedback at self-regulation level*: this level includes self-evaluation strategies and strategies to search for feedback information. It implies a learning capability to create its own feedback and self-assessment processes. This process guides the student in when and where to select and to use the process level and to develop the processes of self-assessment, self-seeking, self-appraisal and self-management.

5.4 Training, Metacognition, and Self-regulation

Considering what we have pointed out above, the instruction process should respond to the characteristics of the learner, the tasks and the teacher. The first of these refer to prior knowledge, the adjustment to the perception of such knowledge and the level of acquisition and development of the learning strategies of the learner, (cognitive, metacognitive and motivational) both for specific and for general dominions.

The second refers to the type of tasks. And the third refers to the type of feedback that the teacher gives, to the tutoring process in the feedback and to the feedback strategies; which guide the development of successful feedback. These strategies are: information on knowledge, information on the type of error, information on procedural knowledge, information on metacognitive procedure, tutorization on errors and their detection, and suitable strategies for their successful solution. In line with Veenman [8], training in metacognitive skills should follow three principles:

- *Position of synthesis*: the teacher has to give the instruction in the context of the task and has to relate the application of the metacognitive skills needed for the solution of the task. The teaching has to connect the task with the conditional metacognitive knowledge and to apply the relevant skill to the task solution routine (procedural knowledge).
- *Give information on the instruction*: students have to be informed of the benefits that the application of strategies and metacognitive skills have for their learning, as those students who do not naturally apply them will need to invest greater effort in analyzing the processes, in order to complete them. Hence, the teacher should transmit the advantages of making such efforts to the learner for successful learning.
- *Prolonged instruction*: a metacognitive intervention which to be effective has to show continuity over time.

Training in metacognitive strategies may focus on different tasks. There are various metacognitive training proposals, among which we may highlight the following:

- Self-instructional training proposals [41, 42] in which the following steps may be distinguished:

  - Definition of the problem: what do I have to do? (What).
  - Focalization of the attention: how will I do it? (How) This implies planning and the selection of satisfactory solution strategies.
  - Follow up of implementation. How am I carrying it out? Am I following the established plan? If I have problems, how can I solve them? (How).
  - How have I done it? (How) Implies the use of self-assessment and self-correction strategies.

- Likewise, the following training strategies are used in self-instruction:

  - Cognitive modeling. The adult undertakes the task, while exercising self-regulation in the steps to arrive at the correct solution.
  - External-hidden guide, in which students complete the task and give themselves instructions on the steps leading to a solution.
  - Self-assessment. On the process and the product in the solution of the task.

Table 5.3 Comparison of the 'Questions' proposed by Hattie; the 'metacognitive self-instructional training procedure' proposed by Meichenbaum and Goodman; and the 'metacognitive skills' proposed by Veenman

Problem solving	Questions (Hattie [33])	Self-instructional training procedure (Meichenbaum and Goodman [41])	Metacognitive skills (Veenman [43])
Relation between identification of the problem and its solution	How am I doing?	What is my problem?	What?
		What are my tasks?	
	Where are you going?	How can I solve them?	What?
			When?
It should offer information on how task solutions can be improved (strategies)	Where to next?	How can I do...?	Why?
		Am I using the best Plan?	How?
		How did I do?	

- The proposal advanced by Veenman WWW&H [43] which distinguishes between:

  - Which strategies to apply (What skills).
  - When to use them (When).
  - Why to use them (Why).
  - How to develop them (How).

All of this in the context of task completion. We recommend this training in university settings, above all in different knowledge domains such as problem-solving in subjects such as physics and Thermodynamics. Table 5.3 presents a comparative list of the feedback components proposed by Hattie, the self-instructional training procedure of Meichenbaum and Goodman and the training methodology of Veenman.

For all these reasons, it is important to translate explanatory models and metacognitive evaluation into concrete experiences [44, 45]. The following section presents a proposal for self-regulation programs applied to the solution of a Thermodynamics problem on an industrial engineering degree course.

5.5 An Example of Intervention in Self-regulation

In the first place, we will analyze the problem in terms of the prior knowledge needed for its solution, after which we propose a training procedure based on self-regulation and, in third place, certain rubrics that will be of use to the teacher for the initial evaluation as well as for student feedback throughout the process of arriving at the solution. We distinguish between the rubric for the analysis of procedural knowledge and the rubric for the analysis of conceptual knowledge.

5.5.1 Analysis of the Problem to Solve

Problem statement. Air enters a compressor that is operating in a steady state at a pressure of 100 kPa, at a temperature of 290 K, and at a speed of 6 m/s in a tube with a cross-section of 0.1 m². The pressure output is 700 kPa, at a temperature of 450 K, and at a ratio of 2 m/s. Heat is transferred from the compressor to the external atmosphere at a speed of 3 kW. Using the ideal gas model, calculate the power that the compressor consumes, in kW. Tables 5.4 and 5.5 present the data on the problem.

  Analysis of prior knowledge. Table 5.6 presents a list of prior conceptual knowledge needed for the successful solution of the problem.

Table 5.4 Entry and output data

Input data		Output data	
Pressure $_1$	100 kPa	Pressure $_2$	700 kPa
T_1	290 K	T_2	450 K
C_1	6 m/s	C_2	2 m/s

Table 5.5 Other data

A_1	0.1 m²
Q_{vc}	−3 kW
W_{vc}	¿?

Table 5.6 Prior conceptual and procedural knowledge to solve the Thermodynamics

Prior conceptual knowledge of the subject matter (Thermodynamics)
Define the system (open or closed)
Define the frontier of the system
Determine possible interactions with the environment in terms of heat and work at the system frontier
Take into account that the material surrounding a system may be an effective insulation, so that heat is not transferred outside the system
Take into account that the temperature difference between the system and the environment is small in which case heat is not transferred
Take into account that the area may not be sufficient to allow the transference of heat to take place
Take into account that when heat Q is unknown, it is calculated through the *balance of energy*
Know the physical laws and relations with which we can describe the behavior of the system
− First Law of Thermodynamics
− Second Law of Thermodynamics
Know the relations between the properties of the substance or substances
Know the second law of Newton on motion
Know Fourier's heat conduction law
Know what a system in a steady state is
Know the International System of Units (SI): pressure, temperature and speed
Know the relation between the difference between system input and output and their relation with the conservation of heat
Understand the concept of energy
Understand the concept of heat transference
Know the International System of Units (SI): pressure, temperature and speed

Table 5.6 (continued)

Prior conceptual knowledge of the subject matter (Thermodynamics)
Know the relation between the difference between system input and output and their relation with the conservation of heat
Understand the concept of energy
Understand the concept of heat transference
Understand the ideal gas model
Understand the concept of consumed power and its measurement in the International System of Units (SI)
Take into account the importance of differentiating between speeds and quantities
Take into account the location of the system frontier when considering a transfer of energy as heat or as work
Know that heat and work represent modes by which energy is transferred
Prior conceptual knowledge related to the problem
The concept of a compressor
The concept of a steady state
The concept of pressure and its measurement (Pa)
The concept of temperature and its measurement (K)
The concept of section
The concept of power and its measurement (*W*)
Prior procedural knowledge
Knowledge of solution heuristics
A. What does the problem require
What prior knowledge do I have?
What knowledge is missing?
Do I have experience in the solution of this type of problem?
B. How can I solve the question? Which strategies should I use to begin its solution?
Self-knowledge strategies, in the case of not knowing some of the concepts involved
Self-reflection strategies
Planning strategies
Self-assessment strategies

Instruments to measure the teaching-learning process. We explain two types of instruments, the first of which refers to the analysis of prior conceptual and procedural knowledge.

There are various options with which to analyze the starting level of the students. It could be from an objective test to a questionnaire on previous knowledge before the instruction begins. In Table 5.7, a questionnaire is presented as an example with which students can analyze their knowledge level before they begin the problem-solving exercise; a Thermodynamics problem, in this case. The questionnaire also allows the teachers to ascertain the level of prior knowledge of their students before beginning the instruction.

We propose an evaluation based on the use of a rubric to evaluate the development of the problem or task. This instrument, as mentioned above, is of use to the student in the analysis process, for feedback on self-learning and for the teacher. In Table 5.8, an example of a rubric is presented with which to study the

Table 5.7 Questionnaire on the student's prior conceptual and procedural knowledge of Thermodynamics

Thermodynamics knowledge questionnaire Prior knowledge level Subject: Thermal engineering					
Professor: Degree in engineering electronics and automation					
Name and surname					
Age	date				
The responses to the questionnaire are rated on a Likert scale of 1 to 5: 1 is 'not at all' and 5 is 'very well'					
Conceptual knowledge	*Assessment*				
1. I understand the concepts of open and closed system in Thermodynamics	1	2	3	4	5
2. I understand the concept of a border system in a Thermodynamics system	1	2	3	4	5
3. I know the laws and physical relationships with which we can describe the behavior of a system:	1	2	3	4	5
• The Law of Conservation of Mass	1	2	3	4	5
• The First Law of Thermodynamics (Energy conservation)	1	2	3	4	5
• The Second Law of Thermodynamics	1	2	3	4	5
5. I know Newton's second law of motion.	1	2	3	4	5
6. I know the heat transfer model of Fourier	1	2	3	4	5
7. I can define what a stationary system is	1	2	3	4	5
8. I know the International System of Units for the measurement of:	1	2	3	4	5
• Pressure	1	2	3	4	5
• Temperature	1	2	3	4	5
• Speed	1	2	3	4	5
9. I know the relationship between the difference (input and output system) and the relation with the conservation of energy	1	2	3	4	5
10. I know the concept of heat transfer	1	2	3	4	5
11. I know the ideal gas model concept					
12. When I solve Thermodynamics problems, I take the border system into account before considering energy transfer as heat or work	1	2	3	4	5
Procedural Knowledge					
13. I look for information to solve practical laboratory tests	1	2	3	4	5
14. I use planning strategies in my laboratory practice	1	2	3	4	5
15. When I am unable to solve the practical laboratory test, I try other methods	1	2	3	4	5
16. I use self-instruction when I try to solve the practical laboratory test	1	2	3	4	5
17. If I fail to understand the practical test, rather than leave it, I try once again to understand it	1	2	3	4	5
18. When I set up the practical laboratory test, I observe it	1	2	3	4	5
19. Even if I am unable to solve a problem, I am motivated to try to solve other similar problems	1	2	3	4	5

Table 5.8 Evaluation rubric for the Thermodynamics problem-solving process

Steps in the solution	Excellent	Very good	Good	Not good enough	Unacceptable
The balance of energy of a volume control in a steady state with only one input flow and one output flow	The student argues the concept of balance of energy in a steady state with only one input and one output flow	The student explains the balance of energy concept in a steady state with only one input and one output flow	The student formulates the balance of energy concept in a steady state with only one input and one output flow	The student has difficulties with the analysis of the balance of energy concept in a steady state with only one input and one output flow	The student fails to identify the balance of energy concept in a steady state with only one input and one output flow
25 %	5	4	3	2	1
The mass flow (m) can be calculated from the input data into the compressor, using the ideal gas thermal equation	The student predicts and explains mass flow, using the compressor input data, applying the ideal gas thermal equation	The student explains the mass concept using the compressor input data, applying the ideal gas thermal equation	The student describes the mass flow using the data from the compressor input, applying the ideal gas thermal equation	The student has difficulties describing and finding the mass flow with the compressor input data, when applying the ideal gas thermal equation	The student fails to find the mass flow and fails to apply the ideal gas thermal equation
25 %	5	4	3	2	1
Specific enthalpies may be calculated with the transformation tables At 290 K, $h_1 = 290.16$ kJ/kg At 450 K, $h_2 = 451.8$ kJ/kg	The student predicts and explains the specific enthalpies using the transformation tables	The student explains the specific enthalpies using the transformation tables	The student calculates the specific enthalpies using the transformation tables	The student has difficulties finding specific enthalpies using the transformation tables	The student does not find the specific enthalpies and has difficulty understanding the concept of enthalpy

(continued)

Table 5.8 (continued)

Steps in the solution	Excellent	Very good	Good	Not good enough	Unacceptable
Conclusions and final solution	The student argues that Q_{vc} and W_{vc} have negative values, such that the heat leaves the compressor and the work is done on the air that flows through it. The consumed power, expressed in terms of kW, is —119.4	The student finds negative values, stating that the heat leaves the compressor and the work is done on the air that flows through it. The consumed power, expressed in terms of kW, is —119.4	The student calculates the power that is consumed in terms of kW	The student has difficulties calculating the consumed power	The student fails to calculate the consumed power
25 %	5	4	3	2	1

Thermodynamics problem explained earlier, which will serve for the analysis of all the problem-solving processes. The rubric in Table 5.9 serves to evaluate the problem-solving process. We followed Bloom's taxonomy to prepare the evaluation levels of the rubric.

Problem solving process Analysis of the text embraces the next steps:

1. The volume control of the figure is in a steady state.
2. Any change of energy between the input and the output is considered insignificant.
3. The ideal gas model is applied for the air.

Solution procedure is stated as follows:

1. Establish the balance of energy of a volume control in a steady state with only one input and output flow (show step 1 and step 2)	$0 = Q_{vc} - W_{vc} + m\left(h_1 + \frac{c_1^2}{2} + gz_1\right) - \left(h_2 + \frac{c_2^2}{2} + gz_2\right)$ Step 1. Whence $W_{vc} = Q_{vc} + m\left[(h_1 - h_2) + \left(\frac{c_1^2-c_2^2}{2}\right)\right]$ Step 2 $m = \frac{A_1 C_1}{v_1} = \frac{A_1 C_1 p_1}{(\bar{R}/M)/T_1}$
2. The change of potential energy is considered insignificant	
3. The mass flow (m) can be calculated from the compressor input data, using the ideal gas thermal equation (show step 3 and step 4)	Step 3 by substitution $= \frac{(0.1\,\text{m}^2)(6\,\text{m}/10^5\,\text{N/m}^2)}{\left(\frac{8314}{28.97}\frac{\text{Nm}}{\text{kgK}}\right)(290\,\text{K})} = 0.72\,\text{kg/s}$ Step 4 Substituting values in the expression of power W_{vc} $W_{vc} = -3\left(\frac{\text{kJ}}{\text{s}}\right) + 0,72\frac{\text{Kg}}{\text{s}}$
4. The specific enthalpies may be calculated in accordance with the transformations table (show Step 5 to Step 9) At 290 K, $h_1 = 290.16$ kJ/kg At 450 K, $h_2 = 451.8$ kJ/kg	Step 5 $(290.16 - 451.8)\,\text{kJ/kg} + \left(\frac{6^2-2^2}{2}\right)\left(\frac{\text{m}^2}{\text{s}^2}\right)$ Step 6 $\left(\frac{1\,\text{N}}{1\,\text{kg·m/s}^2}\right)\left(\frac{1\,\text{kJ}}{10^3\,\text{N·m}}\right)\Big]$ Step 7 $= -3\frac{\text{kJ}}{\text{s}} + 0.72\frac{\text{kg}}{\text{s}}(-161.64 + 0.02)\frac{\text{kJ}}{\text{kg}}$ Step 8 $= -119.4\frac{\text{kJ}}{\text{s}}\left(\frac{1\,\text{kW}}{1\,\text{kJ/s}}\right) = -119.4\,\text{kW}$ Step 9 Conclusions Qvc and Wvc have negative values, such that the heat leaves the compressor and the work is done on the air that flows through it. The consumed power, expressed in terms of kW, is –119.4.

Table 5.9 Rubric to measure appropriate problem-solving procedures

Evaluation criteria	Excellent	Very good	Good	Not good enough	Unacceptable
Understands the problem or task to be solved	The student shows excellent comprehension of the task or the problem to be solved: the wording of the text is analyzed to ensure a coherent understanding. From among the concepts that are presented, the student identifies those that are familiar and those that are not, in order to search for the relevant information. Activities are developed that generalize the solution	The student shows very good comprehension of the task or the problem: the wording of the text is reviewed to ensure a coherent understanding. Among the concepts that are presented, the student identifies those that are familiar and those that are not, in order to search for relevant information	The student shows acceptable comprehension of the task or the problem to be solved: doing little more than following the instructions with no metareflection on the process of arriving at a solution	Understanding of the task or the problem to be solved is not good enough: the student tries to solve the problem using strategies of trial and error with no sign of learning	The student neither understands the problem nor tries to resolve it
20 %	5	4	3	2	1

(continued)

Table 5.9 (continued)

Evaluation criteria	Excellent	Very good	Good	Not good enough	Unacceptable
Analyses which strategies should be used to solve the problem	Before beginning the solution of the problem, the student carries out a reflective analysis on the data from evaluation criteria 1 and the strategies needed to solve it. A connection is established between the steps in the solution and the strategy that will be followed in the solution. Activities are developed that generalize the solution	Before beginning the solution of the problem, a reflective analysis is performed on the basis of the data from Evaluation criteria 1, of the strategies needed for its solution. A connection is established between steps in the solution and strategies that will be developed. Learning from mistakes is evident	Before attempting to solve the problem, the student analyzes the strategies needed to solve it	Before attempting to solve the problem, the student neither establishes the steps in the solution, nor the relation with the strategies needed for its solution	The student fails to take the necessary strategies into account for the solution of the problem
20 %	5	4	3	2	1
Prepares a plan in view of the chosen strategies	The student completes a perfectly designed plan and details the steps and the strategies chosen for the solution in writing	The student prepares a well designed plan and details the steps and strategies selected for the solution in writing	The student prepares a well designed plan of the steps and the strategies selected for the solution	The student presents an unconnected description of the plan to arrive at a solution	The student neither presents an implicit nor an explicit plan for the solution
20 %	5	4	3	2	1

(continued)

Table 5.9 (continued)

Evaluation criteria	Excellent	Very good	Good	Not good enough	Unacceptable
Evaluates the extent to which the steps in the plan are monitored throughout the implementation of the plan	The student completes an exhaustive test of the steps in the solution and the strategies used throughout the implementation of the plan, restructuring the solution, if necessary	The student completes a good test of the steps in the solution and the strategies used throughout its implementation, restructuring the solution, if necessary	The student completes an acceptable test of the steps in the solution and the strategies employed throughout its implementation, restructuring the solution, if necessary	The steps in the solution and the strategies employed throughout its implementation are not tested	The steps in the solution and the strategies employed throughout its implementation are not tested
20 %	5	4	3	2	1
Completes a final evaluation of the results and if they are acceptable restarts the solution of the problem	The student completes an exhaustive test of the results of the solution and compares these with the content of the problem or the task	The student completes a good final test of the results of the solution and compares these with the content of the problem or the task	The student completes an acceptable final test of the results of the test and compares these with the content of the problem or the task	The student completes an unacceptable final test of the results of the solution and compares these with the content of the problem or the task	The student has no final test of the results to compare with the content of the problem or the task
20 %	5	4	3	2	1

5.6 Research and Results

Three research projects are summarized below that incorporate the methodological proposals described in the preceding sections.

Research project 1 aimed to identify self-perception of knowledge among university students. The relation between student and teacher assessments of their self-perceived knowledge was also analyzed. The sample was comprised of 102 students studying for a second cycle academic degree. Differences in knowledge both before and after the instruction period were identified, as well as a few significant correlations between self-knowledge and the assessment of the teacher. The conclusion was to improve the self-perception of the knowledge process in Higher Education to promote safer and more successful learning [46].

Research project 2 analysed the relation between self-regulated learning and the use of rubrics. In the first part, it compared the effects of two different types of feedback on university student learning. Level 1 gave information to students on correct and incorrect learning; level 3 used the rubrics tool and gave information on self-regulated learning. The study involved 72 civil-engineering degree students studying applied physics and materials science. It found no significances between both types of feedback, but it did identify different tendencies. In a second part, different types of assessment (formative and summative) were analysed. Significant differences between all types of assessment were found, except in theoretical areas and with regard to problems of formative and summative assessment in both the experimental and the control groups [47].

Research project 3 looked at the effect of the teaching program on self-regulated learning, the feedback given to students on their learning processes and the use of rubrics facilitate student self-evaluation of their own learning. This program improved the procedural skills of students and the fine-tuning of their self-perceived knowledge, especially with regard to knowledge that relates to the design and estimation of structural elements and the graphic representation of constructive elements. However, this type of methodology requires more homogeneous and longer programs, before we may extend its results (across all subjects of the degree course). Likewise, we have highlighted the importance of the teacher forming an idea of the prior knowledge of students in the subject. Those measures lead to better harmonization of the study plan with the subject matter and the creation of more certain learning outcomes. It is essential for students to become aware of the starting point when they embark on the study of a new subject and of their progress throughout the teaching-learning process. The use of rubrics by the teacher has shown itself to be effective, allowing both teacher and students to measure prior (conceptual and procedural) knowledge. Moreover, rubrics assist the analysis of student progress in the teaching process. The final objective is to increase effective and independent student learning. This aspect is especially relevant for engineering courses, as the work of future engineers has to be autonomous, certain and based on problem-solving and decision-making.

In future research, our intention is to complete longitudinal studies that will provide insight into the relation between work on self-regulation and the efficiency of future graduates in the workplace.

5.7 Conclusions

The use of metacognitive and self-regulation strategies in the teaching-learning process is an important factor in the prediction of successful learning results. This methodology has special links with certain activities such as mathematics, physics and thermodynamics, which have usually presented comprehension problems for students.

All of this acquires special relevance in higher education learning environments, in which the ultimate objective is the effective inclusion of students in future professions. Taking rapid and effective decisions in those working environments is a relevant and necessary factor for employers. Hence, education systems at university level should strengthen the development of teaching-learning strategies that encourage the acquisition and the active use of metacognitive strategies through training in self-regulation. Mediation and feedback from the teacher on the solution processes of the students play a relevant role in this educational process. Teachers should therefore use instruments to assist them in following up the teaching-learning process with their students, on the basis of error analysis and positive correction. In this chapter, we have presented examples that support this approach: training procedures in metacognitive skills, evaluation instruments (rubrics, calibration and feedback) and their application to a real case in higher education involving the evaluation of a Thermodynamics problem and its problem-solving process [48].

Acknowledgments This work was completed with the assistance of funds for educational innovation and pedagogic improvement from the Vice-Rectorate of Teaching Staff and Administrative and Services Personnel of the University of Burgos (Spain) over the period 2012–2015.

References

1. Flavell, J.H.: Metacognition and cognitive monitoring: A new area of cognitive-developmental inquiry. Am. Psychol. **34**(10), 906–911 (1979). doi:10.1037/0003-066X.34.10.906
2. Flavell, J.H.: Cognitive monitoring. In: Dickson, W. (ed.) Children's Oral Communication Skills, pp. 35–60. Academic Press, New York (1981)
3. Brown, A.: Metacognition, executive control, self-regulation, and other more mysterious mechanisms. In: Weinert, F.E., Kluwe, R.H. (eds.) Metacognition, Motivation and Understanding, pp. 65–116. Erlbaum, Hillsdale (1987)
4. Brown, A., DeLoache, J.S.: Skills, plans, and self-regulation. In: Siegel, R.S. (ed.) Children's Thinking: What Develops?, pp. 3–35. Erlbaum, Hillsdale (1978)

5. Van der Stel, M., Veenman, M.V.J.: Relations between intellectual ability and metacognitive skillfulness as predictors of learning performance of young students performing tasks in different domains. Learn. Individ. Differ. **18**, 128–134 (2008). doi:10.1016/j.lindif.2007.08.003
6. Van der Stel, M., Veenman, M.V.J.: Development of metacognitive skillfulness: a longitudinal study. Learn. Individ. Differ. **20**, 220–224 (2010). doi:10.1016/j.lindif.2009.11.005
7. Veenman, M.V.J., Van Hout-Wolters, B.H.A.M., Afflerbach, P.: Metacognition and learning: Conceptual and methodological considerations. Metacognition Learn. **1**, 3–14 (2006). doi:10.1007/s11409-006-6893-0
8. Veenman, M.V.J.: Learning to self-monitor and self-regulate. In: Mayer, R.E., Alexander, P.A. (eds.) Handbook of Research on Learning and Instruction: Educational Psychology Handbook Series, pp. 197–218. Routledge, New York (2011)
9. Efklides, A.: The role of metacognitive experiences in the learning process. Psicothema. **21**(1), 76–82 (2009)
10. Veenman, M.V.J., Elshout, J.J.: Changes in the relation between cognitive and metacognitive skills during the acquisition of expertise. Eur. J. Psychol. Educ. **14**(4), 509–523 (1999)
11. Sáiz, M.C., Montero, E., Bol, A., Carbonero, M.A., Roman, J.M.: Metacognitión y aprendizaje: posibles líneas de intervención educativa en educación superior [metacognitión and learning: possible lines for educational training in higher education]. In: Román, J.M., Carbonero, M.A., Valivieso, J.D. (eds.) Educación, aprendizaje y desarrollo en una sociedad multicultural, pp. 5513–5528. Ediciones de la Asociación Nacional de Psicología y Educación, Valladolid (2011)
12. Kostons, D., Van Gog, T., Paas, F.: Training self-assessment and task selection skills: A cognitive approach to improving self-regulated learning. Learn. Instr. **22**, 121–132 (2012). doi:10.1016/j.learninstruc.2011.08.004
13. Wäschle, K., Allgaier, A., Lachner, A., Fink, F., Nückles, M.: Procrastination and self-efficacy: tracing vicious and virtuous circles in self-regulated learning. Learn. Instr. **29**, 103–114 (2014). doi:10.1016/j.learninstruc.2013.09.005
14. Efklides, A.: Interactions of metacognition with motivation and affect in self-regulated learning: the MASRL model. Educ. Psychol. **46**(1), 6–25 (2011). doi:10.1080/00461520.2011.538645
15. Zimmerman, B.J.: Goal setting: a key proactive source of academic self-regulation. In: Schunk, D.H., Zimmerman, B.J. (eds.), Motivation and Self-regulated Learning: Theory, Research and Applications, pp. 267–295. Lawerence Erlbaum Associates, New York (2008)
16. Zimmerman, B.J., Moylan, A.R.: Self-regulation: where metacognition and motivation intersect. In: Hacker, D.J., Dunlosky, J., Graesser, A.C. (eds.) Handbook of Metacognition in Education, pp. 299–315. Routledge, New York (2009)
17. Schraw, G.: Measuring self-regulation in computer-based learning environments. Educ. Psychol. **45**(4), 258–266 (2010). doi:10.1080/00461520.2010.515936
18. Zimmerman, B.: Investigating self-regulation and motivation: historical background, methodological developments, and future prospects. Am. Educ. Res. J. **45**(1), 166–183 (2008). doi:10.3102/0002831207312909
19. Acevedo, R.: Using hypermedia as a metacognitive tool for enhancing student learning? The role of self-regulated learning. Educ. Psychol. **40**(4), 199–209 (2005)
20. Schraw, G., Kuch, F., Gutierrez, A.P: Measure for measure: Calibrating ten commonly used calibration scores. Learn. Instr. **24**, 48–57. doi:10.1016/j.learninstruc.2012.08.007
21. Veenman, M.V.J.: The assessment and instruction of self-regulation in computer-based environments: a discussion. Metacognitive Learn. **2**(2–3), 177–183 (2007). doi:10.1007/s11409-007-9017-6
22. Sáiz, M.C., Montero, E., Bol, A., Carbonero, M.A.: An analysis of learning-to-learning competences at the university. Electron. J. Res. Educ. Psychol. **10**, 253–270 (2012)
23. Sáiz, M.C., Román, J.M.: Cuatro formas de evaluación en educación superior gestionadas desde la tutoría [Four forms of assessment in higher education managed in a tutoring program]. Revista de Psicodidáctica. **16**(1), 145–161 (2011)

24. Veenman, M.V.J.: Alternative assessment of strategy use with self-report instruments: a discussion. Metacognition Learn. **6**, 205–221 (2011). doi:10.1007/s11409-011-9080-x
25. Veenman, M.V.J., Verheij, J.: Technical students' metacognitive skills: relating general vs. specific metacognitive skills to study success. Learn individ. differ. **13**(3), 259–272 (2003). doi:10.1016/S1041-6080(02)00094-8
26. Panadero, E., Alonso-Tapia, J., Huertas, J.A.: Rubrics and self-assessment scripts on self-regulation, learning and self-efficacy in secondary education. Learn. Individ. Differ. **22**(6), 806–813 (2012). doi:10.1016/j.lindif.2012.04.007
27. Bloom, B.S., Engelhart, M.D., Furst, E.J., Hill, W.H., Krathwohl, D.R.: Taxonomy of Educational Objectives: Handbook I, Cognitive Domain. David McKay, New York (1957)
28. Alonso-Tapia, J., Huertas, J.A., Ruiz, M.A.: On the nature of motivational orientations: implications of assessed goals and gender differences for motivational goal theory. Span. J. Psychol. **13**(1), 232–243 (2010)
29. Panadero, E., Alonso-Tapia, J.: Self-assessment: theoretical and practical connotations. When it happens, how is it acquired and what to do to develop it in our students? Electron. Res. Educ. Psychol. **11**(2), 551–576 (2013). doi:10.14204/ejrep.30.12200
30. Panadero, E., Jonsson, A.: The use of scoring rubrics for formative assessment purposes revisited: a review. Educ. Res. Rev. **9**, 129–144 (2013). doi:10.1016/j.edurev.2013.01.002
31. Panadero, E., Romero, M.: To rubric or not to rubric? The effects of self-assessment on self-regulation, performance and self-efficacy. Assessment in Education: Principles, Policy & Practice, **21**(2), 133–148 (2014). doi:10.1080/0969594X.2013.877872
32. Dinsmore, D.L., Parkinson, M.M.: What are confidence judgments made of? Students' explanations for their confidence ratings and what that means for calibration. Learn. Instr. **24**, 4–14 (2013). doi:10.1016/j.learninstruc.2012.06.001
33. Hattie, J.: Calibration and confidence: where to next? Learn. Instr. **24**, 62–66 (2013). doi:10.1016/j.learninstruc.2012.05.009
34. Hattie, J.A.C.: Visible Learning: a Synthesis of 800 Meta-Analyses on Achievement. Routledge, Abingdon (2009)
35. Hattie, J., Timperley, H.: The power of feedback. Rev. Educ. Res. **77**(1), 81–112 (2007). doi:10.3102/003465430298487
36. Narciss, S.: Feedback strategies for interactive learning tasks. In: Spector, J.M., Merrill, M.D., van Merriënboer, J., Driscoll M.P. (eds.) Handbook of Research on Educational Communications and Technology, (3rd ed.), pp. 125–144. Lawrence Erlbaum Associates, Mahwah (2008)
37. Rakoczy, K., Harks, B., Klieme, E., Blum, W., Hochweber, J.: Written feedback in mathematics: Mediated by students' perception, moderated by goal orientation. Learn. instr. **27**, 63–73 (2013). doi:10.1016/j.learninstruc.2013.03.002
38. Alexander, P.A.: Calibration: what is it and why it matters? An introduction to the special issue on calibrating. Learn. instr. **24**, 1–3 (2013). doi:10.1016/j.learninstruc.2012.10.003
39. Willett, G., Iverson, E.R., Rutz, C., Manduca, C.A.: Measures matter: Evidence of faculty development effects on faculty and student learning. Assessing Writ. **20**, 19–36 (2014). doi:10.1016/j.asw.2013.12.001
40. Hattie, J., Gan, M.: Instructions based on feedback. In: Mayer, R.E., Alexander, P.A. (eds.) Handbook of Research on Learning and Instruction: Educational Psychology Handbook series, pp. 249–271. Routledge, New York (2011)
41. Meichenbaum, D., Goodman, J.: Training impulsive children to talk to themselves: a means of developing self-control. J. Abnorm. Psychol. **77**, 115–1126 (1971)
42. Sáiz, M.C., Román, J.M.: Programa de Entrenamiento Cognitivo para niños pequeños. (7th ed.). [Cognitive Training Program to young children]. CEPE, Madrid (2007)
43. Veenman, M.V.J.: Kennis en Vaardigheden: Soorten Kennis een Vaardigheden die Delevant zijn Voor Rekenwiskunde Taken. [Knowledge and Skills that are Relevant to Math Asks]. In: Duinmaijer, A.F., van Luit, J.E.H., Veenman, M.V.J, Vendel P.C.M. (eds.) Hulp bij leerprobleement: Rekenen-wiskunde, pp. 1–13. Betelgeuze, Zoetermeer (1998)

44. Peña, A., Kayashima, M., Mizoguchi, R., Dominguez, R.: Improving students' meta-cognitive skills within intelligent educational systems: a review. In: Schmorrow, D.D, Fidopiastis, C.M. (eds.) FAC 2011, HCII 2011, LNAI 6780, pp. 442–451. Springer-Verlag, Berlin (2011)
45. Strain, A.C., Azevedo, R., D'Mello, S.K.: Using a false biofeedback methodology to explore relationships between learners' affect, metacognition and performance. Contemp. Educ. Psychol. **38**, 22–39 (2013). doi:10.1016/j.cedpsych.2012.08.001
46. Sáiz, M.C., Payo, R.J.: Auto-percepción del conocimiento en educación superior [knowledge Self-perception in Higher Education]. Revista Iberoamericana de Psicología. **3**(2), 159–174 (2012)
47. Sáiz, M.C., Bol, A.: Aprendizaje Basado en la Evaluación Mediante Rúbricas: Un estudio en Educación [Learning based on rubrics assessment: An educational study]. Suma Psychologica. **21**(1), 28–35 (2014)
48. Sáiz, M.C., Sánchez, M.A., Ortega, V., Manso, M.: Self-regulation and rubrics assessment in structural engineering subjects. Educ. Res. Int. In press (2014)

Chapter 6
Engaging Adolescent Students' Metacognition Through WebQuests: A Case Study of Embedded Metacognition

Hope J. Hartman

Abstract This chapter is a case study which examines how graduate teacher-education students designed WebQuests (WQs) that engaged their students meta-cognitively through embedded activities. As a course requirement, students created their own WQs for their adolescent students after having engaged in one and learning about metacognition from both the perspectives of a teacher and a student. The results showed various types of metacognition were embedded at both the junior high (JH) and high school (HS) levels. Both executive management metacognition (EMM) and strategic knowledge metacognition (SKM) were embedded at both school levels. In addition to individual metacognition, meta-cognitive activities were embedded in a social context, involving pair and group assignments. Sometimes metacognition was required, while other times it was expected. Metacognition was embedded in several WQ components, especially tasks, process, evaluation and conclusions. Conclusion-component metacognitive embeds were generally expected but not required, while metacognitive embeds in the tasks, process and evaluation components were more often required. Affective self-reflections (ASRs) were also embedded in the WQs at both school levels. Implications for future research and designing WQs to maximize metacognitive engagement are discussed.

Keywords Metacognition · E-learning and teaching · Affect · Scaffolding · Social interactions · Multimedia

H.J. Hartman (✉)
The City College of New York, New York, NY, USA
e-mail: hhartman@ccny.cuny.edu

© Springer International Publishing Switzerland 2015
A. Peña-Ayala (ed.), *Metacognition: Fundaments, Applications, and Trends*,
Intelligent Systems Reference Library 76, DOI 10.1007/978-3-319-11062-2_6

135

Abbreviations

A	Let's learn about the smallest thing in the earth (Atom)
AF	If you don't become an actor you'll never be a factor
ASR	Affective self-reflection
BC	Building character: preparing for role
CD	Show me your neighborhood: A quest for cultural diversity
DC	Demystifying the cell
EbM	Embedded metacognition
EMM	Executive management metacognition
F	Life before the rocks: A theatrical journey to Grosse-Ile (Famine)
GM	As Mt. Olympus turns: Greek mythology, a soap opera
GG	Let's get the groove going!
HS	High school
JH	Junior high
M	Machinima: What is it, why you might care and how can it help you in your English lit and theater classes
MLK and MX	Martin Luther King and Malcolm X have something to say to you!
N	Lessons of Nanook from yesterday to today
Nu	Number systems: Does $11 = 11$, 3 or 17?
PWQ	Piaget WebQuest
RT	Right triangles are your friends!
S	Cut it out: printmaking with stencils
SCW	Viva la revolucion: Spanish civil war and better understand the house of Bernarda Alda
SM	Social metacognition
SKM	Strategic knowledge metacognition
WBS	White school—black school: Melba Patillo, integration and the Little Rock 9
WQ/WQs	WebQuest/WebQuests
WS	Writing to show—not tell
WW	Why write?

6.1 Introduction

Since their inception, WQs have been recognized as tools for stimulating students' metacognition, but there has been little research on the topic. For approximately 20 years now, both metacognition and WQs have emerged as topics with important educational implications. This chapter is a case study that focuses on the questions: how did graduate teacher-education students, many of whom were already teachers, design WQs for their students in ways that would engage them metacognitively, to what extent was metacognition embedded in their WQs; what kinds of metacognition were embedded, where, and how?

6.1.1 WebQuests and Metacognition

Dodge [1], developer of the WQ, characterizes it as an inquiry-oriented set of learning activities organized around use of the Internet. Dodge distinguishes between short and long term WQs. Short-term WQs emphasize the goals of acquiring and integrating a considerable amount of material, and the activities last from one to three class periods. Long-term WQs emphasize the goals of deeply analyzing information, and using it to create a product that demonstrates the learner's comprehension of the material. In order to create this product, the learner must transform or reconstruct information in a meaningful way. Activities in a long-term WQ extend from a week to a month. The WQs addressed in this chapter are all long-term.

According to Dodge, WQs have six critical attributes: (1) introduction, provides background information and sets the stage for the learning activities; (2) task, involves activities that are achievable and interesting; (3) information, learners are required to use various resources in order to complete the tasks; (4) process, a series of steps guiding students so they can accomplish the task; (5) guidance, information on how to organize the information acquired; (6) conclusion, to remind students of what they learned and encourage them to extend their learning. The WQs discussed in this chapter all have these attributes because they are embodied in the required Zunal WebQuest template.

Effective WQs are scaffolded structures that use the Web to help learners engage in an authentic learning experience where they transform what they learn into a deeper understanding and reflect on their metacognitive processes [2]. Thus WQs are natural environments for embedding metacognitive activities and developing metacognitive knowledge and skills.

In a review of research on WQs, Abbit and Ophus [3] found that while they may have a positive impact on learner attitudes and collaborative working skills, they do not offer substantial advantages over other instructional approaches for improving student achievement. However, use of WQs for improving student achievement has been found in several subjects. A study in Saudi Arabia [4] found a WQ to be effective for improving the reading comprehension of EFL students compared to control students who did not engage in a WQ, however, they noted, both students and teachers should be trained to maximize the effectiveness of WQs. Teachers need to know how to guide students through WQs in order for them to be effective. A study of WQ use in social studies classrooms in the U.S.A. found them to be more effective than traditional textbook-based classes [5]. The National Council of Teachers of Mathematics (NCTM) in the U.S.A. recommends WQs as tools for teaching topics such as the Fibonacci sequence, because when well-designed, they help students actively engage in the learning process, gather and analyze relevant information, and use higher-order thinking skills [6]. The key to successful WQ use appears to be the quality of their design and the ability of teachers and students to use them effectively.

Metacognition is commonly characterized as awareness and control over one's own thinking processes, knowledge and products, or thinking about one's own thinking. It is often conceptualized with reference to two major types: executive management metacognition (EMM) and strategic knowledge metacognition (SKM) [7]. EMM includes planning, monitoring, evaluating and revising while SKM includes declarative, contextual and procedural knowledge. Research suggests that while some people develop metacognitive knowledge and skills spontaneously, others need direct assistance in developing and using metacognition [8]. Research has documented the positive relationship between metacognition and academic performance [9].

A study of embedded metacognitive strategy training [10] found that students in Turkey who had metacognitive strategy training embedded in their English listening texts performed better on a listening test than control students who did not have embedded metacognition (EbM). However, EbM does not always lead to increased academic success.

Research comparing the combination of cooperative learning with EbM to cooperative learning alone and to traditional instruction in biology with high school (HS) students in Germany found that students who engaged in cooperative learning outperformed students on measures of socioscience decision making compared to those who did not, but there was no significant difference between students who also had EbM in their lessons [11].

Developing students metacognitively can be done through traditional instruction [7] as well as through computer-based learning environments. Azevedo [12], argued for a new paradigm in which computers are used as tools for promoting metacognition, which he viewed as especially important because of the pervasiveness of their use both in and out of school for learning about conceptually rich domains.

Poitras et al. [13] created the MetaHistoReasoning Tool to enhance metacognition when learning history. Students who used metacognitive tools built into this computer-based learning environment had better comprehension, memory, and reasoning about history when engaged in inquiries into historical events than students who did not use these tools.

A comparable study was conducted on teaching biology through use of a web-based tool, Young Researcher, which had prompts for guided reflection built into biology learning experiences. Guided reflection enhanced students' science inquiry skills and their ability to reflect on their own learning [14].

Cho et al. [15] developed a computer-based writing environment, Scaffolded Writing and Revision in the Disciplines, to provide support for students monitoring their own writing. They found that undergraduate and graduate students in the U.S.A. who used this system and improved their self-monitoring skills, also improved their writing more than students who did not develop the self-monitoring skills.

Research on EbM in a problem-solving learning environment for engineering students in Mexico to learn how to solve workplace problems, showed that EbM increased students' metacognitive awareness, helped them solve problems more successfully and earn better grades [16].

The concept of metacognition has evolved from self-regulation to include social metacognition (SM), where students engage in co-regulation [17–21]. Iiskala et al. [17] found that students working collaboratively to solve mathematical problems engaged in "socially-shared metacognition" in which awareness and control over thinking occurred between individuals—not just within them.

Although the concept of SM is relatively new in the literature, pair metacognition has flown beneath the radar for years in the context of pedagogical methods such as Pair Problem Solving [22], Guided Reciprocal Peer Questioning [23] and Think-Pair-Share [24], all of which entail SM in a partner setting. As with pair metacognition, group metacognition has also flown under the radar via use of cooperative learning methods such as Group Investigation [25] and Problem-Based Learning [26]. All of these pair and group teaching methods involve students engaging in SM, although it is usually not discussed explicitly. As Chan [18] noted, there has been little research on how students collaborate and co-regulate in computer-supported learning. A special issue of the journal Metacognition and Learning focused on co-regulation in computer-supported collaborative learning.

Included was a study on task regulation and team regulation by HS students using a computer simulation [27]. Task regulation emphasized comprehension monitoring to ensure students understood the task. Team regulation emphasized how they worked together in order to perform the task successfully. They found that co-regulation, including group planning and monitoring, predicts group performance, so it is important in collaborative inquiry. Computer-supported inquiry, where students work collaboratively on shared tasks, is a common feature of WQs. Another article in that special issue, which also examined adolescents, found that co-regulation on a computer-based historical inquiry task has a positive relationship to group learning outcomes [28].

Affective Self-Regulation. Self-reflections are more than metacognitive activities and should be defined more broadly to include affective self-reflections (ASRs) as well. The affective domain focuses on feelings and includes emotions, attitudes, values and motivation. Affective self-regulation has been described as "meta-affective reflections" and emotional regulation by Chick et al. [29]. It includes sensitivity to one's feelings (awareness) and managing them (control). So awareness and control are the key features of both cognitive and affective self-regulation.

A manifesto on affective learning from the MIT Media Lab made this type of point quite emphatically, especially when dealing with learning in the context of digital technology [30]. Bandura et al. [31] found that adolescents' perceived affective self-regulation efficacy, including regulation of both positive and negative emotions, had an impact in psychosocial situations and was accompanied by regulation of academic and pro-social behavior. A book addressing the importance of affective self-regulation in a wide variety of social situations conceptualizes motivation as a separate, but most important dimension of self-regulation [32].

Considerably earlier, The BACEIS Model of Improving Thinking [33] posited that cognition and affect comprise two separate but interacting subsystems. Awareness and control of one's own thinking is described as metacognition, whereas awareness and control of one's own feelings is described as affective self-regulation (Fig. 6.1). Both must be considered, along with their interactions with each other and with the academic and nonacademic environments, to best enhance intellectual performance. Affective self-regulation includes management of one's own attitudes, values, emotions and motivations.

For example, regulating one's motivation to learn to read can influence one's willingness to use cognitive and metacognitive reading strategies when reading challenging text in the classroom and outside of school, while doing homework.

Thus the BACEIS model is intended to help develop students both cognitively and affectively so that they become independent, self-directed learners who can apply what they learn, as represented in Fig. 6.2 (read clockwise from the top).

A more recent approach to fostering active, meaningful learning, the construction-deconstruction-connectionist process, has much in common with the BACEIS Model. The process approach assumes that classroom learning is a cognitive, psychodynamic and social process [34] and identifies four different metacognitive domains in classroom learning: the self, the professor, classmates and the environment.

A study by Pang and Ross [35] testing this model with college students in Texas, studying criminal justice and British literature, used a four-step process with students who worked in groups and engaged in EbM activities in these two subjects. The results showed that the approach facilitated comprehension of complex constructs, improved students' satisfaction and effectiveness.

Metacognition and WebQuests. There is little literature on the design of WQs or the actual use of metacognition in WQs. Existing studies are primarily about language learning. Work on English for Specific Purposes in Spain [36, 37] which emphasizes use of English for professional and academic contexts, presents goals and guidelines for developing students' metacognition through WQs so they can be autonomous, life-long learners, communicate effectively and develop new literacies for constructing meaning. The English for Specific Purposes WQ includes attention to SKM: declarative—having background knowledge of the discipline, contextual-assessing the specific situation so they can respond appropriately and procedural-knowing how to use a variety of resources to solve problems and answer questions. The ESP WQ also attends to EMM: planning—developing genre awareness, so that they can plan to make specific language choices in order to achieve specific communication goals; monitoring—getting corrective feedback on drafts; and evaluating—reflecting on their learning processes and assessing their progress.

Another study involved a WQ designed to enhance use of EMM strategies for improving oral English skills [38]. College students in China were required to create WQs on the metacognitive strategies of self-planning, self-monitoring and self-evaluating. Each group of students focused on one of the three strategies. Then students shared their WQs. To assess students' reactions, questionnaires were administered and students were interviewed.

The results show that students' attitudes toward oral English and learning through WQ were enhanced by this experience. They also indicate that the WQs

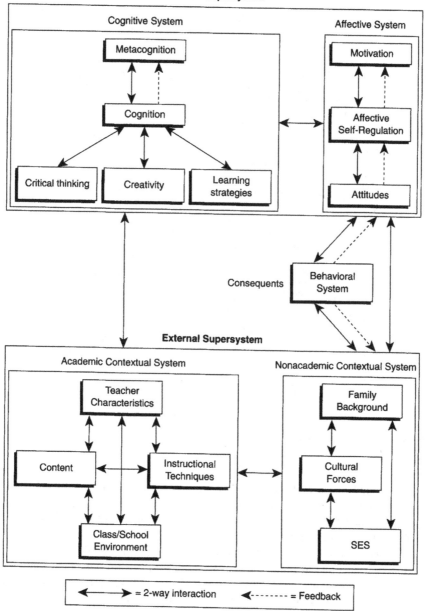

Fig. 6.1 BACEIS model of improving thinking [33]

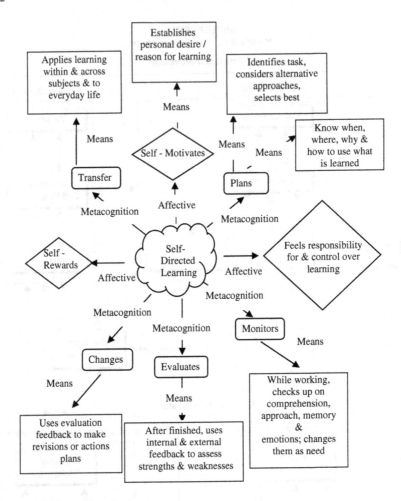

Fig. 6.2 Self-directed learning

improved students' use of metacognitive strategies for speaking English. A similar study on teaching writing in China and found that strategy training for writing English was needed and that WQs improved students' use of EMM metacognitive strategies for writing English [39].

A study of teacher education students in Colombia using WQs for learning English as a Second Language showed WQs to be effective forms of curricula for changing students' views of themselves as learners and future teachers [40]. An exception to this pattern of WQs and metacognition in language learning occurred at an elementary school in Taiwan [41]. Their focus was on using a WQ to learn the science topic of environmental protection soap.

Qualitative and quantitative analyses showed the WQs were effective in helping 6th grade students in the experimental group, where scaffolds built into the WQ assisted students in goal setting, strategic planning, monitoring and self-evaluating, when performing WQ tasks.

Students in the control group, who did not have metacognition scaffolds embedded in the WQs, did not engage in these self-regulatory processes.

Elsewhere [42] I have described rather extensively the teaching metacognition involved in the design and implementation of my Piaget WebQuest: Uncovering and Discovering Piaget [43], which was required of all of the undergraduate and graduate teacher education students in my educational psychology courses. It also describes my teaching metacognition in the follow-up assignment in which students created their individual WQs, and my students' metacognition in these two assignments.

6.1.2 Theoretical Framework

The approach uses a synthesis of several theoretical frameworks, including information processing, constructivism, and situated learning as described in Hartman, 2012 [42]. Mayer's Multimedia Learning Theory [44], an information processing theory, is based on three assumptions: two channels for processing multimedia information are visual and auditory, people have limited capacity for processing information, and our processing system is active. To design effective multimedia tools, such as WQs, which usually involve verbal, pictorial, and auditory processing, these principles must be taken into consideration, as should metacognition and affective variables such as motivation. Metacognition is considered the highest level of thinking in information processing theory [33, 45].

Distributed Learning theory, which focuses on emerging technologies and their use in education [46], posits that new types of communications are needed for conveying content, and advocates using new media as an alternative to the "teaching by telling" approach that has dominated education. WQs were developed around the same time that Dede first proposed Distributed Learning theory, which emphasizes learner control, interactive technologies, and learning-by-doing, all of which typify WQs.

Cognitive constructivism underlies students' experiences as they engage in the learner-centered, inquiry-oriented PWQ, and use what they learn to co-create educational products, instructional activity designs, that require application of the acquired knowledge and skills. The individual, student-created WQ itself is consistent with social constructivist theory, because it is considered a cultural artifact, which Vygotsky [47] depicts as tools, objects, concepts that connect individuals to society and society to individuals. Social constructivism also underlies the scaffolding process of students first engaging in a collaborative, structured WQ experience (PWQ) before creating their own WQ, the scaffolding they received while creating their WQs, as well as the scaffolding they provide for their students while engaging in their WQs.

Situated learning theory is a framework that guides this work. Brown et al. [48] and Lave and Wenger [49] emphasize the importance of the context in which learning activities take place. They argue that learning is situated in and a product of the activity of the learner in the context in which learning occurs. In this case, teacher education students participated in activities which are implemented by members of the educational community: finding, selecting and using a variety of educational resources to create authentic educational products they can use in the real world of their own classrooms with their own students. As members of a community of practice, there was extensive social interaction among them as they developed and shared their WQs.

The second of three parts of the teacher knowledge and learning framework [50], "knowledge-in-practice", depicts this approach in terms of teachers' knowledge about and use of metacognition in their own and their students' practices, as well as the design of learning experiences. This chapter is a case study of metacognitive activities embedded in the WQs designed for adolescents in junior high (JH) schools and high schools (HS). The creation of WQs was a requirement in the course. Major questions posed are: To what extent were WQs embedded with metacognitive activities? What kinds of metacognitive activities were embedded, where and how?

6.2 Method

6.2.1 Participants

Participants were graduate teacher education students enrolled in Adolescent Learning and Development at The City College of New York, a large, urban college in New York City in the Northeastern U.S.A. during Spring, 2011. The 24 students were born in 8 countries: Belgium, Ghana, Haiti, Hong Kong, Jamaica, Taiwan, Turkey and the USA. There were 12 females and 12 males, some of whom were already teachers while others were preparing to be teachers.

Students were allowed to decide whether or not to publish their WQs, and publication was not considered in their course grade. Eighteen of these students still had published WQs as of January 2014. It is these 18 WQs (75 % of the class) that are analyzed in this chapter (Table 6.1).

Eight of these WQs were developed by males (44 %) and 10 by females (56 %). 67 % of the JHWQs were developed by males and 33 % females; at the HS level, 33 % were developed by males and 67 % by females.

6.2.2 Metacognitive Engagement Preparation

Learning about and engaging in metacognition began the first day of class and occurred regularly throughout the course. While teaching I systematically use and explicitly highlight my metacognition to help my students become more aware of what it is (declarative knowledge); why it is important for thinking, learning

Table 6.1 Overview of student-created WebQuests by school level

Level	Title	Subject(s)	URL*
JH	S	Art	88723
JH	AF	Life skills/Careers	93523
JH	GG	Music	90065
JH	A	Science	92839
JH	F	Life skills/Careers	88981
JH	Nu	Mathematics	89904
HS	GM	Social studies	92253
HS	BC	Art/Music	89276
HS	DC	Science	88708
HS	N	Art/Music/Social studies	93333
HS	M	English lit, theater, technology	92968
HS	MLK and MX	Social studies	93527
HS	RT	Mathematics	88992
HS	CD	Art, music	89013
HS	SCW	English, language arts, history	89650
HS	WBS	English, language arts, history	93343
HS	WW	English, language arts	88986
HS	WS	English, language arts	90917

* All begin with http://zunal.com/webquest.php?w

and teaching; when to apply it (contextual knowledge); and how they can assume greater control over their own metacognition, and apply strategies for fostering metacognition in their own students (procedural knowledge).

By explicating the use of metacognition in my own PWQ development, implementation, evaluation and revision, students were provided with a model framework for constructing their own WQs using their own metacognition and embedding it in their WQs for their students.

Metacognition was one of the main topics covered in this course, and it was a theme carried throughout the semester—both enhancing students' metacognition and teaching metacognitively. Metacognition was included in the course textbook and many resources, some required and others optional, were provided at the course's Google website, "Adolescent Learning and Development", and there were many class activities involving metacognition and discussions about it. By engaging in the PWQ, students saw how I embedded and they experienced EbM activities in several components (Table 6.2):

Introduction. Activated prior knowledge through a series of questions regarding what they already knew about Piaget and Constructivism and posed a question for the end, getting them to think about how they could improve their performance on a similar assignment in the future (expected but not required).

Task. Resource Report and an attached template require users to write a summary of and their personal reactions to six resources (3 documents and 3 websites). Instructional Activity Design, with a required template specifying statement

Table 6.2 Model of metacognition embedded in the Piaget WebQuest

WebQuest	Metacognitive activities	Examples
Introduction	Reflection questions	What considerations should you make about teaching based on Piaget's theory of intellectual development and its educational implications?
		How could you improve your performance on a similar assignment in the future?
Tasks	Resource Report summarizing and reacting to Web-based resources (3 websites and 3 documents) using the required template	My summaries of and personal reactions to the information in each of the websites and documents
		Goals/Objectives: What do you want to accomplish? What outcomes do you expect? Your goals/objectives should use concepts consistent with Piaget's theory and its educational implications
	Plan instructional design using the required instructional activity design template	Teaching methods and learning activities: What will the teacher do? What will the students do? Provide detailed descriptions
	Assess individual work, learning and give action plan using the required self-evaluation template	How did you collaborate with your partner/group? What did you contribute? Describe and evaluate your cooperation and contributions on:
		a. Learning from each others' resources
		b. Designing and writing the instructional activity
		What did you learn from the project overall?
		If you could do this whole project over again, what might you do differently?
Evaluation (Rubrics)	Resources report	15 % grade example: Template used. All resources discussed thoroughly (summaries and personal reactions)
	Instructional activity design	15 % grade example: Complete and clear goals/objectives and specifically linked to Piaget. Teaching methods and learning activities and materials clearly and thoroughly specified; consistent with and explicitly connected to Piaget
	Self-evaluation	10 % grade example: Did very thoughtful self-evaluation of all criteria specified on the task page
Conclusion	Reflection questions	How has your knowledge of Piaget's ideas and Cognitive Constructivism changed?
		When, why, how and to what extent might you apply what you learned through this WebQuest to your own teaching?

of goals/objectives, materials/resources, teaching methods and learning activities and connections to Piaget and Constructivism; Self-Evaluation and an attached, required template for assessing their cooperation and contribution to the group process and product, what they learned from the project and an action plan for improving their future performance.

Evaluation rubric. Criteria and percent of grade for each written product using the templates with EbM activities,

Conclusion. Questions expecting students to reflect on: what they learned from engaging in this WQ, what else they would like to know, potential application of what was learned, and improvement of performance.

6.2.3 WebQuests Development

Creation of their WQs required use of the Zunal template with seven main components: Welcome, Introduction, Tasks, Process, Evaluation, Conclusion and Teacher Page, as they had experienced in the PWQ. Graduate students' WQ design and development was scaffolded throughout the semester. First they participated in the long-term PWQ I designed for them, which included EbM activities and lasted throughout the first half of the semester. Their own WQ creation began mid-semester, slightly overlapping their participation in the PWQ.

They were required to design their own long-term WQ for use with adolescents at either the JH or HS level, choosing whatever academic content they wanted. Some of their WQ design and development occurred during class time through scheduled classes in the Multimedia Center where they received feedback from each other and from me, and some of it was created by students during their own time. WQs were also discussed in the course textbook as a way of using 21st century students' interest in and pervasive use of technology.

6.3 Results

Six of the 18 WQs were developed for middle school grades 6–8 (33 %); twelve were developed for HS grades 9–12 (67 %). Subject areas included: mathematics, science, art, music, theater, documentary film making, history/social studies, and English/Language Arts. Following is a qualitative analysis of the metacognitive activities embedded in these WQs. Extensive examples are provided to convey the pervasiveness, richness and wide range of EbM activities.

It is important to keep in mind that many, but not all reflections are metacognitive, and therefore involve thinking about one's own thinking and knowledge (internal). In some cases, reflections embedded in these WQs involve reflecting on external knowledge. For example, in the Conclusion of the GM WQ, students are asked to reflect on: What do these stories tell us about the Greeks themselves? How did the ancient Greeks view their gods? Is this similar to or different from how religions of the modern world conceive of a supreme being? How? Reflections can also be affective in focus, rather than cognitive, as will be discussed later.

6.3.1 Required Versus Expected Metacognition

Two general types of metacognitive activities that emerged were "required", where there was accountability for engaging metacognitively (Table 6.3, See List of Abbreviations and Table 6.1), and "expected", where there was no accountability (Table 6.4). Metacognition was considered to be required when teachers used accountability strategies, such as observing students in class to ensure students engaged in the metacognitive activities or when they graded written answers to metacognitive questions or prompts, such as in homework assignments or journals.

Metacognition was considered to be expected when teachers posed metacognitive questions or prompts for students, but used no accountability strategies such as observation or written documentation. Required metacognition commonly was embedded in the Tasks, Processes and Evaluation pages of the WQs while expected metacognition was often embedded in the Conclusion.

A related distinction is between embedded oral and written metacognitive activities. Oral activities involved students sharing their thinking about a project with others, which will be discussed later in this chapter. Usually oral metacognition was expected but not required.

Written activities most commonly involved writing in a project journal, but also included completing worksheets, and doing homework; usually these metacognitive activities were required (Table 6.3). All JH WQs included both oral and written metacognitive activities.

All HS WQs included written metacognition and 50 % had explicit oral metacognitive activities embedded. In 33 % of the HS WQs, some communication between partners or groups was required, but it wasn't clearly specified whether it was to be oral or written, perhaps through texts or emails. In 17 % of the HS WQs there was no oral metacognition (N and RT).

6.3.2 Executive Management Metacognition

All six JH school WQs included some EMM, although it was minimal in two of them (Table 6.5). The other four (67 %) contained moderate amounts of EMM. EMM was found in several JH WQ components: Introduction, Tasks, Process, Evaluation, and Conclusion.

All of the HS WQs included some EMM (Table 6.6). Although most of it was embedded in the Tasks, Process, Evaluation and Conclusion, occasionally metacognition was embedded the Introduction. Several of the Introductions posed questions to activate students' prior knowledge so they could build on valid information and recognize and overcome misconceptions.

Table 6.3 Metacognition required in WebQuest evaluations

WQ	Evaluation criteria
S	Creativity: Invests time and thought in researching an image that directly reflects something important in their life
	Thorough exploration of how image will work as a stencil and how it will be executed
	Neatness: Uses adequate amount of ink/paint for printing
AF	Journal: Completion of assignments
	Reflections: Contributions to reflections (group and personal journal entries) and the effect they allowed it to have on their choice
GG	Project journal: Lesson summary of each lesson reflecting what you learned and your experience, and progress of your group composition
	Peer evaluation: Answer evaluation questions
A	Reflection journal: How well you learned the scientific skills and terms, how well you participated in your group, what you shared with your group members
F	Ability to reflect: Identified as something to be graded, but no details provided
Nu	Homework: Steps to solve problems demonstrate understanding and accuracy of results
GM	WQ notation guide: questions answered correctly. Thought questions carefully considered. Culminating questions structured as assigned and show deep understanding of Greek mythology. Answers were edited for conventions, spelling and grammar
	Myth creation: Personas of the gods stay true to their depictions in ancient Greek mythology
BC	Individual journal: Creation of a back-story for someone preparing for the role. Shows creativity in character choice, connects research to support character choice, constructs three dimensional character who has a clear objective, takes others' work into consideration so it affects their performance
DC	Individual collaboration journal: Documentation of project collaboration with your partner—typed cell report and actual cell model, both of which are presented to the class and turned in
N	Production proposal: Answer questions completely, proposal has anticipated story Arc including all the parts
	Presentation: Discuss why you chose your task and mention a personal experience
	Production journal: Write answers to sentence prompts about what you never thought about before and what you learned about yourself
	Peer review/Evaluation: Review any aspect of another group's work in any depth
M	Documentation Journal and Group Evaluation both use these criteria: Information is thoughtful and reflective, shows how project was accomplished, what worked well, what could have been better, how it could have been done differently
MLK and MX	Reflections in Pamphlet Created: On articles read, programs watched, and how they changed your mind about the subject (discouraging violence among teenagers using the lives and civil rights philosophies of King and X.)

(continued)

Table 6.3 (continued)

WQ	Evaluation criteria
RT	Article/Video Summaries: Comprehensive and accurate description of all major points, demonstrating understanding of the material
	Task-Specific Criteria: (#2) Clear and complete definition and drawing of a real life problem solved with the Pythagorean theorem
CD	Essay—Metacognitive Development and Assessment: Describe your experience, research methodology, difficulties encountered, memorable moments, misconceptions you had about a particular community that were cleared up, and your interest in participating in future events with that community
SCW	Journal Entries and Monologues show learner has articulated and evaluated his/her performance choices and related these to others' choices in the production
WBS	Self-evaluation: Reflect on your collaboration with your teammates. Voice any concern you may have had, what went well, what went wrong and why. Propose solutions to problems encountered to ensure your next group collaboration works better
WW	Answers to mentor text analysis questions, prewrite of your original piece, peer feedback, and final polished product are all to be graded using rubrics for each, but they're not in the WebQuest
WS	Writing process: clear, concise, well written and edited with no serious errors. Vocabulary used skillfully with precision and purpose
	Pair work: Works toward group goals, sensitive to partner's feelings, helps identify needed changes and encourages pair action for change

Table 6.4 Self-reflections expected in WebQuest conclusions

JH	Self-reflection questions
S	Ask yourself and each other what steps of the process you found easy or difficult?
	What could you have done to make it (your stencil) better?
AF	How did that feel? What have you seen here and what can you learn from it?
	When will you speak up?
	How will you speak up?
GG	What did you know about rhythm before this WebQuest and what did you learn from the WebQuest?
	Does this enhance your appreciation of music?
	What steps may you take in the future to further your music appreciation and learning?
	What would you like to know more about regarding music and rhythm?
A	How did the WebQuest influence your view of learning science? How will this project affect your knowledge and interest in science?
F	How did it feel to get into role as a different person? How did it feel to live through this experience in role?
	How has the experience affected you personally? What is your obligation to protect the weak?
Nu	Do you know what an IP address is? Does a number something like 192.168.1.3 look familiar?

(continued)

Table 6.4   (continued)

HS	Self-reflection questions/Activities
GM	Food for thought. Did you know anything about Greek myths, or mythology in general, before this WQ?
BC	When you first read the play or story what did you think of the character?
	How does some of your own personal life experience help inform your character choices?
DC	What else would you like to learn about the animal cell?
N	How can you bring information from these films into your other classes?
M	Do you think that Machinima is something that might move into the mainstream?
MLK and MX	Explain how your perspective on violence changed after this project
	Think of what King and Malcolm X did not say on violence
RT	What did this lesson teach you about the relative lengths of the sides of a right triangle?
	What about triangles that do not contain a right angle? Did you learn anything about the relative length of their sides?
CD	Have you developed a better understanding of your own ethnicity?
SCW	Does any of this information connect to anything happening in our world today? Connecting the world of the play to the world outside of the play will give you a deeper understanding of your given circumstances and your character!
WBS	What did you understand about the importance and significance of those children's sacrifices? What have they given you with their struggles?
WW	Reflections! Reflect on what you learned about writing and what you learned about yourself as a writer
	Choose the very best piece of work you did and answer the questions on the worksheet attached
WS	How do you think this special training helped you portray a person, a place or an action effectively?
	How might you use what you have learnt in this WebQuest in the future? How does your portfolio reflect what you have learnt?

For example, the introduction to "WBS" began, "What do you know about the Civil Rights Movement?

What do you know about the daily struggles African-Americans went through in their fight for equal rights?" The Introduction to the WS WQ began, "Are you ever hard-pressed on how to describe something?" This use of prior knowledge is a meaningful part of the planning process; it sets the stage for engaging in the task with heightened awareness of whether, the extent to which and how new information fits with existing schemas.

Table 6.5 Executive management metacognition embedded in JH WebQuests

WQEMM	Planning	Monitoring	Evaluating	Revising/Action plan
S	What type of material would you use to cut your template?	To what extent do students consider how much ink/paint they use for printing?	Ask yourself what steps in the process you found easy or difficult?	How would you do it differently if you were to cut another stencil?
	How would you cut the template?			
AF	Create and rehearse a 3–5 min play that expresses the problem you chose but doesn't have a happy ending	Use videos and readings to reconsider the problem you chose	List the issues and solutions that stood out to you	You may decide to edit or change your problem
				Discuss with classmates and write in your journal what you can apply to your life
GG	Compose your own individual rhythm composition by using the rhythm chart on the Process page	Make sure you understand the material well before moving on. Let me know if you have any questions	Reflect on and tell me what you learned from each of the individual lessons on this page	
		Check to make sure you have the correct duration for each measure	Practice with the song and try to listen to see if the rhythm fits the song	
A			How did this WQ influence your view of learning science?	How will the project affect your knowledge and interest in science?
F	Create a character	Please try to stay true to your character when presenting your character to your group	What are the most important things you will take away from this experience?	How might you have changed any prior opinions you had about yourself?
	Write diary entries from different stages of your character's life: I am happy because…, I am sad because…, I am scared because…			
Nu	Pick 3 numbers, one 1–99, one 100–999 and one 1000–9999. Use the table to show how each number is represented in the decimal system	Work cooperatively to help each other fully understand the material	Does a number something like 192.168.1.3 look familiar?	

Table 6.6 Examples: EMM embedded in high school WebQuests

WQ	Planning	Monitoring	Evaluating	Revising/Action plan
GM	Create your own ancient Greek myth	Make sure your myth stays true to the ancient Greek traditions	What was the most interesting thing you learned?	Did your understanding change in any way? How?
BC	Build a character back-story, step by step, using the guided questions		What kind of research was the most powerful to you? Why?	What changed about how you perceived the character?
DC	Testing yourself on what you know about the cell is something you can choose to do or not do. Use your own judgment and discretion	Test yourself on what you know about cells to see how ready you are for tomorrow's Quest	Ask yourself: What are three new things that you learned from doing this Animal Cell project?	Did you have any misconceptions about the Animal Cell at the beginning of the project that have now changed? What are they?
N	Come up with an original documentary idea / Make counter arguments to the director's point of view		After watching these movies I never thought of ____ before. During this process I learned ____ about myself	Make suggestions on how you would do a scene from each movie differently
M	Project 1: Produce a 5 min Machinima based on a sequence from a novel or play you have studied in class over the past year. Choose your software. It must incorporate voice	You will have 4 weeks to finish your projects. Rough drafts of projects are due at the 2 week mark	What criteria would you use to judge if something was a good or successful piece of Machinima?	What would you do differently? Is this something you can see yourself using in the future for other subjects?
MLK and MX	Create a pamphlet which discourages violence among teenagers using the life and civil rights philosophies of Martin Luther King and Malcolm X	Submit drafts of your pamphlets for feedback	How do each of the resources (articles, YouTube videos, documentaries, and movie) change your mind about these men and violence?	Complete your pamphlet using feedback the teacher provides / If you could interview both men what would you want them to say on what they did not say about violence?

(continued)

Table 6.6 (continued)

WQ	Planning	Monitoring	Evaluating	Revising/Action plan
RT	Devise a method of mathematically proving the Pythagorean theorem other than the ones in the five proofs provided		Do an analysis of each of the materials and describe what you learned	How could you apply this knowledge to practical experiences in your everyday life?
CD	Describe your neighborhood using pictures or videos emphasizing community or cultural centers, places of worship, public places, parks and restaurants		Describe the major influences in your family life (food, music, language if other than English etc.) Have you developed a better understanding of your own ethnicity?	In the future, will you reach out and learn as much as possible about other cultures? Has this project contributed to helping you confront your own misconceptions and clear up the confusions you have had about a particular community?
SCW	Are you going to fight?	Before performing, make sure you can answer some key questions: How old are you? Where are you from? What is your occupation?	Would you have fought in the resistance if you had lived in Spain in the 1930s? Why or why not?	How might this experience change the way you approach and read plays in the future?
WBS	Prepare 5 interview questions for someone who was around during the Civil Rights Movement	When writing the answers to your interview questions, try your best to stay true to their personal voice	Use the rubric attached to judge your own work	Each student should turn in one rewrite so there should be three rewrites in every project turned in How will you use the gifts they have fought to give you?
WW	Prewriting your own original piece could be a: brainstorm, web, list, free-write or graphic organizer	Get feedback from your partner using the sentence starters in resource #3	How is this text similar to/different from other types of these texts you've seen?	Using the peer feedback and your mentor text, create a final, polished product
WS	Write first drafts of descriptive: sentences, paragraphs and essays	Peer correction feedback on all individual work during the week	Hand in weekly work to teacher to get feedback	Evaluation criteria: helps identify needed changes; encourages pair action for change

6.3.3 Strategic Knowledge Metacognition

In some cases, the teacher provided SKM knowledge, for example, what is a stencil (declarative), when stencils have been used (contextual), and how to make a stencil (procedural), however there were few cases where students were asked to provide their own strategic metacognitive knowledge.

This same teacher was one of the few who embedded strategic metacognitive questions or tasks for students. He asked: What image would you choose? (declarative) Where would you find it? (contextual) and How would you cut the template? (procedural). In addition to these questions focusing on strategic metacognitive knowledge, they could also be considered executive management prompts for planning the stencil making process, which shows how these two main types of metacognition can overlap. At the JH school level, 3 of the 6 (50 %) of WQs embedded SKM. At the HS level, 6/12 (50 %) of the WQs embedded SKM. See Tables 6.7 and 6.8 for examples.

6.3.4 Social Metacognition

The results support the emerging distinction between personal or individual and social or group metacognition. Although traditionally metacognition has been conceptualized as a person thinking about his or her own thinking, analysis of metacognition embedded in these WQs demonstrates how pairs and groups also engage in metacognitive activities.

Table 6.7 Strategic knowledge metacognition embedded in junior high WebQuests

WQ	Declarative	Contextual	Procedural
S	Discuss what style you might work in	Explain why you chose your image	Discuss ideas and techniques about how to complete the stencil cutting process
	What image would you make?	Where would you find that image?	How would you get your image to stick to the template for cutting?
AF	Write a paragraph describing something in life that is not OK to you	Explain why the problem affects you	How will you speak up?
	What have you seen here and what can you learn from it?	When will you speak up?	
GG	What did you know about rhythm before this WebQuest and what did you learn from the WebQuest?		What steps may you take in the future to further your music appreciation and learning?
	What would you like to know more about regarding music and rhythm?		

Table 6.8 Strategic knowledge metacognition embedded in high school WebQuests

WQ	Declarative	Contextual	Procedural
N	What topics are you interested in?		How can you bring information from these films into your other classes?
M	What is Machinima? What is it not?	Why? (is it Machinima)	How can Machinima help you with your English Lit and/or Theater assignments?
	What software did you choose?		How did you choose it? (software)
MLK and MX	What do you know about Martin Luther King and Malcolm X?		
RT	Describe a scenario where the Pythagorean theorem is needed to calculate a particular length	Explain why other methods of measuring length are not feasible and why knowing the length is important	Describe in detail how the other lengths were measured
SCW	What is the relationship between the characters in your group's scene?	Why is this scene taking place?	
WBS	What might your reaction have solved in Warriors Don't Cry?	Why would you react this way?	How would things go had you been faced with the situation instead of Melba?

SM is defined as working metacognitively with at least one other person. Two forms of SM emerged in the data analysis. One is when students are partnered with another student, which is often referred to as pair or peer learning, and the other is when students work together in groups of three or more.

SM, combining pair and group metacognitive activities, was embedded in 89 % of the WQs. Two of the WQs, one at the JH level (AF) and one at the HS level (SCW) included both pair and group metacognitive activities. One WQ, M, gave students the option of working in pairs or small groups.

At the JH level, five out of six WQs (83 %) embedded SM (Table 6.9). At the HS level, 11 out of 12 (92 %) embedded SM (Table 6.10). Most often SM was embedded in the Process section of the WQS, but it also was found in the Tasks, Conclusion and Evaluation.

Pair Metacognition. Overall, 6 out of 18 (33 %) of WQs embedded pair metacognition. Only one of the JH WQs (17 %), AF, embedded pairs of students working together metacognitively. In Tasks, students were instructed, "with a partner in class …practice how to use your body to display frozen images." In Process, students were instructed "to shake hands and then freeze. One person will step out of the image and walk around it and analyze it, then insert themselves back into the

Table 6.9 Examples of social metacognition: Junior high

WQ	WQ part	SM activities
S	Process	Share your work amongst your classmates. Discuss why you chose your image, what it means to you, what you thought of the process (Ex. It was boring! It was hard! It was better than summer vacation!) and how would you do it differently if you were to cut another stencil
		…talk amongst yourselves about the process of stenciling
	Conclusion	Ask yourself and each other what steps of the process you found easy or difficult? Share your techniques. Do you think your stencil is a good representation of your image? What could you have done to make it better?
AF	Process	…talk with your classmates on how the imaging went. What were the challenges, what was interesting, how did it feel to be frozen and at the mercy of your partner?…the whole group can decide if one member's image accurately represents them all or they can combine their images to create a brand new one that shows the issue…talk with your classmates on what solutions surprised you, which solutions you could try to use it real life
GG	Tasks	You and your partners will compose an 8-measure long piece using 4 percussive instruments…. Start out by deciding the time signature…. Check your work and make sure you have the correct duration for each measure…. Practice with the song and try to listen to see if the rhythm you composed fits the song
F	Process	…as a team, the group must choose a specific movement that represents each moment, then chose a word that represents each moment
		It should symbolize each moment and please continue to stay true to your character. The rhythm and timing must be decided by the group
Nu	Process	Each group will be given a four digit decimal number…you will work together to demonstrate how each number is built. Clearly demonstrate how the value of each place is determined and the number is totaled

Table 6.10 Examples of social metacognition: High school

WQ	WQ part	Social metacognition activities
GM	Process	You will work in groups of three to create a new Greek myth! Each person is assigned a role: facilitator, note taker, or editor. Brainstorm as a group your conceptions of your gods/goddesses and how they relate to one another. Organize your conceptions. Prewrite, Compose a draft, Revise and Edit. The editor should revise and edit the draft, but other groups members are encouraged to help as well. Have a member of another group revise your draft using the peer revising sheet. Print out your final draft and present it to the audience
BC	Process	After they finish journaling, in small groups students share new insights and discoveries. Then they agree on one topic they discussed and create a tableaux
	Conclusion	How did what your cast-mates shared with you about their roles help inform your character's life in the story?
DC	Process	You and your partner will collaborate on: doing a written report on the cell, creating a Cell Model, and presenting them both to the class. Individually you will write a journal documenting your collaboration
N	Process	Peer evaluation on any aspect of another student's work and of any depth of evaluation
M	Process	Groups of 2–4 will choose one of 3 projects. Project 3: produce a Machinima documentary of 5–10 min
		Include a short group report detailing: how you chose the subject matter, what software you chose and how you chose it, your casting process, what material you included and what you decided what not to include and why. Reflection: what you think worked, what didn't and what you would do differently, and why this is Machinima
MLK and MX	Tasks	In groups of at least 5, decide which of the two civil rights philosophies you will use to make your pamphlet. You'll work on it over 3 class periods, must submit a draft and use the feedback to complete it
CD	Tasks and Process	Group Discussion: You will compare your neighborhood to other students' neighborhood emphasizing the major differences and similarities. Share why you think they have similarities with yours
	Process	Presentation: Share with your peers your essay and artwork explaining your thought process and creative process. Reflect your experience
SCW	Task	In groups of 4 you will create scenes with the other characters based on all of your monologues
	Process	Day 1. Read your journal entry to an assigned partner. As listener, tell your partner: what did you notice, what did you like and what would you add? What were the most important parts? How did hearing it make you feel?
		Day 2: Taking the feedback from your partner, turn the journal entry into a monologue to share with the class

(continued)

Table 6.10 (continued)

WQ	WQ part	Social metacognition activities
WBS	Tasks	…groups of three will work on a project related to Warriors Don't Cry with six parts: 1. interviews by each person, 2. letter to Melba Patillo, 3. rewrite a scene from the story, 4. reaction to two websites, 5. presentation, 6. evaluation. Drafts of written work on 1–4 are submitted to the teacher and then revised based on feedback
		Reflect on your collaboration with your teammates. Voice any concern…anything that went well or went wrong. Propose solutions to any problems you may have encountered in order to ensure your next group collaboration works better
WW	Process	Get peer feedback from a partner using the sentence starters provided in resource #3
		Using the peer feedback and your mentor text, create a final, polished product
WS	Tasks and process	Pair Work: write descriptive phrases, sentences, and paragraphs. Interview your partner and write a descriptive essay as a newspaper article
		Peer correction on all individually written work
	Evaluation	Pair Work: helps identify needed changes; encourages pair action for change

image in a different position to create a whole new picture. After a moment, the other person will then step out and do the same." These activities engage students' metacognition, stimulating awareness and control over thinking by removing oneself from an image to analyze it (awareness), and then revising it (control).

At the HS level, 5 out of 12 (42 %) embedded pair metacognition activities. These activities were varied, but mainly focused on obtaining and using peer feedback, or peer evaluation. In peer evaluation one person assesses another, such as one student evaluating another's composition (written or musical).

The person being evaluated is able to use external feedback to self-evaluate his or her product and/or process and make revisions accordingly, and/or plan to improve future products/processes. However, pair metacognition also included reflecting on and evaluating the amount and effectiveness of collaboration with one's partner.

Group Metacognition. Most of the WQs, (12 out of 18, 67 %) included metacognitive group activities (three or more students working together metacognitively); most instances were embedded in the Tasks and Process components (Table 6.9 JH, Table 6.10 HS).

Group metacognition was embedded in and 5 out of 6 (83 %) JH WQs and 7 out of 12 (58 %) of HS WQs.

Group metacognition involved collaborative planning of a written project, getting feedback on it and making revisions before submitting it for a grade. It included a presentation of the project, which required additional planning activities.

Although these EbM activities involved EMM, one example of embedded group SKM is in the WQ M, where the group reported on what material they decided to include and exclude, why, and how they chose their subject matter.

6.3.5 Affective Self-reflections

Affective self-reflections (ASR) were embedded in many of the WQs; 50 % of JH WQs included them while they occurred in 67 % (8 out of 12) of HS WQs. So across school levels, 61 % (11 out of 18) of the WQs embedded ASR.

Although the question, "Have you ever felt neglected or unheard?", was at the beginning of the Introduction section of the AF WQ, ASRs occurred most often in the Conclusion sections, where they were expected rather than required.

The ASRs were commonly geared toward having students assess how they felt about the WQ experience and its impact on their attitudes about and interest in the topic. Examples of ASRs are in Table 6.11.

Table 6.11 Affective self-reflections in junior high and high school WebQuests

WQ	Level	Affective self-reflections
GG	JH	Does this enhance your appreciation of music?
A	JH	How will this project affect your knowledge and interest in science?
F	JH	How did it feel to get into role as a different person? How did it feel to live through this experience in role? How has the experience affected you personally? What is your obligation to protect the weak?
DC	HS	Most importantly, did you have fun?
N	HS	What topics are you interested in?
M	HS	What do you feel are the advantages of Machinima over other story forms?
		Is it something you'd be interested in taking up as a hobby?
MLK and MX	HS	After reading articles, watching YouTube videos, documentaries and a movie on Malcolm X, write a paragraph on each telling me how you feel about each
		What would you like to know about teenage violence?
RT	HS	Write an analysis of each reading describing what you learned and what you liked and disliked
CD	HS	Do you feel more confident in your abilities to thrive in this society?
SCW	HS	How are you (the character you created for the day the war began) feeling? How did hearing your partner's journal entry make you feel?
WBS	HS	After reading the short story Warriors Don't Cry, in your rewrite, take care to describe your feelings and motivations in great detail. Choose 2 links on this WebQuest and write one paragraph on how they made you feel

6.4 Discussion and Conclusion

WQs in this case study embedded several different types of metacognition at both the JH and HS levels, and across academic subjects. All of them included EMM. Whereas SKM was embedded in WQs at both school levels, it was not embedded in all of them. Some of the EbM activities were required while others were expected.

Required metacognitive activities were embedded in the Tasks, Process and Evaluation sections of the WQs most frequently, helping to ensure that the assignments were executed properly.

Although expected metacognition was embedded in some of the Task and Process components, it occurred most often in the Conclusion sections of the WQs in this study. Teachers embedded questions asking students to reflect on their WQ experiences regarding what they learned, how it might impact them in the future, and how they felt about the experience. These are worthwhile and key reflections, so it would make the WQ experience richer, more powerful and fruitful if there were accountability structured into these WQs reflections to help ensure that they actually occur.

Teachers may not appreciate that metacognitive questions and prompts are essentially optional for students if there is no accountability structured in the WQ process, such as through written work, grading and/or observation.

Research is needed to examine variables affecting whether, the extent to which and how students engage in metacognitive activities when they are expected but not required. Research should also address how to design WQs that make required metacognitive engagement more pervasive and effective.

The types of self-reflections examined here can overlap across each other and within a category. EMM can overlap with SKM, for example, as when SKM declarative knowledge, "Describe a scenario where the Pythagorean theorem is needed to calculate a particular length" can also be viewed a step in the planning process (EMM) where devising and solving such a problem is a required academic task. EMM can also overlap within this category.

For example, monitoring can overlap with evaluating when one gets feedback while creating a product and using that feedback to improve the product. By definition, developing an action plan, which is based on evaluation, overlaps with planning, as figuring out how to apply what you have learned from an experience leads to developing a new course of action for the future.

EMM can also overlap with affective-self reflections. For example, evaluating what you liked or didn't like about a project (EMM) involves considering one's own interests, attitudes, values and/or motivations (ASRs).

Cognitive reflections embedded in the WQs in this study include "reflection on action", either before or after a task, as in planning and evaluating and "reflection in action", as monitoring during a task [51]. Friere's [52] concept of praxis is a kind of political and educational metacognition where one reflects on the world and engages in actions to improve it. This concept is similar to EMM, using the results of a self-evaluation to develop an action plan which transforms a product, process or situation so that it is better. Several of the WQs discussed here require or expect students to engage in this type of thinking.

Almost all of the WQs in this case study had resources for individuals, groups and/or pairs to use for learning about the content and designing and implementing their projects attached at the bottom of the Tasks and/or Process pages. These included documents, such as articles to read, as well as design and assessment templates; websites with important information, and YouTube videos. When my students presented their WQs to our class in May, 2011, all these resources were available and I checked them all out subsequently when grading this assignment. However, by January, 2014, these resources were no longer available, and I do not know why. This is clearly a problem with the Zunal WQ site, and one that needs to be considered for others who might consider using it. Resource websites that were embedded directly in the Zunal pages (rather than attached at the bottom) are still available. For some unknown reason, fortunately all of the resources for my PWQ attached at the bottom of the Tasks and Process pages are still there.

This study differs from others examining metacognition in WQs in several ways. Research on metacognition in language-learning related WQs primarily focuses on EMM. The studies of employing metacognitive strategies to enhance speaking and writing English do not address SKM and do not explicitly include affective self-regulation [38, 39] although the Li study included data on attitudes toward oral English and oral English instruction [38].

Questionnaire items such as, "I am satisfied with my oral English" and "I am satisfied with current teaching methods in oral English class" [38] could have been identified as ASRs. While most of the other research on metacognition in WQs specifically sought to see how the WQ format could be used for developing metacognitive approaches to learning, the WQs in this case study focused on learning content, however metacognitive activities were deeply and pervasively embedded in most of these WQs to help achieve the content learning goals and objectives. Although, one of the WQs identified metacognition as a process for development and assessment, most explicitly called for students to reflect on their WQ learning experience thereby emphasizing that students should have *awareness* of their own thoughts, knowledge or feelings.

Almost all of the WQs also asked students to consider revisions or action plans—how they might have done things differently, or better, and/or how they might use what they learned in the future. These reflections emphasize learner *control*—another key dimension of metacognition and affective self-regulation.

One of the limitations of the current study is that it did not investigate whether, how and the extent to which EMM and SKM embedded in WQs resulted in a metacognitive approach to learning that transferred to other academic tasks, in computer-supported learning environments, in the regular classroom, and to non-academic contexts. Also, it is not clear whether or to what extent the EMM and SKM embedded actually affected learning outcomes, such as the products created through the WQ tasks and processes and the ability to transfer what was learned. Research shows that without contextual knowledge, learners are often unable to apply what they have learned, and the declarative and procedural knowledge they have acquired remains inert because of lack of awareness when a situation calls for transfer of what has been learned. These are issues that should be addressed in future research on WQs.

Most, but not all WQs required students to work in pairs or groups and involved SM. These metacognitive activities primarily emphasized executive management processes. There were more group than pair metacognitive activities. The findings here on SM embedded in WQs for adolescents complement Chan et al.'s research on adolescents' work on the simulation, Collisions, where computer scaffolds were designed to support collaborative inquiry and co-regulation. The peer questioning strategies to promote metacognition embedded in several of these WQs are different from the approach developed by Choi et al. [53] because in the WQs in this case study, specific metacognitive questions were assigned to students whereas in the Choi et al. approach, scaffolding was used to help students generate their own metacognitive questions. Future WQ design might include embedding similar SM scaffolding strategies to help students develop self-questioning skills for learning metacognitively.

Future research on SM in WQs might follow the lead of Janssen et al. [28] and look at the impact of co-regulation on the quality of the group product produced through collaboration in a computer-supported environment. Also future research may systematically examine co-regulation by pairs and groups to identify similarities and differences in the cognitive processes used, their impact on affective self-regulation, and their effectiveness in achieving targeted learning outcomes, especially in comparison with WQs that do not use SM.

Additionally, when a group member serves a metacognitive function, such as having a cooperative learning role to revise and edit a group product based on feedback from members, or verify the solution to a problem, research should identify strategies for ensuring and assessing metacognitive contributions by all members of the group.

A few of the WQs had more of an emphasis on affective than cognitive reflections, while most had more cognitive than affective. Some WQs had no affective self-regulation. WQs might be enhanced by embedding and requiring both cognitive and affective self-regulatory activities. Also, WQs might more explicitly explain to and train students in EMM and SKM and affective self-regulation as WQ goals, in addition to, and as a facilitator of, content learning and transfer to new situations.

Finally, WQs might be enhanced by embedding both personal and social required metacognitive activities.

References

1. Dodge, B.: Some thoughts about WebQuests. http://webquest.sdsu.edu/about_webquests.html
2. March, T.: Revisiting WebQuests in a web 2 world: how developments in technology and pedagogy combine to scaffold personal learning. Interact. Educ. Multimedia **15**, 1–17 (2007)

3. Abbit, J., Ophus, J.: What we know about the impacts of Web-Quests: a review of research. AACE J. **16**(4), 441–456 (2008)
4. Alshumaimeri, Y.A., Almasri, M.M.: The effects of using WebQuests on reading comprehension performance of Saudi EFL students. Turk. Online J. Educ. Technol. **11**(4), 295–306 (2012)
5. Kachina, O.: Using WebQuests in the social science classroom. Contemp. Issues Educ. Res. **5**(3), 185–200 (2012)
6. Salsovic, A.: WebQuests—Inquiry-based lesson plans that harness the technology of the Internet—are easy to implement. Math. Teacher **102**(9), 666–671 (2009)
7. Hartman, H.J.: Developing students' metacognitive knowledge and skills. In: Hartman, H.J. (ed.) Metacognition in Learning and Instruction: Theory, Research and Practice. Neuropsychology and Cognition, vol. 19, pp. 33–68. Kluwer Academic Publishers, Dordrecht (2001)
8. Fisher, R.: Thinking about thinking: Developing metacognition in children. Early Childhood Dev. Care **141**, 1–15 (1998)
9. Sternberg, R.J.: Metacognition, abilities and developing expertise: what makes an expert student? Instr. Sci. **26**(1–2), 127–140 (1998)
10. Coskun, A.: The effect of metacognitive strategy training on the listening performance of beginner students. Novitas R: Res. Youth Lang. **4**(1), 35–50 (2010)
11. Eggert, S., Ostermeyer, F., Hasselhorn, M., Bögeholz, S.: Socioscientific decision making in the science classroom: the effect of embedded metacognitive instructions on students' learning outcomes. Educ. Res. Int. **309894**, 12 (2013)
12. Azevedo, R.: Computers as metacognitive tools for enhancing learning. Educ. Psychol. **40**(2), 193–197 (2005)
13. Poitras, E., Lajoie, S., Hong, Y.J.: The design of technology-rich environments as metacognitive tools in history education. Instr. Sci. **40**(6), 1033–1061 (2012)
14. Kori, K., Mäeots, M., Pedaste, M.: Guided reflection to support quality of reflection and inquiry in web-based learning. In: 4th International Conference on Education & Educational Psychology (2013)
15. Cho, K., Cho, M-H., Hacker, D.J.: Self-monitoring support for learning to write. Interact. Learn. Environ. **18**(2), 101–113 (2010)
16. Ramirez-Corona, N., Zaira, R., Lopez-Malo, A., Palau, E.: Assessing metacognitive awareness during problem solving in a kinetics and homogeneous reactor design course. In: 120th American Society for Engineering Education Conference, pp. 1–14, Atlanta (2013)
17. Iiskala, T., Vauras, M., Lehtinen, E.: Socially-shared metacognition in peer learning? Hellenic J. Psychol. **1**, 147–178 (2004)
18. Chan, C.K.K.: Co-regulation of learning in computer-supported collaborative learning environments: A discussion. Metacognition Learn. **7**, 63–73 (2012)
19. Frith, C.D.: The role of metacognition in human social interactions. Philisophical Trans. R Soc B: Biol. Sci. **367**(1599), 2213–2223 (2012)
20. Siegel, M.: Filling in the distance between us: Group metacognition during problem solving in secondary education course. J. Sci. Educ. Technol. **21**(3), 325–341 (2012)
21. Smith, J.: An investigation in the use of collaborative metacognition during mathematical problem solving: A case study with a primary five class in Scotland. PhD Dissertation, University of Glasgow (2013)
22. Whimbey, A., Lochhead, J.: Problem Solving and Comprehension. Franklin Institute Press, Philadelphia (1982)
23. King, A.: From sage on the stage to guide on the side. Coll. Teach. **41**(1), 30–35 (1993)
24. Lyman, F.: The responsive classroom discussion: The inclusion of all students. Mainstreaming Digest. Univ. Maryland, College Park (1981)
25. Sharan, Y., Sharan, S.: Group investigation expands cooperative learning. Educ. Leadersh., 17–21 (1989/1990)

26. Hmelo, C.E., Gotterer, G.S., Bransford, J.D.: A theory driven approach to assessing the cognitive effects of PBL. Instr. Sci. **25**(6), 387–408 (1997)
27. Saab, N., van Joolingan, W., van Hout-Wolters, B.: Support of the collaborative inquiry learning process: influence of support on task and team regulation. Metacognition Learn. **7**(1), 7–23 (2012)
28. Janssen, J., Erkens, G., Kirschner, P.A.: Task-related and social regulation during online collaborative learning. Metacognition Learn. **7**, 25–43 (2012)
29. Chick, N., Karas, T., Kernahan, C.: Learning from their own learning: how metacognitive and meta-affective reflections enhance learning in race-related courses. Int. J. Sch. Teach. Learn. **3**(1), 1–28 (2009)
30. Picard, R.W., Papert, S., Bender, W., Blumberg, B., Breazeal, C., Cavallo, D., Machover, T., Resnick, M., Roy, D., Strohecker, C.: Affective learning—a manifesto. BT Technol. J. **22**(4), 253–269 (2004)
31. Bandura, A., Caprara, G.V., Barbaranelli, C., Gerbino, M., Pastorelli, C.: Role of affective self-regulatory efficacy on diverse spheres of psychosocial functioning. Child Dev. **74**(3), 769–782 (2003)
32. Forgas, J.P., Baumeister, R.F., Tice, D.: Psychology of self-regulation: Cognitive, affective, and motivational processes. Psychology Press, New York (2009)
33. Hartman, H.J., Sternberg, R.J.: A broad BACEIS for improving thinking. Instr. Sci. **21**(5), 401–425 (1993)
34. Illeris, K.: Transformative learning in the perspective of a comprehensive learning theory. J. Transformative Educ. **2**(2), 79–89 (2004)
35. Pang, K., Ross, C.: Assessing the integration of embedded metacognitive strategies in college subjects for improved learning outcomes: a new model of learning activity. J. Effective Teach. **10**(1), 79–97 (2010)
36. Luzon, M.J.: Providing scaffolding and feedback in online learning environments. Mélanges, CRAPEL **28**, 113–122 (2006)
37. Luzon, M.J.: Enhancing Webquest for effective ESP learning. CORELL: Comput. Resour. Lang. Learn. **1**, 1–13 (2007)
38. Li, J., Yue, Y., Yang, M.: On WebQuest-based metacognitive speaking strategy instruction. Theor. Pract. Lang. Stud. **1**(12), 1762–1766 (2011)
39. Lee, J., Liu, W., Wang, L.: On the application of WebQuest in learning of metacognitive writing strategy. In: 5th International Conference on Computational and Information Sciences (ICCIS), pp. 1737–1739. Shiyang, China (2013)
40. Mora, R.A., Martinez, J.D., Alzate-Perez, L., Gomez-Yepes, R., Zapata-Monsalve, L.M.: Rethinking WebQuests in second language teacher education: The case of one Colombian university. In: Wankel, C., Blessinger, P. (eds). Increasing Student Engagement and Retention using Online Learning Activities: Wikis, Blogs & Webquests. Cutting-edge Technologies in Higher Education. vol. 6A. pp. 291–319. Emerald Group Publishing Limited, Bingley, UK (2012)
41. Hsiao, H.S., Tsai, C.C., Lin, C.Y., Lin, C.C.: Implementing a self-regulated WebQuest learning system for Chinese elementary school. Australas. J. Educ. Technol. **28**(2), 315–340 (2012)
42. Hartman, H.J.: Consuming and constructing knowledge through WebQuests. In: Wankel, C., Blessinger, P. (eds). Increasing Student Engagement and Retention using Online Learning Activities: Wikis, Blogs & Webquests. Cutting-edge Technologies in Higher Education. vol. 6A. pp. 255–289. Emerald Group Publishing Limited, Bingley, UK (2012)
43. Hartman, H.J.: Discovering and uncovering Piaget. http://zunal.com/webquest.php?w=22695
44. Mayer, R.E.: Cognitive theory of multimedia learning. In: Mayer, R.E. (ed.) The Cambridge Handbook of Multimedia Learning, pp. 31–48. Cambridge University Press, New York (2005)
45. Sternberg, R.J.: Beyond IQ: A Triarchic Theory of Intelligence. Cambridge University Press, Cambridge (1985)

46. Dede, C.: The evolution of distance education: emerging technologies and distributed learning. Ame J Distance Educ. **10**(2), 4–36 (1996)
47. Vygotsky, L.: Mind in Society: The Development of Higher Psychological Processes. Harvard University Press, Cambridge (1978)
48. Brown, J.S., Collins, A., Duguid, P.: Situated cognition and the culture of learning. Educ. Researcher **18**(1), 32–42 (1989)
49. Lave, J., Wenger, E.: Situated Learning: Legitimate Peripheral Participation. Cambridge University Press, Cambridge (1991)
50. Cochran-Smith, M., Lytle, S.L.: Relationships of knowledge and practice: teacher learning in communities. Rev. Res. Educ. **24**, 249–305 (1999)
51. Schon, D.: The Reflective Practitioner. Basic Books, New York (1983)
52. Friere, P.: Pedagogy of the Oppressed. Herder and Herder, Freiburg (1970)
53. Choi, I., Land, S., Turgeon, A.: Scaffolding peer questioning strategies to facilitate metacognition during online small group discussion. Instr. Sci. **33**(5–6), 483–511 (2005)

Part III
Studies

Chapter 7
The Role of Metacognitive Awareness of Listening Strategies in Listening Proficiency: The Case of Language Learners with Different Levels of Academic Self-regulation

Mehrak Rahimi and Sajad Abedi

Abstract While listeners' metacognitive awareness of listening strategies has been reported to be a significant predictor of listening ability, factors that can moderate this predictive power have been taken for granted in the literature. The present study thus aimed at comparing the relationship between metacognitive awareness of listening strategies and listening proficiency among language learners with different levels of academic self-regulation (low, mid, and high). Three hundred and sixty-nine high-school students participated in the study. To gather the data, Preliminary English Test, Metacognitive Awareness Listening Questionnaire, and Academic Self-Regulation Questionnaire were used. The findings of the study rejected the primary hypothesis that the degree of association between metacognitive awareness and listening comprehension is the strongest among highly self-regulated students. Further, regression analysis demonstrated that metacognitive awareness power value to predict listening proficiency, near to what is reported in the literature, was only gained when the mid self-regulated students were considered in the analysis.

Keywords Metacognition · Listening · Strategies · Self-regulation · Level · Awareness

Abbreviations

ASRQ   Academic self-regulation questionnaire
DA   Directed attention

M. Rahimi (✉) · S. Abedi
English Department, Faculty of Humanities, Shahid Rajaee
Teacher Training University, 1678815811 Lavizan, Tehran, Iran
e-mail: rahimi@srttu.edu

S. Abedi
e-mail: sajjad.abedi@yahoo.com

© Springer International Publishing Switzerland 2015   169
A. Peña-Ayala (ed.), *Metacognition: Fundaments, Applications, and Trends*,
Intelligent Systems Reference Library 76, DOI 10.1007/978-3-319-11062-2_7

EFL	English as a foreign language
ESL	English as a second language
L2	Second language
MALQ	Metacognitive awareness listening questionnaire
MT	Mental translation
PCA	Principal components analysis
PK	Person knowledge
PET	Primary English Test
PE	Planning-evaluation
PS	Problem solving
SLA	Second language acquisition
SRL	Self-regulated learning

7.1 Introduction

What distinguishes cognitive psychology from its predecessor, behaviorism, is the great emphasis it lays on human's mind and the way information is processed, saved and transformed into knowledge by mental processes to direct attention and exhibit behavior. Cognitive psychology brings "internal psychological processes that are involved in making sense of the environment and deciding what action might be appropriate" [1, p. I] into sharp focus. It is a broad field of study that addresses controversial topics about "human memory, perception, attention, pattern recognition, consciousness, neuroscience, representation of knowledge, cognitive development, language, and thinking" [2, p. 2147].

One contentious issue within cognitive psychology is the way control processes regulate different activities the brain is engaged in. These cognitive control processes "configure the cognitive system to process stimuli in a specific manner, and re-configure the cognitive system when certain events tell the observer/actor to treat stimuli in a different way" [3, p. 10]. Cognitive control processes require attention, consciousness, and awareness [4, 5] and include a broad range of mentalistic constructs among which metacognition and self-regulation have been reported to be the most influential ones [6]. While metacognition is simply defined as "cognition about cognition" [7, p. 104], self-regulation refers to "the control of one's present conduct based on motives related to a subsequent goal or ideal that an individual has set for him or herself" [8]. Self-regulation to achieve learning goals and direct one's own learning is labeled self-regulated learning (SRL) that is created as a result of "students' self-generated thoughts and behaviors" [9, p. 125].

Research shows that there is a significant relationship between these two processes as SRL is mostly related to metacognitive awareness of learning strategies that a learner finds most appropriate to use to achieve certain learning goals [10]. It is also known that metacognitive and self-regulated learning theories have some common features, suggesting that these two constructs are somehow integrated and interact with each other in several ways. Learners who know how to integrate cognitive, metacognitive, motivational, and strategic components of learning are good at self-regulation [11]. They are always engaged in self-regulated learning, that is, they have control over their thoughts, emotions, behaviors, and the learning task/situation [12]. Further, the role of both constructs in learning and its related variables has been noted. Language educationists have shown interest in SRL and metacognition as both constructs play a significant role in autonomous learning that guarantees successful language learning [13]. Metacognition plays a key role in many cognitive activities related to language acquisition and language use such as communication, reading/listening comprehension, and writing [14].

Awareness of metacognitive processes is linked to planning, monitoring, and evaluating cognitive processes while doing language tasks and helps learners to "manage, direct, regulate and guide their learning" [13, p. 48]. While research on the role of metacognition in processing the written input dates back to 70 s, research into metacognitive awareness of listening is relatively new.

As listening comprehension is the most difficult language skill for learners of English as Foreign Language (EFL) [15], language researchers and teachers have tried to find effective ways to teach this skill by underscoring the interactive nature of listening and scrutinizing the cognitive and affective factors involved in the process of listening. The studies of successful language listeners and the strategies skillful listeners apply in doing listening tasks drew researchers' attention to metacognitive processes involved in listening comprehension. Follow-up studies showed that metacognitive awareness of listening strategies was worth the attention as it was found to be related to language learners' listening ability [16, 17].

Language learners demonstrate some degree of metacognitive awareness about themselves as second language listeners and the listening processes itself; and this awareness can predict around 13 % of their listening proficiency [18]. The idea of metacognitive listening instruction and promoting awareness of listeners in the process of analyzing the oral input in language learning has its root in making language learners active [13] and self-regulated as metacognitive awareness brings about "a high degree of experienced choice with respect to the initiation and regulation of one's own behavior" [19, p. 9].

What we know about the relationship between metacognition and self-regulation when listening to a foreign language is that, learners' metacognitive knowledge about themselves, the task, and listening strategies are linked to self-regulatory activities such as setting goals; monitoring thoughts, emotions, and behavior; and controlling the learning situation by choosing appropriate listening tasks. Effective use of metacognitive listening strategies plays a critical role in successful listening comprehension and helps listeners to enhance their self-regulation and autonomy in listening [20].

However, we do not still know if the association of metacognitive awareness of listening strategies and listening ability is dependent on academic self-regulation of language learners. As the two constructs are intertwined, it is logical to assume that self-regulated learning helps students to adopt more suitable listening strategies based on the type of listening task they have to perform to learn language. Metacognitive awareness of listening strategies has been found to be related to listening proficiency [21], language learning motivation [20] and listening task/ condition [22], however, the level of learners' self-regulation seems to be taken for granted in the literature. The reason maybe the fact that self-regulation is considered to be the superordinate of metacognition and the two constructs are assumed to have reciprocal impact on each other. The purpose of this study is thus twofold. First, the role of metacognitive awareness of listening strategies in listening proficiency is investigated among EFL learners. Second, the degree of association between metacognitive awareness of listening strategies and listening proficiency is scrutinized among students with different levels of self-regulation.

It is hypothesized that the degree of association between metacognitive awareness of listening strategies and listening proficiency is the strongest when highly self-regulated students are considered in the analysis. The study seeks answers to the following research questions:

- Is there any relationship between metacognitive awareness of listening strategies and listening proficiency?
- Is there any relationship between metacognitive awareness of listening strategies and listening proficiency among low, mid and highly self-regulated EFL learners?
- Is there any difference between the power of metacognitive awareness of listening strategies to predict listening proficiency among low, mid and highly self-regulated EFL learners?

7.1.1 Cognitive Processes Involved in Listening

The construct second language (L2) listening has undergone dramatic changes in the history of language teaching and learning. Once listening was regarded as a passive and receptive activity that did not require considerable effort from the learner's side to be mastered. Form 70 s, however, the way brain processes linguistic input and produces output has been the foci of many studies within the realm of cognitive science and interdisciplinary fields of study such as second language acquisition (SLA) and psycholinguistics. As a result, listening is now believed to be an active process that has a significant role both in language acquisition and comprehension [23] and deserves systematic development and instruction [15]. Listening is:

> A complex active process in which the listener must discriminate between sounds, understand vocabulary and grammatical structures, interpret stress and intonation, retain what was gathered in all of the above and interpret it within the immediate as well as the larger socio-cultural context of the utterance [24, p. 168].

The findings of research on the role of cognition in language learning show that language learners utilize both bottom-up and top-down processes to comprehend the aural text and thus both accuracy and fluency are important features of listening instruction [25]. Bottom-up processing refers to "the decoding of individual words and phrases" [26, p. 166] to achieve higher level meaning [27]. In other words, the meaning is understood when the message is decoded by analyzing and recognizing discrete units of language such as sounds, words, and sentences. Language learners' lexical and grammatical knowledge plays a crucial rule in bottom-up processing as "the input is scanned for familiar words, and grammatical knowledge is used to work out the relationship between elements of sentences" [23, p. 4].

In bottom-up processing the listeners pay "close attention to every details of the language input" [28, p. 74] and "build acoustic features into phonemes, phonemes into syllables, syllables into words, words into syntactic patterns, [and] syntactic patterns into propositional (abstract) meaning" [29, p. 20].

The listeners might use strategies such as translation, identifying single words and clause boundaries, recognizing grammatical relationships between key elements of the text, and using suprasegmental features to identify word and sentence functions to arrive at the meaning [23]. Clark and Clark (1977, cited in [23]) summarize the bottom-up process as:

- [Listeners] take in raw speech and hold a phonological representation of it in working memory.
- They immediately attempt to organize the phonological representation into constituents, identifying their content and function.
- They identify each constituent and then construct underlying propositions, building continually onto a hierarchical representation of propositions.
- Once they have identified the propositions for a constituent, they retain them in working memory and at some point purge memory of the phonological representation. In doing this, they forget the exact wording and retain the meaning (p. 4).

Top-down processing "makes use of higher level, non-sensory information to predict or interpret lower level information that is present in the data" [23, p. 557] and involves the use of prior or background knowledge to understand the information in the written or oral texts.

This prior or background knowledge is called schema. Schema is a set of interrelated features associated with an entity or concept [29] that is created as a result of human's previous experience with the world that makes predictions of future experiences possible [30]. Three types of schema can be used in the process of reading and listening [29]:

- 'World knowledge': including encyclopedic knowledge and previous knowledge of the speaker or writer. This helps us to construct a content schema for a text.
- Knowledge built up from the text so far: a current meaning representation.
- Previous experience of this type of text (a text schema). This can be extended to include: previous experience of the type of task that the listener/reader has to perform (p. 40).

Top-down processing in listening "refers to the use of expectations in order to infer what the speaker may have said or intended to say" [31, p. 53] based on "a bank of prior knowledge" [28, p. 74]; and entails strategies such as elaboration, inferencing, and prediction.

While evidence shows that the use of top-down strategies aids the listeners to process spoken language more quickly [31], it may also lead to "an erroneous conclusion about what a speaker said or meant" [31, p. 53]. Macaro, Graham and Vanderplank [26] believe that the deployment of top-down strategies leads to successful listening if certain criteria are fulfilled:

- The information contained in a text needs to be congruent with the listener's prior knowledge of the topic of the text.
- Learners' lexical knowledge needs to exceed a 'threshold' level.
- Learners need to know how to make effective use of prior knowledge by deploying it flexibly and in combination with linguistic information contained in the input (p. 179).

It is known that listening comprehension involves the integration of both processes or interactive processing [32] when "top-down and bottom-up processes interact, so that lack of information at one level can be compensated for by checking against information at the other level" [32, p. 88]. In this way both understanding the linguistic message and activating the previous knowledge are utilized to process the spoken language. The listeners' preference to use one process depends on both the characteristics of the listener (e.g., prior background knowledge) and the text (e.g., its type) [23]. This preference is also linked to the tactics listeners use to respond to the challenges or difficulties they face while they are listening, as difficulty is synonymous with the text "cognitive load" [31, p. 49] or its linguistic and non-linguistic characteristics [33]. In other words, "text difficulty is a reflection of the cognitive processes required for an adequate understanding of a text" [31, p. 49].

How to control cognitive processes to overcome difficulties of listening has been the foci of many studies that aimed to describe the characteristic features of more and less successful listeners. The way language learners use language learning strategies and the type of strategies they use have become the main theme of research on successful listening for more than three decades.

7.1.2 Listening Strategies

Strategy is generally defined as "a planned series of actions for achieving something" [34]. In language education, language learning strategies are defined as "operations employed by the learner to aid the acquisition, storage, retrieval and use of information, specific actions taken by the learner to make learning easier, faster, more enjoyable, more self-directed, more effective and more transferable to new situations" [35, p. 167]. Language learning strategies are of three basic types, that is, cognitive, metacognitive and socio-affective [36].

Cognitive strategies are those strategies that help learners "make and strengthen association between new and already known information" [35, p. 167] and include strategies such as repetition, resourcing, translation, and note-taking [37].

Metacognitive strategies are strategies that learners use to manage and regulate their learning and entail strategies such as self-monitoring, directed attention, and self-evaluation [37]. Socio-affective strategies are strategies learners use to interact with others [38] and include strategies such as cooperation and question for clarification [37]. Language learners use these strategies consciously [39] both directly and indirectly in the process of their language learning [40].

Language learning strategy research emerged and evolved with studies that focused on finding the characteristics of good language learners [37] in order to help other language learners make their learning more effective and less challenging. Five aspects of successful language learning have been surveyed by strategy research [41, p. 708]:

- A concern for language form.
- A concern for communication (functional practice).
- An active task approach.
- An awareness of the learning process.
- A capacity to use strategies flexibly in accordance with task requirements.

Strategy research has focused on core themes such as the influence of individual differences and situational variables on strategy use [41], models of strategy instruction and their impact on both strategy use and language learning [42], and the type of strategy successful language learners use [35].

Recent studies on learning strategies have paid more attention to the strategies language learners use "in the context of specific tasks and skills" [43, p. 27] such as learning grammar [39], reading comprehension [44], writing [45], vocabulary learning [46], and listening comprehension [18].

Within this framework listening strategies are defined as the ways listeners approach and manage their listening task [23]. Different experts have suggested varied types of listening strategies, mostly based on and similar to O'Malley and Chamot's framework of language learning strategies. Oxford [40] gave a long list of strategies listeners use such as associating, elaborating, and creating mental linkages.

Vandergrift [47] proposed a framework of listening strategies consisting of three main types of strategies including metacognitive, cognitive, and socio-affective strategies. Similarly, Buck [48] suggested that successful listeners use cognitive and metacognitive strategies. Recently, however, Rost [31] "collectively, using introspection and retrospection methodologies, and coupled with measures of actual effects of strategy use on comprehension and retention" (p. 64) has identified five listening strategies that listeners use in the process of listening including:

- Predicting speaker intentions and activating ideas.
- Monitoring one's own comprehension.
- Asking for clarification (with increasingly focused informational requests).
- Making inferences from incomplete information.
- Providing personal responses about content (p. 64).

It is evident that using listening strategies is related to listening task type [22], language learning motivation [20], listening anxiety [49] and listening ability [50].

The findings of a string of studies done on metacognitive awareness of listening strategies have shown that listeners' metacognitive awareness of the strategies they use in the process of listening can be a contributory factor to successful listening and thus deserve more serious attention.

7.1.3 Metacognitive Awareness of Listening Strategies

Metacognitive knowledge is defined as "the knowledge concerning one's own cognitive processes and products or anything related to them, e.g., the learning-relevant properties of information or data" [14, p. 232].

Metacognition has two basic components: knowledge of cognition and regulation of cognition [51]. Knowledge of cognition entails a set of beliefs about "the way cognitive processes work and about what cognitive states are" [52, p. 125].

This knowledge consists of three types of knowledge including person knowledge, task knowledge, and strategic knowledge [53].

Person knowledge is a learner's knowledge about his/her cognitive ability and motivational beliefs [54]. Task knowledge is about how learners can process information in doing tasks and solving problems [16] and what type of strategies are more useful in doing certain tasks based on task condition [54].

Strategy knowledge is the knowledge of when to deploy a certain strategy that is more useful and effective in doing the learning task.

Regulation of cognition, on the other hand, means how learners control and manage their mental processes and more specifically their learning. This component has four basic functions of:

- Becoming aware of the existence of a cognitive problem.
- Planning and activating the appropriate strategies to address it.
- Predicting one's performance.
- Monitoring and regulating the ongoing cognitive activity [52, p. 126].

The two components of metacognition have been found to be related, although the nature of this relationship is still open to question [55]. Needless to say, metacognitive strategies have a great role in both components of metacognition.

Metacognitive strategies are "higher order executive skills that may entail of planning for, monitoring, or evaluating the success of activity" [36, p. 44] by the help of which learners manage, direct, regulate, and guide their learning [56].

As "all language learning strategies are related to the features of control, goal-directedness, autonomy, and self-efficacy", [35, p. 167] learners' understanding and awareness of these strategies can help them to become more conscious of their learning processes and thus more efficiently control these processes.

Metacognitive awareness of listening strategies is "students' perceptions of themselves as listeners, their perceptions of the requirements of listening tasks, and their awareness of the strategies they deploy to achieve comprehension" [18, p. 438].

Table 7.1 Examples of metacognitive listening strategies [18] cited in [22, p. 1155]

Metacognitive listening strategies	Examples
Problem solving (PS)	Using known words to deduce the meaning of unknown words; using the general idea of a text to deduce unknown words; using one's experience and general knowledge in interpreting the text; adjusting one's interpretation upon realizing that it is not correct; monitoring the accuracy of one's inferences for congruency with the developing interpretation; comparing the developing interpretation with one's knowledge of the topic
Planning-evaluation (PE)	Having a plan for listening; thinking about similar texts as a guide for listening; having a goal in mind while listening; periodically checking one's satisfaction with the ongoing interpretation while listening; evaluating the strategic effectiveness of one's listening efforts
Mental translation (MT)	Avoiding translation in one's head while listening; avoiding translation of the key words; avoiding word for word translation
Person knowledge (PK)	Assessing the perceived difficulty of listening compared with the three other language skills; assessing learners' linguistic confidence in L2 listening; assessing the level of anxiety experienced in L2 listening
Directed attention (DA)	Getting back on track when losing concentration; focusing harder when having difficulty understanding; recovering concentration when one's mind wanders; not giving up when one experiences difficulties understanding

The learners who are more metacognitively aware, process and store new information more efficiently, can manage their learning skillfully [18], and may result in better learning outcome [16].

Five types of listening strategies have been proposed including problem solving (inferencing and monitoring), planning-evaluation (preparation and self-evaluation), mental translation (translating), person knowledge (self-efficacy), and directed attention (concentrating and staying on task) [18]. The types and examples of listening strategies are summarized in Table 7.1.

7.1.4 Metacognition and Self-regulation

Self-regulation is the "self-generated thoughts, feelings, and actions that are planned and cyclically adapted to the attainment of personal goals" [57, p. 14]. In the academic setting, self-regulation is considered an indispensable characteristic of successful learners as self-regulated students are "proactive in their efforts to learn because they are aware of their strengths and limitations and because they are guided by personally set goals and task-related strategies" [57, p. 14).

Self-regulated learning is assumed to subsume some interrelated psychological constructs such as metacognition, motivation, strategies, and beliefs [55]. Self-regulated learners are intrinsically motivated [58], are aware of how to regulate their learning and managing the learning task, can deploy learning strategies to reach their goals [59], and thus are likely to do better in their learning [60]. Metacognitive and self-regulated learning theories aim at understanding how people learn and explore the acquisition, evaluation and regulation of knowledge. They view learners as being able to monitor their own learning and agree that the learners benefit from metacognitive activities such as setting goals and evaluating/regulating one's progress.

This view is reflected in Zimmerman and Moylan's social cognitive model of self-regulated learning [11], where self-regulation integrates with metacognitive processes and key measures of motivation. Based on this model, self-regulation consists of a cycle with three phases of forethought, performance, and self-regulation (Fig. 7.1).

The forethought phase consists of two parts, that is, task analysis and self-motivation beliefs. The first component includes the way the learners analyze the learning task into some components and plan to do the task. The second component

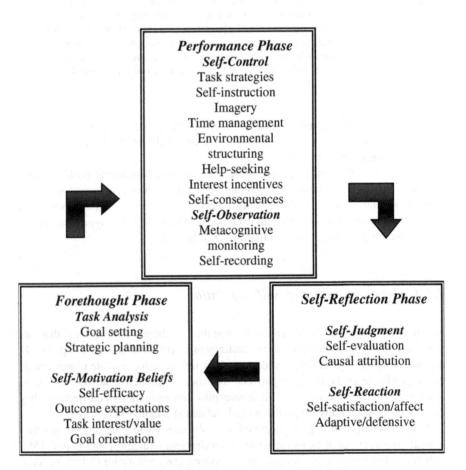

Fig. 7.1 A cyclical phase model of self-regulation, taken form [11, p. 300]

consists of the beliefs learners have about themselves as learners, their ultimate performance, and interest/goal in doing a task.

The performance phase consists of two basic parts, that is, self-control and self-observation. Self-control "involves a variety of task-specific as well as general strategies" [11, p. 302]. These strategies are selected based on learners' outcomes and thus self-observation and its two components, i.e., metacognitive monitoring and self-recording have a key role in students' self-control of their performance.

The self-reflection phase consists of the way the learners evaluate their performance (self-judgment) and the way they react to their self-judgment or their enthusiasm to continue doing a certain task (self-reaction).

Language learning strategy frameworks are considered to be a part of self-regulation models [39], like the aforementioned one, as their goal is "equipping learners with the skills necessary to plan, execute, monitor, and evaluate their learning processes" [59, p. 58]. They help learners become self-regulated, autonomous, and goal-directed learners who can consciously control the process of their learning [35]. It is suggested that "the adequate use of learning strategies is considered crucial for successful self-regulation" [59, p. 58].

While metacognition is a subpart of self-regulation, the two constructs have been found to have a large degree of overlap. Metacognition and self-regulation share two basic functions, namely monitoring and control [61, p. 14].

The former monitors and controls the cognition, the latter monitors and controls behavior to reach a specific goal [62]. Research shows that the role of metacognition within the superordinate self-regulation model is still open to question [55].

It is also unknown if the self-regulatory ability of learners has a significant role in their metacognitive awareness of language learning strategies in general, and the strategies they use to manage specific language tasks such as listening comprehension in particular.

7.2 Method

7.2.1 The Context of the Study

The study has been carried out in an EFL setting where language learners basically learn English in the classroom and under instruction with a limited chance of communicating in English out of the classroom. EFL setting is contrasted with English as a second language (ESL) setting where a language learner "learns English in a setting in which the language is necessary for everyday life ... or in a country in which English plays an important role in education, business, and government..." [27, p. 180]. Therefore, in this context, all activities and tasks to learn English out of the classroom are regulated by students themselves.

As listening comprehension is the least important skill in the current EFL curriculum of Iran, those students who are more interested in this skill have to manage their learning themselves to reach higher skillfulness in processing the oral input.

This may include setting goals to do a variety of language tasks in general and listening tasks in particular; applying different types of learning strategies in the process of doing these tasks; monitoring and self-evaluating the listening performance; seeking help from other sources, classmates, and peers; controlling environmental variables such as the place and time of doing the tasks; planning and organizing the listening tasks; and accepting one's learning responsibilities.

Since self-regulation eases the difficulty, tension, and worry of listening comprehension in a foreign language, it is expected that highly self-regulated language learners become more aware of the listening strategies they use in listening and thus they show a better listening performance.

It can therefore be hypothesized that metacognitive awareness of listening strategies has a significantly contributing role in listening ability of highly self-regulated learners when they are compared with low and mid self-regulated learners. This study has been carried out on the basis of this assumption.

7.2.2 Participants

Three hundred and sixty nine high-school students of one small city in Iran took part in the study. The participants included all grade 3 and 4 high-school students of the city. The students were majored in humanities (43 %), natural sciences (26.6 %), and mathematics (30.4 %). Both male (n = 138, 37.4 %) and female (n = 231, 62.6 %) students were sampled.

7.2.3 Instruments

To collect data for the current study three instruments were used including:

- Preliminary English Test (PET)
- Academic Self-Regulation Questionnaire (ASRQ)
- Metacognitive Awareness Listening Questionnaire (MALQ)

Preliminary English Test (PET). PET is a part of Cambridge Main Suite, a group of examinations developed by Cambridge ESOL at Cambridge University. There are five tests (levels) in The Main Suite and PET is at Level B1, above Key English Test (KET) and below First Certificate in English (FCE). PET has been accredited as an Entry Level 3 ESOL certificate in the UK.

PET has three main sections including reading (5 parts)/writing (3 parts), listening (4 parts), and speaking (4 parts). The objective of the listening part is "the assessment of candidates' ability to understand dialogues and monologues in both informal and neutral settings on a range of everyday topics" [63, p. 5].

The listening part includes four parts and 25 items. Part 1 consists of 7 discrete three-option multiple choice items with visuals; and asks the examinees to listen to neutral or informal monologues or dialogues and identify key information from short exchanges. Part 2 has 6 three-option multiple-choice items; and asks the examinees to listen to a longer monologue or interview and identify specific information and detailed meaning.

Part 3 has 6 gaps to be filled with one or more words; and asks the examinees to listen to a longer monologue to identify, understand and interpret information. Part 4 has 6 items, in the form of statements to be decided by the examinees to be correct or incorrect; and asks the examinees to listen to a longer informal dialogue for detailed meaning and identify the attitudes and/or opinions of the speakers.

PET was used for two purposes in this study: (a) to provide students with listening tasks so that their metacognitive awareness of listening strategies can be subsequently tested, and (b) to assess participants' English listening proficiency.

Academic Self-Regulation Questionnaire (ASRQ). To assess participants' academic self-regulation, the Persian version of Academic Self-Regulation Questionnaire (ASRQ) originally developed by Magno [64] was used.

ASRQ consists of 55 items, categorized into 7 factors including memory strategies (14 items), goal setting (5 items), self-evaluation (12 items), asking for help (8 items), managing the surrounding (5 items), learning responsibility (5 items), and planning (6 items). The respondents were asked to rate themselves regarding self-regulatory items on a four-point Likert scale including 1 'completely disagree', 2 'disagree', 3 'agree', and 4 'completely agree'. A sample item of ASRQ is "I use note cards to write information I need to remember".

The result of the principal components analysis (PCA) has shown that the seven factors of the scale explain 42.54 % of the total variance of the construct [64]. The reliability coefficients of the components of the scale reported to be 0.73–0.87 [64].

The overall reliability of ASRQ was found to be 0.91 in the current study. The reliability coefficients of the seven factors were found to be 0.82 for memory strategies, 0.87 for goal setting, 0.83 for self-evaluation, 0.74 for asking for help, 0.73 for managing the surrounding, 0.75 for learning responsibility, and 0.78 for planning.

Metacognitive Awareness Listening Questionnaire (MALQ). The Persian version of Metacognitive Awareness Listening Questionnaire (MALQ) originally developed and validated by Vandergrift et al. [18] was used to assess participants' metacognitive listening strategies awareness and perceived use while they listen to English texts [22]. MALQ was immediately administered after PET.

MALQ contains 21 items consisting of five factors including problem solving (6 items), planning-evaluation (5 items), mental translation (3 items), person knowledge (3 items), and directed attention (4 items). The respondents were asked to rate themselves regarding their perceived use of listening strategies on a six-point Likert scale ranging from 1 (completely disagree) to 6 (completely agree). A sample item of MALQ is "As I listen, I compare what I understand with what I know about the topic".

Factor structure of the Persian version of the scale has been investigated by PCA and it has been found that the five factors explain more than 61 % of the variance of MALQ [22]. The reliability coefficients of 0.74–0.85 were reported for the scale [22].

The overall reliability of MALQ was estimated to be 0.84 using Cronbach's alpha in the current study. The reliability coefficients of five factors of the scale were found to be 0.74 for problem solving, 0.75 for planning-evaluation, 0.78 for mental translation, 0.74 for person knowledge, and 0.68 for directed attention respectively.

7.2.4 The Procedure

Prior to the study, the bureaucratic procedure of getting the permission of the educational office, school principals, and teachers were followed. Having gained the permissions, the researchers attended the schools and started collecting the required data. Three instruments of the study were administered in a single session in each class. At the beginning of each session the listening part of PET was administered in order to assess the participants' level of listening comprehension.

Before test administration, a clear explanation on how participants had to go through the test was given and sample questions were shown. Immediately after the administration of PET, MALQ was distributed among the participants. Prior to the administration, a quick and clear explanation was given about the goal of the questionnaire and how to answer it. The participants were asked to fill in the questionnaire based on the listening test they had just had. In the last phase, ASRQ was administered. The whole procedure of data gathering from 369 students lasted around 2 months.

7.3 Results

7.3.1 Metacognitive Awareness of Listening Strategies and Listening Comprehension

In order to answer question number 1 and find the relationship between metacognitive awareness of listening strategies and listening proficiency, Pearson correlation method was used (Table 7.1). As Table 7.2 shows, there is no significant relationship between metacognitive awareness of listening strategies and listening ability. When the subscales of MALQ are considered, listening ability is just correlated with one strategy, that is *problem solving* ($r = 0.170, p < 0.01$).

Table 7.2 Correlation coefficients of PET and MALQ (and its subsection)

	MALQ	PE	DA	PK	MT	PS
PET	0.090	0.063	0.057	−0.091	0.069	0.170[**]

** $P < 0.01$

7.3.2 Metacognitive Awareness of Listening Strategies and Listening Comprehension Among Groups of Learners with Different Levels of SR

In order to answer question number 2 and find the relationship between metacognitive awareness of listening strategies and listening proficiency among low, mid and highly self-regulated language learners, three steps were taken.

Step 1. The participants were divided into three groups of low, mid, and highly self-regulated learners based on their scores on ASRQ. First the mean and standard deviation of participants on ASRQ were calculated (mean = 3.04, SD = 0.369).

Then, based on the mean and half of standard deviation three groups were formed with respect to their academic self-regulation: highly self-regulated learners based on mean +1/2 SD, low self-regulated learners based on mean −1/2 SD, and average self-regulated learners based on the interval between highly self-regulated learners and low self-regulated learners.

As Table 7.3 shows around one-third of the participants were highly self-regulated, while two-third of them were low and mid self-regulated learners.

Step 2. Pearson correlation method was used to find the relationship between MALQ and PET considering three groups of language learners based on their level of self-regulation (low, mid and highly self-regulated).

As Table 7.4 shows, there is no significant relationship between PET and MALQ (and its subsections) when the analysis is done considering highly self-regulated learners. In other words, there is no relationship between English listening proficiency and metacognitive awareness and perceived use of listening strategies among highly self-regulated learners.

Similarly, when low self-regulated learners were considered in the analysis, metacognitive awareness of listening strategies and listening ability were not found to be significantly related. However, among the subscales of MALQ just

Table 7.3 Levels of self-regulation

Level	Frequency	Percent
Low	99	26.8
Average	151	41
High	119	32.2
Total	369	100.0

Table 7.4 Correlation coefficients of PET and MALQ (and its subsections) among three groups of learners with different levels of SR

Groups		MALQ	PE	DA	PK	MT	PS
Highly SR	PET	0.004	0.005	0.044	−0.146	−0.082	0.106
Low SR	PET	0.034	−0.035	−0.066	−0.149	0.198*	0.122
Mid SR	PET	0.272**	0.259**	0.197*	0.011	0.104	0.322**

$** \ P < 0.01 \ * \ p < 0.05$

Table 7.5 The values of Z_{obs}

Z_{obs}						
Comparisons	PET/ MALQ	PET/ PE	PET/ DA	PET/ PK	PET/ MT	PET/ PS
Mid-High	2.20*	2.07*	1.26	1.24	1.47	1.84
Mid-Low	2.04*	2.39*	2.07*	1.28	0.74	1.58

one type of strategy, that is mental translation, was found to be significantly associated with listening ability ($r = 0.198$, $p < 0.05$).

By contrast, when mid self-regulated language learners were included in the analysis, a significant relationship was found between listening ability and metacognitive awareness of listening strategies ($r = 0.272$, $p < 0.01$). Also, metacognitive awareness of three kinds of listening strategies were found to be significantly related to listening proficiency, that is planning-evaluation ($r = 0.259$, $p < 0.01$), directed attention ($r = 0.197$, $p < 0.05$), and problem solving ($r = 0.322$, $p < 0.01$).

Step 3. Fisher's Z was used to statistically compare the strength of correlations between metacognitive awareness of listening strategies and listening proficiency among three groups of learners with different levels of self-regulation (Table 7.5).

As the result of fisher's Z shows, there is a significant difference between the correlation of listening proficiency with metacognitive awareness of listening strategies on the whole and planning-evaluation strategies when mid self-regulated group was compared with highly self-regulated group. Further, there is a significant difference between the correlation of listening proficiency with metacognitive awareness of listening strategies on the whole and both planning-evaluation and directed attention strategies when mid self-regulated group is compared with low self-regulated group.

7.3.3 MALQ as the Predictor of PET Among Three Groups of Learners with Different Levels of SR

In order to answer research question 3 and find the power of metacognitive awareness of listening strategies to predict listening proficiency among three groups of learners with different levels of self-regulation, multiple regressions method was used.

The result of regression analysis showed that this model was not significant among highly self-regulated [$F = (113, 5) = 1.655$, $p = 0.151$] and low self-regulated learners [$F (93, 5) = 1.947$, $p = 0.094$]. In other words, metacognitive awareness of listening strategies does not have any role in the listening ability of these two groups of language learners. However, the regression model was significant among mid self-regulated learners. As Table 7.6 shows, metacognitive awareness of listening strategies can predict more than 10 percent of the variance of listening proficiency among mid self-regulated learners (Table 7.6).

Table 7.6 Model summary

Model	R	R square	Adjusted R square	Std. error of the estimate
1	0.364	0.132	0.102	3.039

Table 7.7 ANOVA

Model	Sum of squares	df	Mean square	F	Sig.
1 Regression	202.935	5	40.587	4.393	0.001

Table 7.8 Standardized and unstandardized coefficients

Model	Unstandardized coefficients		Standardized coefficients	t	Sig.
	B	Std. error	Beta		
1 (Constant)	7.874	1.655		4.758	0.000
PE	0.090	0.078	0.118	1.156	0.249
DA	0.080	0.090	0.085	0.894	0.373
PK	−0.150	0.094	−0.138	−1.598	0.112
MT	−0.075	0.106	−0.063	−0.709	0.480
PS	0.194	0.069	0.289	2.821	0.005*

* $p<0.05$

Table 7.7 shows that the result of ANOVA has reached the statistical significance (F = 4.39; df = 5; p = 0.001 <0.05). To find out which independent variable (metacognitive listening strategies) significantly predicted the dependent variable (listening proficiency) the standardized and unstandardized coefficients were considered (Table 7.8). According to Table 7.8 there is only one variable, problem solving, that makes a statistically significant contribution to the construct listening proficiency (β = 0.289, t = 2.82, p = 0.05). In other words only problem solving strategies have a major role in listening proficiency of mid self-regulated students.

7.4 Discussion

The purpose of this study was investigating the relationship between metacognitive awareness of listening strategies and listening proficiency among language learners with different levels of self-regulation (low, mid, and high). It was primarily hypothesized that metacognitive awareness of listening strategies and listening proficiency has the strongest degree of association when highly self-regulated language learners are considered in the analysis.

The findings of the study revealed that generally there is no relationship between Iranian language learners' metacognitive awareness of listening strategies and their listening ability. The gained result is in contrast with studies done on the relationship between these two variables [18], as most of them have reported a significant relationship between all types of metacognitive listening strategies and listening proficiency. The reason can be related to the fact that Iranian high-school students generally have an average level of metacognitive awareness of listening strategies and thus they are not very skillful in using these strategies in managing listening tasks [22].

The lack of relationship between these two variables can also be attributed to the type of English listening instruction Iranian students receive. As listening has a limited role in Iranian EFL curriculum and is mostly taught based on comprehension-based teaching rather than interactive listening or strategy-based instruction, it cannot lead to successful listening [65]. It is known that these types of listening instruction cannot contribute to shaping language learners' listening self-efficacy [66] and the way they have to effectively perform listening tasks.

Further, the role of language teachers and their procedural and declarative knowledge in improving their students' skills can be a contributing factor in the limited use of listening strategies by students [67].

However, a significant, weak and positive relationship between metacognitive awareness of problem solving strategy and listening ability was found. Problem-solving represents a group of strategies listeners use to infer and monitor inferences [16]. This finding partially corroborates the findings of other studies that skilled listeners are more successful in applying listening strategies. The reason why just one strategy, problem solving, is related to listening ability can be explained by limited listening experience of Iranian EFL learners.

As listening is a challenging and difficult task for EFL learners, it needs careful instruction to help learners acquire required skills to manage their listening task. When listening is neglected in the curriculum or is taught through unsuitable methods (such as just bottom-up approaches) students are not aware of their own abilities as listeners and what strategies they can use to facilitate the difficult task of listening comprehension. It is worth mentioning that as problem-solving strategies are related to learners' autonomy [68], it is possible that the participants of this study managed to use them in their listening tasks without any formal instruction.

Surprisingly, the study revealed that metacognitive awareness of listening strategies and listening comprehension were not related when highly self-regulated learners were included in the analysis. This finding seems to be in contrast with previous research on the relationship between metacognitive strategies and successful listening comprehension. Empirical evidence shows that an important difference between more skilled and less skilled listeners lies in their use of metacognitive strategies [16, 37] and skilled listeners use twice as many metacognitive strategies as their less skilled counterparts [21].

Furthermore, evidence shows that the use of metacognitive strategies helps learners manage, regulate and control their learning processes [24]. Although the findings of the current study are in full contrast with findings of the previous studies, the research base is not large enough to rely on, as in none of the previous studies the effect of academic self-regulation was considered as an affective variable in the relationship between metacognitive strategy use while listening.

One possible reason of this finding, though, can be related to distinctive characteristics of self-regulated students who are proactive in managing their learning on the basis of the set goals [57]. The ultimate goal of many students in Iran is passing the national university entrance exam whose English section basically focuses on reading, grammar and vocabulary. Therefore, it is possible that highly self-regulated students are not motivated to improve their listening skill and thus do not care about how to do listening tasks.

Further, it was found that the relationship between metacognitive awareness of listening strategies and listening proficiency among low self-regulated learners was not significant. This finding is fully consistent with previous studies providing evidence for the influence of self-regulatory strategies on listening proficiency and metacognitive listening strategies use. Research has shown that self-regulatory strategies are critical to the development of metacognitive knowledge and strategy use [16]. Therefore, those listeners who are low self-regulated naturally are not fully aware of their metacognitive listening strategies and employ less metacognitive strategies; and thus will have unsuccessful listening performance.

The findings of the study also revealed a significant relationship between metacognitive awareness of listening strategies and listening ability in general when mid self-regulated learners were considered in the analysis. Also, listening proficiency was found to be related to planning-evaluation strategy, directed attention strategy and problem-solving strategy among this group of learners.

These findings are in full agreement with previous studies as research shows that the most effective metacognitive strategies include the combination of planning and monitoring and the combination of planning and evaluation [69].

Metacognitive strategies help learners regulate and oversee learning activities like taking conscious control of learning, planning, utilizing strategies, and monitoring the process of learning [39]. According to Wenden [56], metacognitive knowledge affects self-regulation of learning in planning, monitoring and evaluating skills. In addition, it is believed that metacognitive knowledge is a prerequisite to self-regulated learning [11] and provides the knowledge base for planning, monitoring and evaluation [70].

Further, as fisher's Z showed the correlation of metacognitive awareness of listening strategies and listening proficiency was significantly stronger when mid self-regulated group was compared with low and highly self-regulated groups. This finding is partly contradictory with the findings of previous studies as research on self-regulation shows that metacognitive strategies are among those features that differentiate students who self-regulate their learning from those who do not [11, 57].

Research has also indicated that learners with higher levels of overall self-regulation generally achieve higher levels of language achievement. In addition, empirical studies show that the effective use of metacognitive awareness of listening strategies plays a critical role in successful listening comprehension and both help listeners to enhance their self-regulation and autonomy in listening [20, 21].

In other words, those learners who employ more effective metacognitive strategies are more likely to attain higher levels of self-regulation. The contradictory results about the mid self-regulated students may have different reasons such as learners' self-efficacy beliefs, motivation, attitudes towards listening tasks, and listening anxiety. Therefore, further research is required to shed more light on this issue.

The study confirmed that metacognitive awareness of listening strategies can predict more than 10 % of the variance of listening proficiency among mid self-regulated language learners. This value is near to what has been reported in the literature, that is, 13 % [18]. The results of regression analysis suggest that an optimum level of self-regulation is required to make language learners able to use their metacognitive strategies of listening in the process of listening comprehension.

However, there is not enough evidence to fully support this idea. Further research is required to explore the influence of self-regulation strategies on learners' metacognitive awareness and use of learning strategies and its impact on language skills specially listening comprehension.

7.5 Conclusions

This study attempted to provide a glimpse of the relationship between EFL learners' metacognitive awareness of listening and their listening comprehension, as well as investigating the possible role of academic self-regulation in this regard. The sample was selected from EFL learners based on the assumption that their academic self-regulation assists them to use listening strategies more skillfully in the process of listening and thus better listening proficiency is guaranteed.

The findings of the study demonstrated a close relationship between listening proficiency and metacognitive awareness of listening strategies among mid self-regulated students, while no significant relationship between the two constructs was found in low and highly self-regulated groups and the whole sample.

Also the findings of the current study indicated that students' metacognitive knowledge functions as a good predictor of listening proficiency when students have average level of academic self-regulation.

This provides a surprising conclusion that metacognitive awareness, which is believed to be a key characteristic of successful listeners, is not a predictor of listening proficiency for low and highly self-regulated students, at least among the participants of this study. The findings of this study support other findings of strategy research that strategy use is highly situational and context-specific and it is related to language learners' individual differences [41].

7.6 Suggestions for Further Studies

The findings of this study revealed surprising results about the relationship between self-regulation and metacognition indicating that higher academic self-regulation does not necessarily guarantee metacognitive awareness and/or adoption of learning strategies in doing language tasks. What seems to be worth of further corroboration is the underlying reasons of this finding, that is what intrinsic and extrinsic variables may cause self-regulated learners limit their use of learning strategies.

Among the intrinsic factors one may focus on the emotional side of the learners such as motivation and desire to do certain types of tasks and the contribution of these tasks to their expected outcomes. Self-efficacy, anxiety, and autonomy are among other learners' characteristics that can have a key role in mediating the relationship between self-regulation, strategy use, and task performance.

Among the extrinsic factors, the role of different types of instruction, teaching materials, and the context of teaching can be investigated in this regard. For instance, experimental studies can be carried out within the framework of strategy-based instruction by implementing teaching materials that focus on oral language skills in private schools or language institutes to see if the same results are gained.

One may also focus on alternative research methods to scrutinize the role of self-regulation in using learning strategies while doing language tasks. Qualitative methods of data gathering can reveal more facts about cognitive control processes of self-regulation and metacognition while learners are supposed to do different types of tasks. Also, longitudinal studies rather than single-shot designs or case studies may yield more convincing reasons about the findings of this study.

References

1. Eysenck, M.: Principles of Cognitive Psychology, 2nd edn. Psychology Press, Hove (2001)
2. Frensch, P.A.: Cognitive psychology: Overview. In: Smelser, N.J., Baltes, P.B. (eds.) International Encyclopedia of the Social and Behavioral Sciences, pp. 2147–2154. Elsevier, Amsterdam (2001)
3. Kunde, W., Reuss, H., Kiesel, A.: Consciousness and cognitive control. Adv. Cogn. Psychol. 8(1), 9–18 (2012)
4. Dehaene, S., Naccache, L.: Towards a cognitive neuroscience of consciousness: evidence and a workspace framework. Cognition 79(1–2), 1–37 (2001)
5. Jack, A.I., Shallice, T.: Introspective physicalism as an approach to the science of consciousness. Cognition 79, 161–196 (2001)
6. Schunk, D.H.: Metacognition, self-regulation, and self-regulated learning: research recommendations. Educ. Psychol. Rev. 20(4), 463–467 (2008)
7. Flavell, J.H.: Cognitive Development, 2nd edn. Prentice Hall, Englewood Cliffs, NJ (1985)
8. Zimmerman, B.J., Schunk, D.H.: Motivation: an essential dimension of self-regulated learning. In: Schunk, D.H., Zimmerman, B.J. (eds.) Motivation and Self-Regulated Learning: Theory, Research and Applications, pp. 1–30. Lawrence Erlbaum, New York (2008)
9. Schunk, D.H.: Social cognitive theory and self-regulated learning. In: Schunk, H., Zimmerman, B.J. (eds.) Self-Regulated Learning and Academic Achievement: Theoretical Perspectives, 2nd edn, pp. 125–149. Erlbaum, Hillsdale, NJ (2001)
10. Ziegler, N., Moeller, A.: Increasing self-regulated learning through the linguafolio. Foreign Lang Annals. 45(3), 330–348 (2012)
11. Zimmerman, B.J., Moylan, A.R.: Self-regulation: where metacognition and motivation intersect. In: Hacker, D.J., Dunlosky, J., Graesser, A.C. (eds.) Handbook of Metacognition in Education, pp. 299–315. Routledge, NY (2009)

12. Kormos, J., Csizér, K.: The interaction of motivation, self-regulatory strategies, and autonomous learning behavior in different learner groups. TESOL Q. **48**(2), 275–299 (2013)
13. Wang, J., Spencer, K., Xing, M.: Metacognitive beliefs and strategies in learning Chinese as a foreign language. System **37**(1), 46–56 (2009)
14. Flavell, J.H.: Metacognitive aspects of problem solving. In: Resnick, L.B. (ed.) The Nature of Intelligence, vol. 12, pp. 231–236. Erlbaum, England (1976)
15. Vandergrift, L.: Listening to learn or learning to listen? Annu. Rev. Appl. Linguist. **24**, 3–25 (2004)
16. Vandergrift, L., Tafaghodtari, M.H.: Teaching L2 learners how to listen does make a difference: an empirical study. Lang. Learn. **60**(2), 470–497 (2010)
17. Zhang, W., Liu, M.: Investigating cognitive and metacognitive strategy use during an English proficiency test. Indonesian J. Engl. Lang. Teach. **4**(2), 122–139 (2008)
18. Vandergrift, L., Goh, C.M.C., Mareschal, C.J., Tafaghodtari, M.H.: The metacognitive awareness listening questionnaire (MALQ): development and validation. Lang. Learn. **56**(3), 431–462 (2006)
19. Sinclair, B.: Learner autonomy: the next phase? In: Sinclair, B., McGrath, I., Lamb, T. (eds.) Learner Autonomy, Teacher Autonomy: Future Directions pp. 4–14. Longman, UK (2000)
20. Vandergrift, L.: Relationships among motivation orientations, metacognitive awareness and proficiency in L2 listening. Appl. Linguist. **26**(1), 70–89 (2005)
21. Vandergrift, L.: Orchestrating strategy use: Toward a model of the skilled second language listener. Lang. Learn. **53**(3), 463–496 (2003)
22. Rahimi, M., Katal, M.: The role of metacognitive listening strategies awareness and podcast-use readiness in using podcasting for learning English as a foreign language. Comput. Hum. Behav. **28**(4), 1153–1161 (2012)
23. Richards, J.C.: Teaching Listening and Speaking: from Theory to Practice. Cambridge University Press, Cambridge (2008)
24. Vandergrift, L.: Facilitating second language listening comprehension: acquiring successful strategies. ELT J. **53**(3), 168–176 (1999)
25. Hinkel, E.: Current perspectives on teaching the four skills. TESOL Q. **40**(1), 109–131 (2006)
26. Macaro, E., Graham, S., Vanderplank, R.: A review of listening strategies: focus on sources of knowledge and on success. In: Cohen, A.D., Macaro, E. (eds.) Language Learners Strategies: 30 years of Research and Practice, pp. 165–185. Oxford University Press, Oxford (2007)
27. Richards, J.C., Schmidts, R.: Longman Dictionary of Language Teaching and Applied Linguistics, 3rd edn. Pearson Education Limited, UK (2002)
28. Moley, J.: Aural comprehension instruction: Principles and practices. In: Celce-Murcia, M. (ed.) Teaching English as a Second or Foreign Language, pp. 69–85. Heinle and Heinle, US (2001)
29. Field, J.: Psycholinguistics: A Resource Book for Students. Routledge Taylor & Francis Group, London (2006)
30. Nunan, D.: Introducing Discourse Analysis. Penguin, London (1993)
31. Rost, M.: Areas of research that influence L2 listening instruction. In: Usó-Juan, E., Martínez-Flor, A. (eds.) Current Trends in the Development and Teaching of the Four Language Skills, pp. 47–74. Mouton de Gruyter, NY (2006)
32. Peterson, P.W.: Skills and strategies for proficient listening. In: Celce-Murcia, M. (ed.) Teaching English as a Second or Foreign Language pp. 87–101. Heinle and Heinle, US (2001)
33. Wilson, J.: How to Teach Listening. Pearson Longman, UK (2008)
34. Longman Dictionary of Contemporary English. Longman Pearson, UK (2011)
35. Oxford, R.L.: Language learning strategies. In: Carter, R., Nunan, D. (eds.) The Cambridge Guide to Teaching English to Speakers of Other Languages, pp. 166–172. Cambridge University Press, Cambridge (2001)

36. O'Malley, J.M., Chamot, A.U.: Learning Strategies in Second Language Acquisition. Cambridge University Press, Cambridge (1990)
37. O'Malley, J.M., Chamot, A.U., Stewer-Manzanare, G., Russo, R.P., Kupper, L.: Learning strategy application with students of English as a second language. TESOL Q. **19**(3), 557–584 (1985)
38. Brown, H.D.: Teaching by Principles: An Interactive Approach to Language Learning. Longman Pearson, US (2007)
39. Oxford, R., Lee, K.R., Park, G.: L2 grammar strategies: the second cinderella and beyond. In: Cohen, A., Macaro, E. (eds.) Language Learners Strategies: 30 years of Research and Practice, pp. 117–139. Oxford University Press, Oxford (2007)
40. Oxford, R.: Language Learning Strategies: What Every Teacher Should Know. Newbury House, NY (1990)
41. Ellis, R.: The Study of Second Language Acquisition, 2nd edn. Oxford University Press, Oxford (2008)
42. Goh, C.: Metacognitive instruction for second language listening development theory, practice and research implications. RELC J. **39**(2), 188–213 (2008)
43. Grenfell, M., Macaro, E.: Language learner strategies: Claims and critiques. In: Cohen, A.D., Macaro, E. (eds.) Language Learners Strategies: 30 years of Research and Practice, pp. 9–28. Oxford University Press, Oxford (2007)
44. Mokhtari, K., Reichard, C.: Investigating the strategic reading processes of first and second language readers in two different cultural contexts. System **32**(3), 379–394 (2004)
45. Khaldieh, S.A.: Learning strategies and writing process of proficient vs. less-proficient learners of Arabic. Foreign Language Annuals. **33**(5), 522–533 (2000)
46. Schmitt, N.: Vocabulary in Language Teaching. Cambridge University Press, Cambridge (2000)
47. Vandergrift, L.: The comprehension strategies of second language (French) listeners: a descriptive study. Foreign Lang. Annals. **30**(3), 387–409 (1997)
48. Buck, G.: Assessing Listening. Cambridge University Press, Cambridge (2001)
49. Golchi, M.M.: Listening anxiety and its relationship with listening strategy use and listening comprehension among Iranian IELTS learners. Int. J. Engl. Linguist. **2**(4), 115–128 (2012)
50. Bidabadi, F.S., Yamat, H.: The relationship between listening strategies used by Iranian EFL freshmen university students and their listening proficiency level. Engl. Lang. Teach. **4**(1), 26–32 (2011)
51. Brown, A.L.: The development of memory: knowing, knowing about knowing and knowing how to know. In: Reese, W. (ed.) Advances in Child Development and Behavior, vol. 10, pp. 103–152. Academic Press, New York (1975)
52. Mazzoni, G., Kirsch, I.: Autobiographical memories and beliefs: a preliminary metacognitive model. In: Perfect, T.J., Schwartz, B.L. (eds.) Applied Metacognition, pp. 121–145. Cambridge University Press, New York (2002)
53. Pintrich, P.R.: The role of metacognitive knowledge in learning, teaching, and assessing. Theory Pract. **41**(4), 219–225 (2002)
54. Bartimote-Aufflick, K., Brew, A., Ainley, M.: University teachers engaged in critical self-regulation: How may they influence their students? In: Efklides, A., Misailidi, P. (eds.) Trends and Prospects in Metacognition Research, pp. 427–444. Springer, US (2010)
55. Sperling, R.A., Howard, B.C., Staley, R., DuBois, N.: Metacognition and self–regulated learning constructs. Educ. Res. Eval.: Int. J. Theory and Pract. **10**(2), 117–139 (2004)
56. Wenden, A.L.: An introduction to metacognitive knowledge and beliefs in language learning: beyond the basis. System **27**(4), 435–441 (1999)
57. Zimmerman, B.J.: Attainment of self–regulation: a social cognitive perspective. In: Boekaerts, M., Pintrich, P.R., Zeidner, M. (eds.) Handbook of Self-Regulation, pp. 13–19. Academic Press, San Diego, CA (2000)
58. Nakata, Y.: Toward a framework for self-regulated language-learning. TESL Can. J./Revnue TESL Du Can. **27**(2), 1–10 (2010)

59. Finkbeiner, C., Knierim, M., Smasal, M., Ludwig, P.: Self-regulated cooperative EFL reading tasks: students' strategy use and teachers' support. In: Special issue Awareness Matters: Language, culture, literacy. Lang. Awareness. **21**(1–2), 57–83 (2012)
60. Thompson, D.: promoting metacognitive skills in intermediate Spanish: report of a classroom research project. Foreign Lang. Annals. **45**, 447–462 (2012)
61. Efklides, A., Misailidi, P.: Introduction: The present and the future in metacognition. In: Efklides, A., Misailidi, P. (eds.) Trends and Prospects in Metacognition Research, pp. 1–18. Springer, US (2010)
62. Carver, C.S., Scheier, M.F.: On the Self-Regulation of Behavior. Cambridge University Press, Cambridge, UK (1998)
63. Preliminary English Test. Cambridge University Press, Cambridge, UK (2010)
64. Magno, C.: Assessing academic self-regulated learning among Filipino college students: the factor structure and item fit. Int. J. Educ. Psychol. Assess. **5**, 61–76 (2010)
65. Field, J.: Listening in the Language Classroom. Cambridge University Press, Cambridge (2008)
66. Graham, S.: Self-efficacy and academic listening. J. Engl. Acad. Purp. **10**(2), 113–117 (2011)
67. Sangster, P., Anderson, C., O'Hara, P.: Perceived and actual levels of knowledge about language amongst primary and secondary student teachers: do they know what they think they know? Lang. Awareness **22**(4), 293–319 (2013)
68. Chamot, A., Dale, M., O'Malley, J., Spanos, G.: Learning and problem solving strategies of ESL students. Bilingual Res. J. **16**(3–4), 1–28 (1992)
69. Lai, E.: Metacognition: A Literature Review. Technical report. Pearson Research Report Series (2011)
70. Perkins, D.N., Salomon, G.: Are cognitive skills context bound? Educ. Res. **18**(1), 16–25 (1989)

Chapter 8
Fostering Student Metacognition and Motivation in STEM through Citizen Science Programs

Suzanne E. Hiller and Anastasia Kitsantas

Abstract Current educational trends center on enhancing science, technology, engineering, and mathematics (STEM) programs within formal and informal learning settings to motivate students to enter science related careers. The purpose of the chapter is to examine the role of informal learning setting activities, namely citizen science programs, in promoting student metacognition and STEM career motivation from a social cognitive career perspective. Citizen science programs involve data collection on the part of hobbyists to contribute to the work of professional scientists on large scale natural studies. In order to assemble data that is scientifically rigorous, hobbyists receive training on data collection. According to research studies, the interaction with field experts in developing scientific observation skills has far reaching benefits in terms of student development of metacognitive skills, motivation, achievement, and STEM career pathways. In this chapter, we describe citizen science programs as a type of informal science learning approach that offers significant contributions in enhancing student metacognition and STEM career paths. We also provide information on essential characteristics of effective citizen science programs. Third, we focus on metacognition and its role on student achievement in science from a social cognitive perspective. Specifically, we address the development of scientific observation skills as a metacognitive process of student self-regulation, an integral component of citizen science programs. Fourth, we review research on the impact of citizen science programs on metacognition, motivational processes, and career planning. Fifth, we describe the design of effective citizen science programs within an educational context. Finally, we will discuss implications for educators, students,

S.E. Hiller (✉)
College of Education and Human Development, George Mason University, West Building, Suite 2100 4400 University Drive, Fairfax, VA 22030, USA
e-mail: shiller@gmu.edu

A. Kitsantas
College of Education and Human Development, George Mason University, West Building, Room 2001, MS6D2, 4400 University Drive, Fairfax, VA 22030, USA
e-mail: akitsant@gmu.edu

© Springer International Publishing Switzerland 2015
A. Peña-Ayala (ed.), *Metacognition: Fundaments, Applications, and Trends*,
Intelligent Systems Reference Library 76, DOI 10.1007/978-3-319-11062-2_8

parents, and policymakers as citizen science programs are pertinent pedagogical activities which reinforce student metacognition, science achievement, and STEM career motivation. Future directions and recommendations of research are also discussed.

Keywords Citizen   science   programs · Metacognition · Self-regulation · Self-efficacy · Scientific observation skills

Abbreviations

CBW   Classroom birdwatch
CFW   Classroom feederwatch
CSSS   Citizen science self-efficacy scale
LAL   Linulus amebocyte lysate
MLMP   Monarch larva monitoring project
SPT   Seed preference test
STEM   Science, technology, engineering mathematics

8.1 Introduction

Transcending national boundaries and cultural contexts, a commonality in educational goals is to provide a foundational support for children to flourish in terms of career pathways. Science education initiatives focus on the development of integrative science, technology, engineering, and mathematics (STEM) instruction to foster a work force with high levels of expertise in prospective science oriented occupations. Effective practices for strengthening student knowledge and standardized test performance center on inquiry based instructional delivery, professional development which targets content and pedagogical knowledge, and integrative STEM technologies [1].

Furthermore, recent research supports informal natural science learning contexts which engage students in authentic activities while promoting high levels of interest and firsthand experience with the habits and behaviors of scientists [2–4]. Citizen science programs are one type of informal activity in which individuals volunteer to collect data to support scientific research in a host of domains including bird surveying, invasive plant species, and vernal pools and can augment formal classroom studies while encouraging student performance and motivation. The goal of the chapter is to review the literature and examine the impact of citizen science programs particularly on student self-regulation, with an emphasis on metacognition and motivation.

A prominent aspect of establishing proficiency in science achievement rests in the development of metacognitive skills through self-regulation practices [5–9]. In essence, students actively engage in behaviors centered on achieving instructional goals through planning, reflection, readjustments, and repeated attempts until achieving mastery levels of performance [9, 10]. In fact, self-regulation is more than metacognition. It refers to the degree to which learners are metacognitively, motivationally and behaviorally proactive participants of their own learning process [11]. For adolescent students, training programs which support metacognitive development are essential for academic success and ultimately influence vocational interests [12]. A focus on metacognition is particularly potent in the secondary school years which are fundamental in the development of career motivation [13, 14].

Career pathways emerge over the course of many experiences and strongly relate to self-efficacy and interest [15, 16]. Effective practices which stem from student interests and authentic applications of science skills support long term retention of information and performance achievement [4]. Typically, researchers have examined these constructs and their interaction in formal school settings. As influential learning platforms, citizen science programs offer an alternative perspective on advancing middle school student achievement and career motivation for STEM career trajectories [2, 17].

In the second section we describe citizen science programs as a type of informal natural science learning approach that offers significant contributions in promoting student metacognition and STEM career paths [2] and provide information on essential characteristics of effective citizen science programs. In the third section, we focus on metacognition and its role on student achievement in science from a social cognitive perspective. Specifically, we address the development of scientific observation skills as a metacognitive process of student self-regulation [2, 18], an integral component of citizen science programs. In the fourth section, we examine literature related to the impact of citizen science programs on metacognition, motivational processes, and career planning. Although empirical research in this area is limited, we summarize some recent relevant studies, and we discuss how to develop effective citizen science programs in the fifth section. Finally, we discuss implications for practice as citizen science programs are relevant pedagogical activities which reinforce student metacognition, science achievement, and STEM career motivation.

8.2 A Description of Citizen Science Programs

Well-defined citizen science monitoring programs depend on scientific processes via collaborative efforts between hobbyists, researchers, and scientists [19]. The goals of the programs generally center on the need for assistance in data collection over widespread areas, large volumes of data points often related to the study of biotic or abiotic factors, and over an extended time duration. These parameters are often too difficult for a scientist to collect on an individual basis.

For instance, the Southern Africa Bird Atlas program originating in 1986 and spanning over 20 years focused on the record of bird species across Botswana, Lesotho, Namibia, South Africa, Swaziland, and Zimbabwe. In this context, a primary skill set was the ability to differentiate between bird species and to assist in assessing population trends. Subsequent databases focused on frogs, reptiles, spiders, and butterflies [20]. In other cases, the goals of the program may be to assess environmental conditions. Contrail cirrus clouds (initially formed behind airplane exhausts) have the potential to create a radiative imbalance which ultimately may play a role in climate changes [21]. From this perspective, citizen science data from ground observations identifying contrail types offers scientists insight into the extent of incoming short wave radiation and outgoing long wave radiation. By way of this information, the goal for scientists is to analyze climatology changes related to anthropogenic (pollution related to human activity) cirrus cloud formation [22].

Regardless of the topic of study, common characteristics of citizen science programs involve a training component, opportunities for preliminary data collection experiences, and independent field work. Typically, citizen science programs provide resources for data collection, data submission, manuals, online tutorials, and on site instruction for identifying the subject under study which could range from birds, wildflowers, leaf damage, or impervious surfaces.

An essential characteristic of these citizen science programs is that hobbyists receive specific training and follow a designated protocol in order to collect viable data for scientific study [22, 23]. The coordination of scientific processes to study widespread biological and environmental trends are context specific and may fluctuate in terms of the age of participants, the time of year for data collection, protocol criteria, and goals of the program. For example one of the oldest, established citizen science programs, the National Audubon's Christmas Bird Count occurs in December and January and centers on identifying bird species across large geographical areas [24].

In contrast, the Great Whale Count, a citizen science program monitoring the Hawaiian humpback whale, occurs from January 31 to March 11 in 20 min increments between 8:30 a.m. and 12:00 p.m. during the breeding and calving seasons. Teams of volunteers and a site leader record the proportion of humpback whales three miles off the island of Maui rather than determining population estimates [25].

Despite variations in goals, subject matter, and training protocol, field experts serve as facilitators in instructing citizen scientists on how to collect data through the use of scientific observation skills with pre survey training programs. For example the Delaware Bay Horseshoe Crab Spawning Survey holds training programs to prepare individuals for the spring survey during the horseshoe crab spawning season [26]. Through this program, citizen scientists learn how to collect data on horseshoe crab characteristics using a grid approach.

In order to follow a standardized process, the role of the scientist is to instruct volunteers on handling procedures, measurements, interpreting variations of color, shape, and size, pacing procedures, and recording measurements. More recently, citizen science programs are including cell phones which contain GPS receivers. This technology provides geographical information about organisms and the environment. In addition, there is the potential to incorporate environmental factors such as weather observations [27]. Depending on the program goals, scientists may instruct individuals on how to handle equipment in terms of data collection, how to make classifications across species, and how to submit information.

In another instance, one online citizen science program known as Old Weather, directs volunteers in reading handwritten log books from old ships to compile a database of weather systems from the past. In this type of forum, volunteers may not have access to a scientist but receive guidelines on how to interpret and record online data through instructions and/or tutorials [28]. Citizen science programs have limitless applications in terms of topics of study and educational implications particularly with the advent of technological tools such as monitoring devices.

Although a variety of programs both online and onsite provide opportunities for individuals to become involved in data collection, from a social cognitive perspective, it is the role of the field expert as mentor that has extensive implications for metacognitive development and STEM career motivation. However, a recent question that has emerged in regards to these programs is how to maintain accuracy of observations. One of the controversies surrounding citizen science programs is the reliability of data collection by hobbyists [23], particularly by younger children who may not have mastery skills in terms of scientific observation [3].

This concern may deter scientists from including students in data collection due to the perception that professional collaborations based on hobbyist contributions will not be considered credible by the scientific community. Yet in a study of naturalists' perceptions of effective environmental practices, many individuals viewed children's work products as an asset to professional scientific databases [29].

8.3 Student Metacognition and Motivation in Science Contexts

From a social cognitive perspective, a central feature of self-regulation is a cyclical feedback loop which supports the interaction between self-regulatory processes, motivational constructs, and learning accounts. Learners' cyclical feedback loops involve three phases: forethought, performance, and self-reflection [30].

In the forethought phase, an individual sets goals and plans steps in order to acquire a skill or master a concept. Student sources of self-motivation including self-efficacy beliefs, outcome expectations, goal-orientation and task interest, all which play an influential role in setting goals and engaging in *strategic planning*.

During the performance phase, the plan is enacted. At this point, the sub processes of task strategies and metacognition steer an individual's observations.

In the self-reflection stage, an individual makes judgments about their performance to prepare for another attempt [6] with the goal of becoming independent of the mentor in terms of metacognitive development, motivation, and behavior [31].

For adolescents, social support from mentors during this process is a central motivational force [2, 32]. As with many subject domains, a self-regulated feedback loop in which an individual is responsible for shaping their learning [33] is particularly crucial in developing scientific observation skills, the driving cognitive process in citizen science programs. Through field expert guidance, individuals gain new knowledge by coordinating observational information with content knowledge to extend learning.

This type of cyclical self-regulatory process to strengthen metacognitive skills is fundamental in advanced performance in biological studies [18, 34]. As part of the self-regulation process, metacognition centers on an individual's ability to monitor their cognitive functioning based on new or prior knowledge [35]. Due to inexperience, children are often unable to assess their metacognitive functioning clearly [36]. Metacognitive monitoring as part of the performance phase of self-regulation assists an individual in evaluating their progress in reaching an academic goal [37].

However, being self-motivated to engage in metacognitive processes is critical. Figure 8.1 shows the process of developing scientific observation skills in terms of both metacognitive and motivational processes of self-regulated learning. In terms of science development, guiding students through metacognitive processes has positive ramifications for science achievement. For example, DiBenedetto and Zimmerman [6] used a series of microanalytic measures in a science context related to tornados. Fifty-one ($N = 51$) eleventh grade students participated in a series of metacognitive prompts through each of the three phases of self-regulation (forethought, performance, and self-reflection), while studying tornadoes.

The purpose of a microanalytic approach is to capture self-regulation processes in the moment rather than to use survey results as representations of individual characteristics. For example, during each phase of self-regulation, researchers asked participants microanalytic prompts which typified aspects of forethought,

Fig. 8.1 A cyclical self-regulatory feedback loop and citizen science

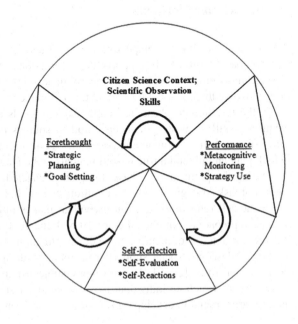

performance, and self-reflection. Participant answers were recorded for each phase.

In the forethought phase, researchers asked questions related to plans students had made to read tornado text and to take a test. Metacognitive prompts assessed the performance phase and included questions related to how confident participants were about their answers for each question and the approach that was used to create conceptual models. To measure student capabilities in the self-reflection stage, individuals summarized how well they had learned the concept [6].

All components of the three phases of self-regulation through the microanalytic approach had positive correlations with academic achievement. Findings indicated that the microanalytic measures served as predictors of student performance to a greater extent than teacher ratings and that those students with high self-regulatory processes outperformed their peers. These findings were particularly relevant for students who engaged more fully in metacognitive monitoring during the performance phase of self-regulation [6]. Similarly, Peters and Kitsantas [8] examined the impact of explicit metacognitive processes on student content knowledge, nature of science understanding, and self-efficacy for metacognitive and self-regulatory strategies.

Findings indicated that individuals exposed to metacognitive prompts through the four phases of progressing from novice to mastery levels (observation, emulation, self-control, and self-regulation) demonstrated greater proficiency in terms of content knowledge and nature of science understanding although there were not significant differences for self-efficacy for metacognitive and self-regulatory processes.

Moreover, students engaged in self-regulation strategies within science inquiry based contexts have less anxiety and greater career motivation than their counterparts [2, 38]. Within an outdoor context, Houseal et al. [3] analyzed the impact of metacognitive processes during a teacher-student-scientist partnership entitled Expedition: Yellowstone! on academic achievement.

The purpose of this study was to examine the influence of an authentic inquiry based program on fourth through eighth grade students and teachers. The program targeted teacher professional development, student data collection, and engagement in field science by way of incorporating metacognitive prompts.

The emphasis of the program was to engage students in aspects of scientific research and data interpretation apart from data collection. Findings indicated that the use of metacognitive strategies while conducting geobiological field work positively influenced student and teacher content knowledge and science attitudes [3].

For citizen science programs, a fundamental metacognitive process involves the coordination of sensory observations with content knowledge through the use of scientific observation skills. Notably different from general observation, scientific observation skills stem from activities including counting, collecting, measuring, and classifying and rely heavily on the synthesis of physically collected information and prior knowledge [18, 39]. This type of unification requires systematic training via modeling and guidance from trained field experts [2, 18].

The development of the metacognitive process of scientific observation skills has substantial influence on student achievement, self-efficacy, and career motivation. Problems, however, exist with current assessments of these constructs through the use of self-report surveys. These types of measures may create limitations due to the accuracy of an individual's perceptions of their capabilities based on past experiences.

It may be difficult for individuals to recall or quantify aspects of their learning based on the amount, quality, or applications of a perception or behavior [40]. Most recently, researchers have shifted their attention to using a microanalytic methodology to assess metacognitive and motivational processes in science education.

Microanalysis involves assessing a learner's response to questions while he or she is engaged in a learning task before, during and after learning [6]. For example, in a horseshoe crab context, students met with field experts to receive protocol training for data collection on measuring the interocular distance, determining gender, and identifying relative age. Students at key points of their learning are asked questions about their understanding and motivation to pursue this task.

Once participants had practiced collecting data, the researcher and field expert monitored how accurate students were in collecting data. Based on these event measures which are brief and task-specific, participants have the opportunity to accurately monitor, reflect, and identify horseshoe crabs from other organisms and measure in centimeters consistently.

In addition, another benefit of incorporating these event measures with citizen science studies is to triangulate findings related to self-efficacy of scientific observations along with quantitative measures and qualitative data sources [41]. Furthermore, combining microanalytic assessment with essential components of scientific observation skills such as Noticing, Expectations, creating Observations Records, and Productive Dispositions, which have been described as foundational aspects of scientific observational skill development by Eberbach and Crawley [18] may help students engage in a self-regulatory cyclical loop.

The first component, Noticing, which requires an integration of perceptual and cognitive processes, refers to the transition of observation skills as a scientific process. For example, if students have access to various skull types, they may initially notice skull differences and similarities. Next, students may begin to distinguish that the shape of teeth in the skulls may vary by organism. In the final stage, students learn to categorize animals as carnivores, herbivores, or omnivores by the shape of the teeth. In addition, students may begin to classify organisms based on defining characteristics such as the number of incisors or the presence of only canines while excluding extraneous information for classification purposes [18].

Below, taking into consideration these foundational skills, we present Tables 8.1, 8.2, 8.3 and 8.4 with suggestions on how to use a microanalytic approach to develop scientific observation skills using the three phases of self-regulation including, forethought (strategic planning), performance (metacognitive monitoring), and self-reflection (self-evaluation).

Table 8.1 Microanalytic assessment for supporting use of metacognitive and motivational processes of self-regulated learning in a citizen science activity across levels of learning for noticing

Self-regulatory phases	Example assessments
Novice	
Noticing: notice a horseshoe crab is different from other organisms	
Forethought Strategic planning	How will you tell if a horseshoe crab is different from other animals?
Performance Metacognitive monitoring	How confident are you that this is a horseshoe crab?
Self-reflection Self-evaluation	How well have you learned to tell a horseshoe crab apart from other animals?
Transition	
Noticing: notice more relevant features and identify patterns of features of horseshoe crabs	
Forethought Strategic planning	Which characteristics will you use to tell a horseshoe crab apart from other animals?
Performance Metacognitive monitoring	How confident are you that the features you have selected tell that a specimen is a horseshoe crab rather than another animal?
Self-reflection Self-evaluation	How well have the features you selected help to tell a horseshoe crab apart from another animal?
Expert	
Noticing: notice and describe relevant features of horseshoe crabs and ignore irrelevant features using disciplinary structure (e.g., taxonomy)	
Forethought Strategic planning	Which characteristics will you include to classify the animal as a horseshoe crab? Which characteristics will you exclude? (For example, why is the horseshoe crab more closely related to arachnids rather than crustaceans?)
Performance Metacognitive monitoring	How confident are you that the features you are including and not including will help determine the classification of an animal as a horseshoe crab?
Self-reflection Self-evaluation	How well have you learned to classify a horseshoe crab apart from other animals?

The goal is to assist students as they progress in their skill sets to approach expert levels. Specifically, we outline one citizen science example where each microanalytic assessment is representative of the three phases of the self-regulation model to assist students as they engage in self-regulated learning. We start with Table 8.1 which highlights the use of microanalytic prompts to guide students through Noticing with horseshoe crabs as an example.

A second underlying component of the development of scientific observation skills is Expectations, in which individuals coordinate observations with scientific ideas [18]. For example, students may have vague expectations about observations related to bird data collection. Individuals may consider that all birds have feathers on their heads. As individuals transcend through the development of Expectations,

Table 8.2 Microanalytic assessments for supporting use of metacognitive and motivational processes of self-regulated learning in a citizen science activity across levels of learning for expectations

Self-regulatory phases	Example assessments
Novice	
Expectations: vague expectations about observations related to horseshoe crabs	
Forethought Strategic planning	How will the size of the horseshoe crab determine the gender?
Performance Metacognitive monitoring	How confident are you that the size will accurately identify the horseshoe crab's gender?
Self-reflection Self-evaluation	How well have you learned to identify a horseshoe crab's gender by size?
Transition	
Expectations: more explicit expectations about horseshoe crabs that reflect observations	
Forethought Strategic planning	What size variations are possible between male and female horseshoe crabs?
Performance Metacognitive monitoring	How confident are you that the size range you have selected accurately identifies the gender of the horseshoe crab?
Self-reflection Self-evaluation	How well have the size ranges you selected help to tell a male and female horseshoe crab apart?
Expert	
Expectations: explicit hypothesis consistent with a theoretical framework shapes observations	
Forethought Strategic planning	What hypothesis have you formed about identifying a horseshoe crab's gender by size?
Performance Metacognitive monitoring	How confident are you that the hypothesis you have formed relates to previous studies on horseshoe crabs?
Self-reflection Self-evaluation	How well have you learned to form a hypothesis on determining horseshoe crab gender?

they may encounter the absence of feathers on the head of vultures and begin to explain this absence. In the final stage, individuals develop a hypothesis about the absence of the feathers based on the coordination of previous literature and observations. As described in Table 8.2 with horseshoe crabs as an example, instructors may facilitate the process of Expectations through microanalytic prompts.

Furthermore, creating Observation Records is a skill that does not occur inherently to most individuals. In order for individuals to acquire the ability to generate viable data sources, mentoring and guidance is required [18]. In a citizen science context, individuals typically follow a protocol to answer a specific question. Contributors submit data to a well-defined database. In this instance, students require an understanding of how data collection systems design is central as a tool to answering questions. For example, young students may be asked to make observations informally about moths.

Table 8.3 Microanalytic assessments for supporting use of metacognitive and motivational processes of self-regulated learning in a citizen science activity across levels of learning for observation records

Self-regulatory phases	Example assessments
Novice	
Observation records: observe without collecting or recording observations	
Forethought Strategic planning	What observations have you collected?
Performance Metacognitive monitoring	How confident are you that the way you collected data will help you to understand horseshoe crabs?
Self-reflection Self-evaluation	How well have you learned to collect data on horseshoe crabs?
Transition	
Observation records: record observations	
Forethought Strategic planning	How did you collect and record observations?
Performance Metacognitive monitoring	How confident are you that the system you created for recording observations will help to answer questions?
Self-reflection Self-evaluation	How well did you record observations?
Expert	
Observation records: record observations using established disciplinary procedures	
Forethought Strategic planning	How did you collect and record observations?
Performance Metacognitive monitoring	How confident are you that you collected and recorded observations based on a standard procedure?
Self-reflection Self-evaluation	How well have you learned to collect and re-cord observations using a standard procedure?

Next, students could begin to create an organizational system about their observations by making a list of questions and then designing a data table. Finally, students would be able to participate in National Moth Week [42] by submitting photographs and data. Table 8.3 serves as a way of transitioning students in developing and understanding the purpose of data collection and record keeping which will assist them in engaging in self-reflection with horseshoe crabs as an example.

Productive Dispositions relates to the level of interest and commitment of an individual over a prolonged period of time. Identity formation and interest are driving forces in the development of scientific observation skills and career paths [18, 41]. Thus, the development of Productive Dispositions requires repeated exposure to field work with scientists. Students benefit from encouragement to study objects and/or organisms of interest, to use creative skills, to

Table 8.4 Microanalytic assessments for supporting use of metacognitive and motivational processes of self-regulated learning in a citizen science activity across levels of learning for productive dispositions

Self-regulatory phases	Example assessments
Novice	
Productive dispositions: opportunistic and incidental observations	
Forethought	What additional observations have you made about horseshoe crabs?
Strategic planning	
Performance	How confident are you that these observations will help to learn about horseshoe crabs?
Metacognitive monitoring	
Self-reflection	How well have you learned to collect additional observations about horseshoe crabs?
Self-evaluation	
Transition	
Productive dispositions: intentionally seek information and observations about horseshoe crabs	
Forethought	What other questions do you have about horseshoe crabs?
Strategic planning	
Performance	How confident are you that you will be able to find information on these questions?
Metacognitive monitoring	
Self-reflection	How well have you found information on these questions?
Self-evaluation	
Expert	
Productive dispositions: persistent, sustained engagement; high level of interest	
Forethought	How interested are you in studying horseshoe crabs in the future?
Strategic planning	
Performance	How confident are you that you will continue studying horseshoe crabs?
Metacognitive monitoring	
Self-reflection	What steps have you taken to continue studying horseshoe crabs?
Self-evaluation	

collect additional information, to ask new questions, and to form new hypotheses. Table 8.4 offers insight into the use of the microanalytic approach to encourage the development of Productive Dispositions.

8.4 The Impact of Citizen Science Programs on Metacognition, Motivational Processes, and Career Planning

The latest research in citizen science literature illuminates the positive benefits of citizen science participation. As researchers continues to refine current understandings of well-formulated citizen science structures, studies support the influence of data collection for scientific research on cognitive development related to scientific inquiry [39, 40]. More recent attention focuses on the influence of these types of activities on individual development in terms of metacognitive science process skills, knowledge gains, self-efficacy, and motivation [2, 43–48].

Focusing on the impact of citizen science programs on student metacognition is critical as authentic real world experiences influence student achievement in STEM disciplines [1]. In the effort to promote STEM careers, informal learning settings afford exceptional experiences to strengthen student academic achievement and career motivation. Specifically, the development of science skills and content knowledge through metacognitive processes and expert modeling has a significant impact on student achievement and STEM career motivation [2]. For example, determining the gender or relative age of horseshoe crabs requires more than simple observation skills.

A student citizen scientist identifying these characteristics must examine a multitude of features in terms of color distinctions, arc structure, size comparisons, and anatomical variations to categorize organisms by gender [41]. These distinctions require training through modeling and preliminary experiences with a field expert. Based on metacognitive processes involving science process skills and content mastery, sensory observations coupled with specialized content experience require extensive development in order to approach a level of expertise [18]. Citizen science activities embedded within a social cognitive perspective guide students to become metacognitively aware. The use of scientific observation skills is a metacognitive process essential to field work data collection. Citizen science programs rely on the application of scientific observations skills.

The distinction between scientific observation skills and general observation skills is that an individual uses metacognitive skills to integrate sensory observations with content knowledge [18, 34]. Through social cognitive principles such as modeling, self-monitoring, repeated practice, self-reflection, and with mentor guidance, these processes shape a refined mastery of understanding. This type of scaffolding positively influences participants' metacognition, self-efficacy for science process skills, science literacy, mastery experiences, content knowledge, and motivation [2, 46, 47].

The general public has informally engaged in scientific inquiry for centuries. Largely based out of interest, individuals involved in scientific research as early as the 1,600s recruited assistants with less experience to collect observations on natural specimens. Since the onset of professional scientists, joint efforts between accomplished researchers and voluntary partnerships in citizen science programs emerged over 100 years ago [49]. Only within the last two decades, researchers began examining the impact of citizen science programs on volunteer contributors' scientific literacy, knowledge acquisition, and environmental attitudes [49]. Much of this type of original citizen science literature focused on pedagogical advancements for volunteers who were primarily adults. These first studies were instrumental in terms of learning how to measure effects, identify the role of scientists as mentors, and to frame effective programs including curriculum materials and training protocol.

Trumbull et al. [50] conducted one of the preliminary studies which attempted to examine the relationship between participants' understanding of scientific processes in conjunction with a Seed Preference Test (SPT). The program, in affiliation with the Cornell Lab of Ornithology and the National Science Foundation, requested thousands of volunteers to assist with determining seed preferences for ground feeders. Following the program, researchers analyzed 750 letters which were sent to Cornell unexpectedly by program enthusiasts. The purpose of this analysis was to assess the impact of the program on factors related to scientific inquiry including problem formulation, the development of hypotheses, designing studies, planning or conducting an investigation, data interpretation, and knowledge synthesis. As a data source, researchers categorized the science processes described in these letters by applying a coding scheme. Of these factors, writings most often referenced information related to general observations and hypotheses formation. Although early in the development of citizen science literature, these findings tentatively highlighted that individuals engaged in citizen science programs participated in a range of cognitive processes based in scientific inquiry [50].

In a subsequent study affiliated with the Cornell Lab of Ornithology, Brossard et al. [45] assessed the impact of the ornithological citizen science program known as Birdhouse Network on content and scientific process gains as well as changes in attitude related to the environment. Using a repeated measure design with adults ($N = 798$), participants completed a research developed content knowledge scale and an adjusted Attitude Toward Organized Science Scale based on the Science and Engineering Indicators [51]. Further, participants completed the New Environmental Paradigm Scale developed by Dunlap and Van Liere [52] which measured participants' attitudes towards protecting the environment. Findings indicated that the citizen science programs based in ornithological studies had a positive effect on content knowledge but minimal impact on environmental attitudes or science process understanding [45].

In contrast to minimal science literacy findings, Cronje et al. [46] subsequent study revealed positive impacts on science content and process development. The purpose of this pilot study was to examine the impact of an invasive species citizen science program on adult science content knowledge and as a newly developed science literacy measure. During a two day workshop, treatment participants ($N = 57$) received training on invasive plant identification, methods for collecting samples, and instruction on the use of technology for data collection.

Treatment adult participants completed newly developed measure framed around four open ended items on science process knowledge. The open ended items were then coded to represent elements of science literacy. In addition, a comparison group ($N = 90$) completed the measures online without intervention exposure. Using an independent sample t-test to compare the groups, results indicated that although there were no significant main effects between the treatment and comparison groups prior to the training program, the treatment group engaged in the invasive plant citizen science program outperformed the comparison group on measures centered on science literacy [46].

Building on prior research, a later study conducted by Crall et al. [53] examined citizen scientist development in terms of content knowledge, science processes, and environmental attitudes following another plant monitoring program in which volunteers studied the abundance of invasive plant species in a specific area. Measures included the Attitude Toward Organized Science Scale [45], the Science and Engineering Indicators related to the habits and cognitive processes of scientists [54], the New Environmental Paradigm [52, 55], and response items centered on content knowledge and science processes measure [53]. Pre and post measures were completed by an adult treatment group ($N = 166$) and comparison group ($N = 48$). As in past works, the citizen science program positively influenced gains on content specific measures whereas overall science literacy and environmental attitudes were not significantly impacted. Researchers hypothesized that these consistent findings reflected time restraints in that participants required more exposure with field work to develop cognitive domains [53].

Some researchers focused efforts on the influence of citizen science programs with adolescent students. Specifically, researchers have examined citizen science programs, with scientific observation skills as a cornerstone, in terms of the positive influence on metacognition, academic achievement, and career motivation [2, 47]. Moreover, guidance and mentoring from field experts has become a focal point when examining the effects of these types of programs on student development.

In terms of motivation, Tomasek [47] researched the impact of a subsequent version of the ornithological citizen science program called ebird and the associated curriculum known as Classroom BirdWatch (CBW), both developed through the Cornell Lab of Ornithology and the National Audubon Society, on middle school academic performance, cognitive processes, and motivation. The study measured the impact of the citizen science program in terms of a sense of autonomy (level of responsibility), competence (mastery of content), relatedness (connection to scientists), and intrinsic motivation; factors associated with self-regulatory skills and motivation. In addition, the research design addressed the impact of the curriculum on the development of cognitive skills through authentic inquiry [47].

The intervention included a program in which students and teachers collaborated with scientists with specific training on setting up feeders, identifying species, and submitting data online. Students extended learning by asking questions, analyzing data, and writing findings to a newsletter [47]. Using a mixed method design with one hundred twenty-two ($N = 122$) students in grades fifth, seventh, and eighth participated in the study. Data sources ranged from surveys, focus groups, and student developed research reports. Based on the citizen science intervention, there were correlations among autonomy, competence, relatedness, and interest levels. Further, the implications of this study were that the cognitive processes centered on observation skills were reflective of authentic scientific inquiry [47].

These findings, mirrored in a subsequent study on middle school student involvement in plant studies, showed that the program increased interest for science and application of scientific processes. With the premise that interest forms the foundation for motivation within the context of the Sequoia and Kings Canyon National Park, Sutton [48] highlighted the influence of the citizen science program with fifth grade students ($N = 49$). The purpose of the study was to examine the impact of a citizen science program related to the study of climate change and plant studies. Students received classroom training on the identification of plant, insect interaction, and data collection and observation tools.

The participants then worked with a field expert collecting observations for two months. Findings indicated that the citizen science program via field expert modeling resulted in significant positive main effects for student literacy, interest, and affect particularly in terms of scientific process understanding [48].

The positive benefits of citizen science extend to a variety of topic domains apart from ornithological studies. For example, Jeanpierre et al. [44] focused on a Minnesota/Texas based study which examined the collaborative partnership between high school students, teachers, and researchers through the Monarch Larva Monitoring Project (MLMP) through the use of inquiry in teacher developed lessons. The program, sponsored by the Museum of Minnesota and a National Science Foundation grant, included teams of teachers and students who attended a one week training program in the spring in Minnesota and one week in Texas in the fall. After receiving training and guidance on the migratory patterns of butterflies from field experts, the teams designed research studies based on the workshops. Working within their local regions, students and teachers monitored monarch butterflies initially with the support of scientists and applied techniques at home based surveys [44].

Using a mixed methodology approach, researchers analyzed the use of inquiry in subsequent teacher developed lessons following the monarch butterfly training. Based on the activity, findings indicated that both student-teacher generated reports were reflective of effective practices related to inquiry learning. Further the teachers demonstrated heightened content knowledge and application of scientific inquiry in lesson plans [44].

In an effort to extend current understandings of the benefits of a citizen science program on student career motivation with social cognitive career theory as the basis of interpretation, Hiller and Kitsantas [2] examined the impact of a horseshoe crab citizen science program on student achievement and STEM career motivation for eighth grade students. Within the quasi-experimental framework, the treatment group ($N = 45$) conducted field work with naturalists to contribute data to an investigative research study on potential speciation effects as well as to experience the protocol training of a survey usually conducted by adults.

The comparison group ($N = 41$) studied horseshoe crabs as part of school instruction using a PowerPoint developed by a field expert. Pre and post measures were given to participants and centered on social cognitive career theory constructs including academic achievement, self-efficacy, interest, outcome expectations, choice goals, and content knowledge were given to participants before and after the intervention. Furthermore two citizen science scales were incorporated in the study.

The first measure, a Citizen Science Self-Efficacy Scale (CSSS), [41] developed based on the perceptions of naturalists on effective environmental education practices, [29] assessed self-efficacy for scientific observation skills. The Citizen Science Outcome Expectations Scale [30] addressed students' judgments about the impact of the citizen science program on reaching their career aspirations. In accordance with the hypothesis that the treatment group would outperform the comparison group on all variables, findings indicated significant gains on measures related to metacognition, self-regulation, motivation, and academic achievement. Further, path analyses revealed that interest and self-efficacy for scientific observation skills in conjunction positively impacted content knowledge and outcome expectations. Ultimately, through path analysis, these relationships had a positive influence on choice goals related to science career paths [2]. In accordance with these findings, the interaction of self-efficacy, interest, and outcome expectations during collaborative work with scientists positively affected student science achievement and career motivation.

A qualitative component of a mixed method study conducted by Hiller [41] further supported the influence of the metacognitive development of scientific observation skills, self-efficacy, academic achievement, and career motivation. Twenty ($N = 20$) eighth grade students participated in interviews following a citizen science horseshoe crab program. Through the use of a version of constant comparative analysis approach, student perceptions of a horseshoe crab citizen science program revealed that this type of activity with expert modeling had positive benefits in terms of bolstering student achievement and interest for science processes and content knowledge. The phased coding scheme of open, axial, and emergent theme development revealed three emergent themes: (a) student perceptions of

Table 8.5 Outcomes of qualitative data sources [41]

Theme	Quotation
Science skills, learning, and fun	...It was pretty fun. You get to spend a day like out on the beach, looking at horseshoe crabs, collecting data...and learning something new that you hadn't learned in like awhile
Training and collaboration	Well I felt like I was helping mankind. I was helping people because my work would help the scientist and her studies and her data
Future careers	If being a lawyer doesn't work out I know I can always do something else...Because I always thought I didn't know that much about science. I wasn't any good at science...Well not anymore 'cause I learned a lot of things I thought I didn't know

Note From The Impact of a Citizen Science Program on Student Achievement and Motivation: A Social Cognitive Career Perspective, by S.E. Hiller, 2012, Proquest Information and Learning Company, p. 104. Reprinted with permission.

science skills, learning, and fun, (b) the influence of student and field expert training and collaboration, (c) future careers as seen in Table 8.5 with participants' quotes as data sources reflective of each theme [41]. The overall findings from these data sources were that students perceived their science skills had improved based on collaborative efforts with scientists.

A distinct pattern which occurred across emergent themes was the level of importance participants placed on the role of the field expert apart from their experiences with horseshoe crabs. When describing their capabilities in terms of scientific observation skills, adolescents incorporated information about horseshoe crabs to substantiate their perceptions of their skill level mastery. However, students emphatically described the way scientists engaged participants through modeling and scaffolding.

Participants centered on the level of responsibility required to support professional research. Student self-efficacy and identity formation as a scientist developed based on joint efforts. In general children noted that the experience was influential in exploring career options related to STEM. As one participant stated, "I was really kind of shocked that I got taken with a professional scientist…like he would actually pick me to do it…I felt good about today. I learned a lot I never did before" pp. 110–111 [41].

The latest literature on the benefits of citizen science, albeit limited due to the recent onset of the inclusion of this type of activity in educational settings, extends to a variety of contexts and highlights positive benefits in the motivation and metacognitive development of adolescents and children. As a support of formal education curriculum, citizen science programs provide students with tangible experiences to develop science processes skills and content knowledge through collaborative, mentoring partnerships with scientists. In particular, field experts as models and facilitators, are foundational strongholds in terms of coordinating scientific observation skills with content knowledge and protocol requirements. The outcome of student exposure to collaborative professional efforts fosters metacognitive processes, achievement, and career motivation [2]. Based on the research findings, we present some guidelines on how to structure citizen science programs for school aged students.

8.5 Creating Effective Citizen Science Programs

The development of effective citizen science programs, appropriate for school based applications, requires collaborative efforts between scientists, teachers, administrators, students, and parents. Below we describe components of well-designed citizen science programs within a school context. First we outline a sequence of events to prepare students for protocol training in such a program. Secondly, we offer a narrative description of an example of a school based citizen science program. Third, we provide suggestions for developing a structure for incorporating citizen science within school programs based on recent research.

8.5.1 Citizen Science Programs and Training Protocol

Citizen science programs have the potential to foster self-efficacy, academic achievement, and career motivation for adolescents [2, 47, 48]. The development of a citizen science program for children, particularly as an extension of formal classroom curriculum, may greatly enhance overall student progress with specific components as shown in Fig. 8.2.

In order to provide students with authentic applications of science process and content knowledge skills, encouraging students to engage in collaborative efforts which include interactions with science professions as a mentor may be a highly beneficial experience. Developing a team effort between the teachers, students, and scientist is preferable, particularly when students have exposure to modeling and guidance in terms of safety procedures, scientific observation, data collection, and recording. Further, the citizen science program centers around a specific scientific research purpose with well-defined protocol for data collection and systematic process for submitting data observations which are clear and accessible to individuals [56]. Throughout the school year, teachers may prepare students to develop scientific observation skills and content knowledge as part of the science curriculum.

Training in scientific observation skills may be done with in class instruction, and before and after school training programs. In addition, exposure to a variety of objects to develop scientific observation skills prepares students for the experience. Natural specimens such as seashells, sea stars, fossils, skulls, leaves, and pine cones are viable for developing science process skills as outlined by Eberbach and Crawley [18].

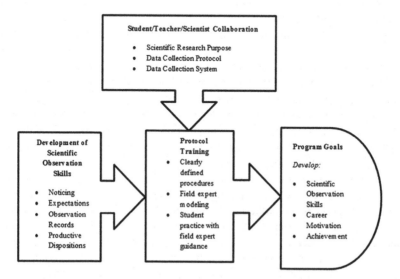

Fig. 8.2 A conceptual model of a well-designed citizen science program

In conjunction with the activities of professional scientists, standardized procedures and criteria for data collection which are piloted with hobbyists in outdoor contexts is an advisable way to ensure that students have maximized experience in collecting and contributing valid data [57]. Once the citizen science program is in place, participants should be well-trained in data collection protocol and submission as volunteer error may have repercussions for valid scientific interpretations [58].

Options for initial instruction include classroom discussion, PowerPoints, films, and in class demonstrations. Some programs may have online tutorials, classification guides, instruction manuals, or deliver materials to the students' location. Following first exposure, individuals should observe field experts model appropriate procedures and then an opportunity to practice in front of field experts and teachers prior to data collection to ensure the use of proper techniques. The end goal of the approach is to heighten science process skills such as scientific observation skills, develop interest intrinsic, provide career options, and increase academic learning.

8.5.2 An Example of Citizen Science Training

There are countless opportunities for students to participate in citizen science programs. Training and protocol criteria vary depending on the needs of the study, the topic of study, data collection procedures, and data contributions [56]. The following description contains excerpts from training procedures reflective of Fig. 8.2 within a horseshoe crab citizen science context from both a pilot study and a quasi-experimental study.

In the morning, students assembled at a park preserve to receive training from a field expert. Students had exposure to content information about horseshoe crabs during the school year as part of their science curriculum while others had also attended two, 12 week, one hour programs to learn more about horseshoe crabs as part of a career training program at the school. The students would have the opportunity to collect data for a professional biological research study as well as to mimic a survey which required participants to be 18 years old [41, 59].

During the morning session, students received instruction from a naturalist led discussion through a PowerPoint, and an activity to test for Linulus amebocyte lysate (LAL), a protein derivative extracted from horseshoe crab blood which is used by the biomedical industry to test for bacteria. To meet the needs of the research study, the discussion centered on differentiating between males and females, the relative age, reproductive processes of horseshoe crabs, and the form and function of horseshoe crab anatomy, by directly observing the horseshoe crabs. This background information was necessary for participants in order to learn how to collect data. Participants learned that they would be collecting data for a scientist to examine size changes of female horseshoe crabs along the east coast of the United States. Other information would be used for research based in exploration [41, 59].

The second item of the training session involved some guidelines for taking measurements and recording observations. Students were given instruction on how to measure the distance between the eyes (known as the interocular) with a tape measure in metric units. The naturalists modeled how to take measurements with a male and female pair. Subjects were to identify the male and female, the relative ages of both, and the interocular measurement. The findings were recorded on a data table [41, 59].

Following the morning training, students visited a naturalist center to speak with a naturalist, observe some live specimens, and participate in a scavenger hunt to further refine their knowledge base. At this point, some students had initial exposure to handling a horseshoe crab [41, 59]. Once students returned to the park facility, field experts reviewed handling procedures and measurement requirements again through discussion and demonstrations. Students learned how to pace off and then drop a frame made of PVC in order to count horseshoe crabs for the mock survey. First the field expert demonstrated and then had the students practice several times. Throughout the activity, the instructor emphasized the role of accuracy in order to provide acceptable data [41, 59]. Following the practice session, students entered a beach to meet three field experts. The overall spectacle was dramatic. Students were met with the sights and sounds of waves, thousands of offshore birds and horseshoe crabs which densely lined the beach.

Each of the field experts worked with smaller teams of two and three students to first review handling procedures, measurement skills, and data collection protocol with live horseshoe crabs. For a second data table, participants were to examine a cluster of horseshoe crabs. On this table, participants indicated whether males were attached to females or if they were satellite males; those horseshoe crabs surrounding the females but not attached. Participants were to record gender, relative age, and the interocular for each. Following modeling from the field experts, students began to take measurements. The naturalist observed their technique to check protocol specifications [41, 59].

During the training process, students were keenly interested in the animals but were tentative. The group stayed back as a whole when the naturalists held up the horseshoe crabs. The participants had their arms crossed or hands clasped. Jessica, who was very inquisitive, kept her hands clasped around her book back and then would motion towards the horseshoe crab as the naturalists described the animal's structure. As the training continued, students in general would crouch near the naturalists and motion towards the animals. If a horseshoe crab approached the students, they would laugh nervously [59]. At one point Jessica interjected and said, "If we were to touch it, would it hurt?" When the naturalists indicated that it would be safe to handle the animal, Jessica turned to a peer and said, "You can touch it." Group members tentatively poked the tail spine and jerked back. Another participant, Pat, nervously peered over a friend's shoulder during the training [59].

As the morning session progressed, students began to ask more questions and demonstrated a basic understanding of the organisms through responses to questions. Field experts would select individual horseshoe crabs and ask students to identify the gender and make age approximations. Once the initial training ended, students gently proceeded. Connie approached the horseshoe crabs and began to wipe the sand off of them [59]. As students worked in teams, they meticulously took interocular measurements and helped record data. The field experts alternated between teams to oversee data collection by making corrections and giving verbal encouragement. The field experts would direct students to observe a variety of clues to determine age and gender and emphasized precise measurements for the interocular (distance between eyes). Once teams demonstrated sound procedures, they were encouraged to begin data collection. Team members began to refine their teamwork by distributing roles; one holding the horseshoe crab, one measuring the interocular, one recording the data. Then individuals would double check the measurements [41, 59]. Field experts continued to monitor the teams. Excited students would run to the field experts to ask additional questions and to consult the experts when anomalies occurred. For example, one team located a one eyed horseshoe crab and checked to see if it was possible to use the specimen since there was no interocular measurement [41].

Participants in a pilot study had the opportunity to collect additional data in the evening due to the tide cycles. The comfort level of the students in working with horseshoe crabs was a stark contrast in the evening as compared to the morning. The conditions were much more difficult as there was less light and a higher tide. Participants worked efficiently to collect data on the horseshoe crabs.

Audio tapes from the evening session were filled with excitement. Shouts rang out such as, "That's a male!" "No, no, no. Wait. That's a female. So now we have at least one!" "Sweet!" "It's a male. Do you want me to hold it?" "Yes please. It's awesome" [59]. During the experience, Connie's remark echoed the field observations of the participants. "It is very difficult because they keep moving, and the waves are coming in, and it's dark out. And it's just like very, very hard but it's challenging. But I think it's fun" [59].

Once students returned to the park facility, students combined the data for analysis within the classroom as well as for the biological study. Students graphed their data, discussed the overall experience in terms of observations, challenges for collecting data, and reviewed how the data would be useful for data interpretations in the future [41, 59].

8.5.3 The Structure of Citizen Science Programs for Students

Scientific observations are the impetus for scientific processes centered on hypotheses formation, data interpretation, and theory development [39]. Although citizen scientists have made it possible for researchers to analyze large population trends, as with any research study, accuracy for both volunteers and professionals requires that research designers test data points depending on the contextual requirements once accuracy specificity has been established [60].

In some cases, education level has been shown to be a predictor of hobbyist capabilities in data collection. For example, seventh graders at 95 % accuracy in identifying crustaceans outscored third graders by 15 %. As a result, Delaney et al. [57] used data from children and adolescents for aspects of biological research analysis while incorporating data from individuals with at least two years of college experience for further analysis.

Due to errors during scientific investigations, even with specific protocol criteria, inexperience may lead to restrictions for volunteer participation in citizen science programs [57]. Professional researchers may raise concern that children and adolescents may lack the cognitive skills needed for data collection [25]. Nevertheless, scientific researchers are encouraging volunteer data input from children. For example, children in England have participated in a citizen science program to analyze leaf damage of the horse-chestnut tree citizen science program due to parasitic moth larvae known as Cameraria ohridella [61]. Findings revealed a high level of accordance for most variables between observations collected by experts and children.

For the detection of parasitoids, children's estimates were low in the sample. A series of statistical adjustments enabled the scientists to account for this variation in data interpretation. Based on these contributions, scientists were able to establish the rate of leaf damage in the horse-chestnut tree with the presence of C. ohridella [61]. Similarly, volunteer contributions in a lady beetle study revealed accuracy levels between 81–100 % on species identification [58].

Further, in the horseshoe crab citizen science context, with the support of field experts who checked teams of students' work, findings revealed high levels of correctness on compiling variables under study [41]. As the advantages of engaging volunteers in large scale studies has become more apparent, researchers continue to improve efforts to gauge the accuracy of data collection for citizen scientists. One way to bolster the use of validity of volunteer contributions is to incorporate well defined citizen science structures and tested standardized protocol requirements.

Citizen science is a cost effective option for researchers [58] that enhances scientific understanding particularly for large scale studies while enabling interested individuals to participate in science investigations which promote metacognitive functions and knowledge acquisition. The facets of typical citizen science programs which are pivotal in successful programs are important when designing programs for children and adolescents. These types of activities should center on an authentic research problem, a team mentality among school staff, students, and professional researchers, with a standardized data collection protocol which steers the action of the group [56].

Components of strong citizen science programs often include opportunities to collaborate with experts, systematic methodologies, and standardized training programs [19]. Distinct from citizen science programs open to the general public, citizen science programs incorporated as part of a formal classroom structure requires coordination between administrators, teachers, scientists and students. Figure 8.3 highlights the roles of each type of school based participant in a citizen science program.

Fig. 8.3 The role of citizen
science participants within
a school context

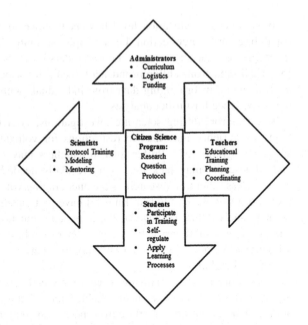

Administrators play a key role in the dynamic of engaging students in informal
science activities as they are essential in providing a support ranging from curriculum development, funding opportunities, and logistics [62]. Thus, without administrative support, minimal citizen science engagement is possible within a school
framework.

Scientists involved in the program often are responsible for developing and testing corresponding curriculum materials, protocol requirements, training, and mentoring [56]. Furthermore, for scientifically based studies, professional scientists
should regularly asses the accuracy for observations and establish parameters for
acceptable ranges in errors [60]. For adolescents, the role of the scientist as mentor
in developing metacognitive skills is a particularly strong component in enhancing
academic achievement and career motivation [2, 41].

In order to establish effective citizen science programs, teachers may provide
preliminary training of science process skills such as scientific observation skills
and content knowledge as part of the standard curriculum throughout the year as
well as planning and coordinating activities, trips, and materials. Students involved
in citizen science programs are responsible for engaging in training with full effort
as their contributions have an impact on the overall scientific study [57, 61]. In
order to benefit from the activities, students should use self-regulatory skills to
improve their skill.

Once the program logistics have been established, students may receive specific protocol training in school prior to the activity with leaflets, online materials, and classroom instruction. With basic skills in place, and the protocol tested,

students should receive training in data collection and recording techniques while reviewing the overall goals of the program so that students are aware of how researchers will incorporate the data into the study [56]. After data collection and submission, students benefit from opportunities such as graphing, observation discussions, and writings related to the activity. A natural progression in this type of activity is for students to engage in further questioning, hypotheses development, scientific study designing, and conducting investigations.

8.6 Implications for Educators, Students, Parents, and Policymakers

There are several implications of citizen science programs based on this research. Although educators are bound by instructional time restraints and limited resources, incorporating citizen science programs within the formal school curriculum supports science achievement and STEM career pathways. As such, informal natural science learning opportunities which encourage collaboration between professional scientists, teachers, and children are relevant educational funding choices.

Further, the use of online citizen science programs may provide schools with less costly alternatives to engage students in citizen science programs such as whale communications, light pollution, and human cell analysis. Although these options engage students in authentic applications of science content knowledge and the development of scientific observation skills, the role of a field expert within this context is irreplaceable. The mentoring relationship in the development of scientific observation skills has great potential for establishing the sense of identity as a scientist within children [17, 18, 41]. It is the guidance and reflection through the phases of self-regulation with the ongoing support of an expert which bolsters metacognitive functioning, science achievement, and career goal setting. For educators, attending workshops and forming partnerships with scientists through citizen science programs provides students with opportunities unique from traditional classroom settings.

Social cognitive career researchers often consider barriers and supports within the development of career pathways. The role of scientists in field work settings, whose presence typifies social supports, is crucial to the academic development of students [2]. Policymakers have a critical role in advocating for informal learning opportunities as part of the school curriculum by securing monetary support, instructional resources, and access to scientists.

For children and parents, volunteering in citizen science programs through community venues, school centered activities, and online resources is one way to enhance student achievement and career motivation. Regardless of the context or subject of study, both field oriented and web-based citizen science programs are an emerging option to promote student self-efficacy, motivation, and competence for STEM careers.

8.7 Conclusion and Future Directions

In this chapter, we described citizen science as a type of informal natural science learning approach and reviewed research on the impact of these types of programs on student metacognition and interest in STEM career paths. A rising body of research shows that one of the most compelling reasons to incorporate citizen science programs as part of formal educational training is the influence this type of activity has on student self-efficacy, achievement, and career motivation. Specifically research shows that during authentic field work, instructors through synthesis of scientific observation skills and content knowledge can teach students to become metacognitively aware of their own learning.

We recommend that in an effort to promote STEM careers, informal natural science learning opportunities should become part of the school curriculum to strengthen student academic achievement and career interest. Future research should include longitudinal studies of students from elementary to high school in terms of metacognitive development and STEM career motivation in relation to ongoing citizen science programs. In addition, the use of microanalytic approaches to measure metacognitive and motivational processes of self-regulated learning may assist in a better understanding of the development of effective practices during citizen science programs.

Fundamental aspects of effective programs for metacognitive development rest in basic structural components. It is imperative that students receive specific training in protocol requirements and the use of science process skills such as scientific observation skills and content information. In this capacity, scientists serve as a key factor in establishing a science oriented work force in the future.

Research studies and citizen science programs which address the accuracy of student observations may help to substantiate support and funding for continued experiences. Moreover, forming a connection with a scientist during citizen science experiences affords a profound opportunity to engage in memorable science activities, while creating the perception of self a s "scientist," increasing metacognitive functioning, and the potential for motivation in a career trajectory.

Freitag and Pfeffer [19] view the focus on cognitive processes within citizen science programs as a way of altering the divide between the general public and the realm of the scientists. The result is the establishment of well-informed individuals who ultimately participate in decisions related to managing and protecting the environment. For students, the opportunity to engage in real world, collaborative, scientific studies has limitless applications in terms of extending metacognitive skills, academic achievement, and STEM career motivation.

References

1. Hansen, M., Gonzalez, T.: Investigating the relationship between STEM learning principles and student achievement in math and science. Am. J. Educ. **120**(2), 139–171 (2014)
2. Hiller, S.E., Kitsantas, A.: The effect of a horseshoe crab citizen science program on student science performance and STEM career motivation. Sch. Sci. Math. J. In Press (2014)

3. Houseal, A.K., Abd-El-Khalick, F., Destefano, L.: Impact of a student–teacher–scientist partnership on students' and teachers' content knowledge, attitudes toward science, and pedagogical practices. J. Res. Sci. Teach. **51**(1), 84–115 (2014)
4. Johnson, C.C., Zhang, D., Kahle, J.B.: Effective science instruction: impact on high-stakes assessment performance. Res. Middle Level Educ. Online **35**(9), 1–12 (2012)
5. Bembenutty, H., Clearly, T., Kitsantas, A.: Applications of self-regulated learning applied across diverse disciplines. A Tribute to Barry J. Zimmerman. Information Age Publishing, Charlotte (2013)
6. DiBenedetto, M.K., Zimmerman, B.J.: Construct and predictive validity of microanalytic measures of students' self-regulation of science learning. Learn. Individ. Differ. **26**, 30–41 (2013)
7. Kitsantas, A., Steen, S., Huie, F.: The role of self-regulated strategies and goal orientation in predicting achievement in elementary school children. Int. Electron. J. Elementary Educ. **2**(1), 65–81 (2009)
8. Peters, E., Kitsantas, A.: The effect of nature of science metacognitive prompts on science students' content and nature of science knowledge, metacognition, and self-regulatory efficacy. School Sci. Math. **110**(8), 382–396 (2010)
9. Zimmerman, B.J.: From cognitive modeling to self-regulation: a social cognitive career path. Educ. Psychol. **48**(3), 135–147 (2013)
10. Schunk, D.H.: Social cognitive theory and self-regulated learning. In: Zimmerman, B.J., Schunk's, D.H. (eds.) Self-Regulated Learning and Academic Achievement: Theoretical Perspectives, 2nd edn, pp. 191–225. Routledge, New York (2009)
11. Zimmerman, B.J.: A social cognitive view of self-regulated academic learning. J. Educ. Psychol. **81**(3), 329–339 (1989)
12. Sperling, R.A., Richmond, A.S., Ramsay, C.M., Klapp, M.: The measurement and predictive ability of metacognition in middle school learners. J. Educ. Res. **105**(1), 1–7 (2012)
13. Quinn, G., Lyons, T.: High school students' perceptions of school science and science careers: a critical look at a critical issue. Sci. Educ. Int. **22**(4), 225–238 (2011)
14. Rogers, M.E., Creed, P.A.: A longitudinal examination of adolescent career planning and exploration using a social cognitive career theory framework. J. Adolesc. **34**(1), 163–172 (2011)
15. Buccheri, G., Gürber, N.A., Brühwiler, C.: The impact of gender on interest in science topics and the choice of scientific and technical vocations. In: Special issue students' interest in science across the world: Findings from the PISA study. Int. J. Sci. Educ. **33**(1), 159–178
16. Patrick, L., Care, E., Ainley, M.: The relationship between vocational interests, self-efficacy, and achievement in the prediction of educational pathways. J. Career Assess. **19**(1), 61–74 (2011)
17. Bombaugh, R.: From citizen scientist to engineers. J. Prof. Issues Eng. Educ. Pract. **126**(2), 64–68 (2000)
18. Eberbach, C., Crowley, K.: From everyday to scientific observation: how children learn to observe the biologist's world. Rev. Educ. Res. **79**(1), 39–68 (2009)
19. Freitag, A., Pfeffer, M.J.: Process, not product: investigating recommendations for improving citizen science "success". PLoS One **8**(5), 1–5 (2013)
20. Harrison, J.A., Underhill, L.G., Barnard, P.: The seminal legacy of the southern african bird atlas project. S. Afr. J. Sci. **104**(3–4), 82–83 (2008)
21. Burkhardt, U., Kärcher, B.: Global radiative forcing from contrail cirrus. Nature Climate Change **1**, 54–55 (2011)
22. Fowler, A., Whyatt, J.D., Davies, G., Ellis, R.: How reliable are citizen-derived scientific data? Assessing the quality of contrail observations made by the general public. Trans. GIS **17**(4), 488–506 (2013)
23. Snäll, T., Kindvall, O., Nilsson, J., Pärt, T.: Evaluating citizen-based presence data for bird monitoring. Biol. Conserv. **144**(2), 804–810 (2011)
24. Christmas Bird Count. National Audubon. http://birds.audubon.org/christmas-bird-count
25. Great Whale Count. Pacific Whale Foundation. http://www.pacificwhale.org/content/great-whale-count-3

26. Delaware Bay Horseshoe Crab Spawning Survey. Delaware Department of Natural Resources and Environmental Control. http://www.dnrec.delaware.gov/coastal/DNERR/Page s/DNERRHSCSpawningSurvey.aspx

27. Citizen Science National Geographic Education. http://education.nationalgeographic.com/ education/encyclopedia/citizen-science/?ar_a=1

28. Old Weather. Zooinverse Project. www.oldweather.org

29. Hiller, S.E., Reybold, L.E.: Field expert's perceptions of scientific observations in the natural world. In: Annual Meeting of the North American Association for Environmental Education. Convention Center (October), Raleigh, NC (2011)

30. Zimmerman, B.J.: Attaining self-regulation: a social cognitive perspective. In: Boekart, M., Pintrich, P., Zeidner, M. (eds.) Handbook of Self-Regulation, pp. 13–39. Academic Press, New York (2000)

31. Zimmerman, B.J., Kitsantas, A.: Developmental phases in self-regulation: Shifting from process goals to outcome goals. J. Educ. Psychol. **89**(1), 29–36 (1997)

32. Rice, L., Barth, J.M., Guadagno, R.E., Smith, G.P.A., McCallum, D.M.: The role of social support in students' perceived abilities and attitudes toward math and science. J. Youth Adolesc. **42**(7), 1028–1040 (2013)

33. Zimmerman, B.J., Schunk, D.H.: A reflection of theories of self-regulated learning and academic achievement. In: Zimmerman, B.J., Schunk's, D.H. (eds.) Self-Regulated Learning and Academic Achievement: Theoretical Perspectives 2nd edn, pp. 289–307. Routledge, New York (2009)

34. Kohlhauf, L., Rutke, U., Neuhaus, B.: Influence of previous knowledge, language skills and domain-specific interest on observation competency. J. Sci. Educ. Technol. **20**(5), 667–678 (2011)

35. Dinsmore, D.L., Alexander, P.A., Loughlin, S.M.: Focusing the conceptual lens on metacognition, self-regulation, and self-regulated learning. Educ. Psychol. Rev. **20**(4), 391–409 (2008)

36. Bandura, A.: Self-Efficacy: The Exercise of Control. Freeman and Company, New York (1997)

37. Zimmerman, B.J.: Goal setting: a key proactive source of academic self-regulation. In: Schunk, D.H., Zimmerman's, B.J. (eds.) Motivation and Self-Regulated Learning: Theory Research and Applications, pp. 267–295. Routledge, New York (2008)

38. Moote, J.K., Williams, J.M., Sproule, J.: When students take control: Investigating the impact of the crest inquiry-based learning program on self-regulated processes and related motivations in young science students. J. Cogn. Educ. Psychol. **12**(2), 178–196 (2013)

39. Cartwright, N.: Nature's Capacities and Their Measurements. Clarendon Press, Oxford (1989)

40. Perry, N.E., Winne, P.H.: Learning from learning kits: gStudy traces of students' self-regulated engagements with computerized content. Educ. Psychol. Rev. **18**(3), 211–228 (2006)

41. Hiller, S.E.: The impact of a citizen science program on student achievement and motivation: A social cognitive career perspective. PhD Dissertation, George Mason University (2012)

42. National Moth Week. Scistarter. http://scistarter.com/project/965-National%20Moth%20 Week

43. Crall, A.W.: Developing and Evaluating a National Citizen Science Program for Invasive Species. PhD Dissertation, retrieved from Proquest Information and Learning Company. (UMI No. 3421961) (2010)

44. Jeanpierre, B., Oberhauser, K., Freeman, C.: Characteristics of professional development that effect change in secondary science teachers' classroom practices. J. Res. Sci. Teach. **42**(6), 668–690 (2005)

45. Brossard, D., Lewenstein, B., Bonney, R.: Scientific knowledge and attitude change: the impact of a citizen science project. Int. J. Sci. Educ. **27**(9), 1099–1121 (2005)

46. Cronje, R., Rohlinger, S., Crall, A., Newman, G.: Does participation in citizen science improve scientific literacy? A study to compare assessment methods. Appl. Environ. Educ. Commun. **10**(3), 135–145 (2011)
47. Tomasek, T.M.: Student cognition and motivation during the classroom birdwatch citizen science project. Master Dissertation. University of North Carolina (2006)
48. Sutton, S.: Increasing science literacy among English language learners through plant phenological monitoring: A citizen science program at Sequoia and Kings Canyon National Parks. Master Dissertation, Prescott College (2009)
49. Miller-Rushing, A., Primack, R., Bonney, R.: The history of public participation in ecological research. Front. Ecol. Environ. **10**(6), 285–290 (2012)
50. Trumbull, D.J., Bonney, R., Bascom, D., Cabral, A.: Thinking scientifically during participation in a citizen-science project. Sci. Educ. **84**(2), 265–275 (2000)
51. Science & Engineering Indicators. Technical Report, National Science Board (NSB 96-21). Washington, DC: U. S. Government Printing Office (1996)
52. Dunlap, R.E., Van Liere, K.D.: The new environmental paradigm: a proposed measuring instrument and preliminary results. J. Environ. Educ. **9**(4), 10–19 (1978)
53. Crall, A.W., Jordan, R., Holfelder, K., Newman, G.J., Graham, J., Waller, D.M.: The impacts of an invasive species citizen science training program on participant attitudes, behavior, and science literacy. Public Underst. Sci. **22**(6), 745–764 (2012)
54. Science & Engineering Indicators. Technical Report, National Science Board (NSB 01A). Washington, DC: U. S. Government Printing Office (2008)
55. Dunlap, R.E., Van Liere, K.D.: The new environmental paradigm: a proposed measuring instrument and preliminary results. J. Environ. Educ. **40**(1), 19–28 (2008)
56. Bonney, R., Cooper, C.B., Dickinson, J., Kelling, S., Phillips, T., Rosenberg, K.V., Shirk, J.: Citizen Science: a developing tool for expanding science knowledge and scientific literacy. Bioscience **59**(11), 977–984 (2009)
57. Delaney, D.G., Sperling, C.D., Adams, C.S., Leung, B.: Marine invasive species: validation of citizen science and implications for national monitoring networks. Biol. Invasions **10**(1), 117–128 (2008)
58. Gardiner, M.M., Allee, L.L., Brown, P.M.J., Losey, J.E., Roy, H.E., Smyth, R.R.: Lessons from lady beetles: accuracy of monitoring data from us and UK citizen-science programs. Front. Ecol. Environ. **10**(9), 471–476 (2012)
59. Hiller, S.E.: The Effect of a Horseshoe Crab Citizen Science Project on Student Self-efficacy and Career Trajectories. Round table session at the annual research conference of the American Educational Research Association, Vancouver (2012)
60. Crall, A.W., Newman, G.J., Stohlgren, T.J., Holfelder, K.A., Graham, J., Waller, D.M.: Assessing citizen science data quality: An invasive species case study. Conserv. Lett. **4**(6), 433–442 (2011)
61. Pocock, M.J.O., Evans, D.M.: The success of the Horse-Chestnut leaf-miner, Cameraria ohridella, in the UK revealed with hypothesis-led citizen science. Plos One. **9**(1), e86226 (2014)
62. Ernst, J.: Influences on and obstacles to K-12 administrators' support for environment-based education. J. Environ. Educ. **43**(2), 73–92 (2012)

Chapter 9
Personal Self-regulation, Self-regulated Learning and Coping Strategies, in University Context with Stress

Jesús de la Fuente, Lucía Zapata, José Manuel Martínez-Vicente, Paul Sander and Dave Putwain

Abstract Personal self-regulation is an important variable in education and research, but self-regulated learning is the construct seen most often in the educational context. Existing studies do not seek to establish relationships between personal self-regulation and other educational variables. We define conceptual characteristics and relationships of personal self-regulation (personal presage variable), self-regulated learning (meta-cognitive, process variable) and coping strategies (meta-motivational, meta-affective process variable), establishing the importance of these variables in future meta-cognition research. These relationships have been established conceptually and empirically within the 3P and DEDEPRO Models, and are confirmed in recent research: namely, the importance of personal self-regulation in determining the degree of cognitive self-regulation during the process of university learning with stress; the relationship between personal self-regulation and the type and quantity of coping strategies,

J. de la Fuente (✉) · J.M. Martínez-Vicente
Department of Psychology, University of Almería, Carretera de Sacramento s/n, La Cañada de San Urbano 04120 Almería, Spain
e-mail: jfuente@ual.es

J.M. Martínez-Vicente
e-mail: jvicente@ual.es

L. Zapata
Education & Psychology I+D+i. Science and Technology Spin-Off, University of Almería, Costa Azul No 68, 04009 Almería, Spain
e-mail: lucia@epsychology.com

P. Sander
Department of Psychology, Cardiff Metropolitan University UWIC, Western Avenue, Llandaff, Cardiff CF5 2YG, UK
e-mail: paul.sander@uclmail.net

D. Putwain
Faculty of Education, Edge Hill University, Ormskirk L39 4QP, UK
e-mail: putwaind@edgehill.ac.uk

© Springer International Publishing Switzerland 2015
A. Peña-Ayala (ed.), *Metacognition: Fundaments, Applications, and Trends*,
Intelligent Systems Reference Library 76, DOI 10.1007/978-3-319-11062-2_9

and the relationship between self-regulated learning and coping. We conclude by discussing our experience with an online self-help system designed for university students.

Keywords 3P model · DEDEPRO model · Personal self-regulation · Self-regulated learning · Coping strategies

Abbreviations

ANOVA   Analysis of variance
BAC   Blood alcohol content
CAR   Cuestionario de autorregulacion personal
CFI   Comparative fit index
DEDEPRO   Design, development, product
ICT   Information and communication technologies
IFI   Incremental fit index
MANOVA   Multivariate analysis of variance
NFI   Normed fit index
RFI   Relative fit index
RMSEA   Root mean square error of approximation
SEM   Structural equation modeling
SRL   Self-regulated learning
SRQ   Self regulation questionnaire
TLI   Tucker Lewis index
3P   Presage, process, product

9.1 Introduction

As a psychological variable inherent to the competencies of an individual's personal development, personal self-regulation is presently the object of much interest in education and research. However, there is still a scarcity of studies that seek to establish relationships between personal self-regulation and other educational variables. The construct of self-regulation is found in educational contexts, but normally in reference to self-regulated learning [1–4] which is the name given to applying general self-regulation (or the self-regulation used by persons in their daily life) to the specific conditions of learning situations.

Self-regulation has been used with different shades of meaning in different contexts. In the field of health or substance abuse, and in educational contexts that deal with regulating the teaching-learning process, the concept of "personal self-regulation" has been used [5]. The present chapter has four aims:

(1) To delimit the conceptual characteristics of three different constructs, each with theoretical potential: personal self-regulation, self-regulated learning and coping strategies. (2) To show the importance of these variables in future research in meta-cognition, since they refer to different general aspects of self-regulation, of meta-cognition and of specific meta-motivation while learning (self-regulated learning) and to meta-affective control in situations of academic stress (coping strategies). (3) To demonstrate the relationships between these variables, as a research hypothesis based on prior evidence and empirical dates. (4) To illustrate intervention strategies for improving self-regulation and coping in university students.

9.2 The Process of Teaching-Learning as a Source of Stress

In Higher Education, teaching and learning processes form part of a single binomial for the purpose of preparing university students and ensuring their success. Currently, higher education is undergoing changes due to the need for quality education, with a view to increased employment.

This new system is based on teaching for competencies, meaning new demands for both students and teachers, and restructuring the teaching-learning process itself [6–9]. It becomes essential for students to have an active role in constructing their own learning, while the teacher becomes responsible for advising and assisting students throughout the process [10]. This context of competency-based learning means greater formative knowledge requirements, whether conceptual (knowing), procedural (knowing how), or attitudinal (wanting to do). So it is that, within this new scenario, students have a bigger workload, they must be more responsible and they must be consistently more independent in their learning process. These changes affect how they ought to approach the educational situation, taking into account affective-motivational variables, cognitive variables and strategic variables alike. This new scenario can become a stressful context for students, due to its novelty and to the demands of competency-based learning [11–13].

It is within this teaching-learning context that we study the different variables that make up the present study, working from two different heuristics: Biggs' 3P Model (Presage, Process and Product) [14] and the DEDEPRO Model [10, 15]. The combination of these two models offers a framework for analyzing teaching-learning situations and for a better understanding of the structure of existing research and the variables that are being studied. Another reason for adopting both models is their complementary nature. Recently, relationships between personal self-regulation and other educational variables have been established conceptually and empirically in the framework of the 3P Model [14] and the DEDEPRO Model [10, 16] see Fig. 9.1.

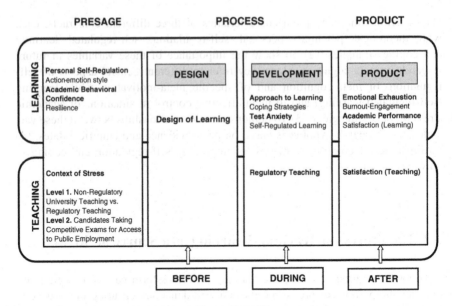

Fig. 9.1 Relationships between the variables studied [65]

9.3 Personal Self-regulation as a Student Meta-Cognitive and Meta Motivational Presage Variable

Personal self-regulation refers to the capacity or ability to control our own thoughts, emotions and actions. Through self-regulation we are able to consciously control the amount that we eat, whether to act on an impulse, our task execution, obsessive thoughts, and even the extent that we allow ourselves to listen to our own emotions.

We can therefore affirm that personal self-regulation is a vital process that allows people to behave adequately, carry out tasks properly, and abstain from activities that may be harmful to their own well-being. Self-regulation is used in a number of processes including the regulation of emotions, thoughts and actions for physical or behavioral control or restraint [17].

Different theoretical models have outlined the characteristics of this psychological construct. From a sequential approach, Kanfer [18] proposed a model within the so-called open-loop conception [19].

Self-regulation is conceived as a self-correcting procedure when faced with discrepancies, indications of imminent danger, or conflictive motivational states that activate the system of observation. The present study adopts this conception. Miller and Brown [20] modify postulates of the Kanfer [18] model, providing a better explanation for changes in addictions. Within Miller and Brown's theoretical model for addictive behaviors [20], it is assumed that self-regulation is developed through seven successive processes:

1. Informational input (self-observation) is the first process that occurs in self-regulation, where persons obtain information about their own behavior, especially about a potentially problematic behavior. In this process, persons increase their understanding of the nature and impact of the behavior to be changed.
2. In Self-evaluation, one looks for consistency between expected performance and actual performance, and this includes becoming aware of the negative consequences of a behavior. In other words, this process is produced when a person becomes aware that a behavior may be problematic. The observed behavior is compared to some personal criterion, which may be: (1) internal, where the actual behavior is compared to the ideal; or (2) external, comparing the behavior to social norms. If one discovers that the behavior does not meet a certain standard or norm, a negative feeling may result. When these reactions (whether cognitive, affective or behavioral) are sufficiently strong, they may lead us to the next process.
3. Instigation to change is triggered by perceptions of discrepancy and dissatisfaction in the above evaluation. According to this model, this impetus from discrepancies is essential for advancement to further stages of self-regulation [21].
4. Searching for options to reduce discrepancies that have been detected above.
5. Formulating a plan where one sets down a schedule, activities to be pursued, places and any other aspects to be considered in the attainment of one's goals.
6. Implementing the plan, the stage where one executes all that was planned in the prior phase.
7. The final phase is addressed through a *comprehensive assessment*, addressing both the effectiveness of one's planning and the attainment of goals.

If there is a deficit in any of these self-regulation processes, one's behavior regulation will suffer. Within this theoretical framework, Brown [21] defines self-regulation as a person's ability to "plan, monitor and direct his or her behavior in changing situations" (p. 62). In essence, this model adopts the self-regulation postulates of Zimmerman [22], by defining moments of planning, control and thoughtful evaluation of one's action.

Hoyle [23] also speaks of these discrepancies and of the actions that we carry out in order to obtain our objectives and what we desire. He calls these actions self-regulation, actions that are natural and often are automatic responses of a healthy person in order to cope with the day-to-day discrepancies that are found between one's expectations or desires and one's reality. This self-regulation takes the qualifier "personal" in order to differentiate it from "academic", and has been studied in both adolescents [5, 24] and university students [25].

Personal self-regulation is a construct that has been used to a greater extent in the field of health [21, 26, 27]. However, after Zimmerman [22] showed the existence of processes that are common to different domains, experts have begun to show interest in analyzing the self-regulating components that are common to different spheres of life, such as education and work.

Brown et al. [28] constructed the Self-Regulation Questionnaire (SRQ) to measure self-regulation based on their theoretical model. Later, after performing further analyses, they developed an abbreviated version, the Short Self-Regulation Questionaire (SSRQ), which was validated in a Spanish sample by Pichardo et al. [29].

The data show good fit to the structure of seventeen items grouped under four factors (goal setting-planning, perseverance, decision making and learning from mistakes). These factors are adopted in the present chapter and are seen in Fig. 9.2, which establishes the moments at which each phase takes place.

This instrument has been used mainly in connection with substance abuse, and has been submitted to an examination of its psychometric characteristics on several occasions [30, 31]. Its use has also been extended beyond substance abuse to address aspects such as psychological well-being, disposition to happiness [32], depression symptoms [33] and career adaptability [34], and is in demand in other areas such as education [5]. In 2005, a monograph of *Applied Psychology: An International Review* (vol. 54, no 2) [35] presents different studies that inquired into the similarities and differences of self-regulation as used in several domains of psychology, such as education and health. This monograph represents an advance in the study of self-regulation in the main areas of applied psychology: work and organizations, health and education [35]. Karoly et al. [36] reviewed the papers published in this monograph and sought to establish the similarities and differences in self-regulation activities: academic, health-related and work-related.

One of their conclusions [36] states that there is a "meta-theoretical convergence" among the areas of psychology. They identified differences and similarities in aspects pertaining to conceptions, methodologies, assessment and intervention. Among the similarities, they found components that were common to all the areas, such as "goal selection, goal setting, feedback sensitivity, discrepancy (error) monitoring, self-evaluative judgment, self-corrective instrumental action, and the emergence of self-efficacy beliefs" [36].

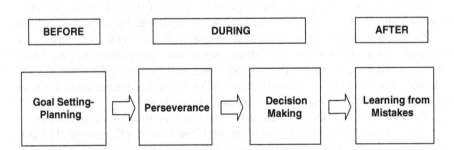

Fig. 9.2 Factors of personal self-regulation [16, p. 24]

9.3.1 Prior Evidence on Personal Self-regulation

Personal self-regulation, as a psychological variable that is closely tied to subjects' personal development competencies, has attracted interest in the sphere of educational psychology. Prior studies have shown that self-regulation has a significant role in health as well as in success, whether academic or work-related [36, 37]. We can think of the process of self-regulation as having a personal, behavioral and contextual nature [4, 19] adding goals as a key factor [38, 39].

Taking personal regulation as a *presage variable* in the sphere of educational psychology, de la Fuente and Cardelle-Elawar [40] define it as a student variable "that determines the level of effort that students will sustain in the process of active learning for the completion of a given task". It is widely recognized as the means by which students transform their mental skills into problem solving survival skills [40].

As we have stated earlier, there are many studies from the sphere of healthcare that incorporate personal self-regulation as a study variable. Within this broad field, addictions have been most often related to this variable, since they represent a highly important topic to today's society.

From these studies, we are able to affirm that personal self-regulation plays a very important role in substance abuse or abstinence [21, 26, 27]. Muraven et al. [26] discovered greater blood alcohol content (BAC) in persons with less self-regulation, and a lower BAC in persons with higher self-regulation. Muraven et al. [27] examined whether there was a relationship between alcohol consumption and distress over time in two samples of social drinkers.

They found that less self-regulation in alcohol use implied a greater alcohol intake and greater feelings of distress. Tangney et al. [41] found that higher self-regulation scores correlated with less alcohol abuse, a higher grade point average, better psychological and emotional adjustment as well as optimal responses. Ferrari et al. [42] revealed that self-regulation scores were positively related to the length of abstinence. As self-regulation increased, so did the length of abstinence. Their study examined the relations between changes in self-regulation and self-efficacy as predictors of abstaining from substances.

They found that changes in self-regulation and in self-efficacy were significantly predictive of the probability of abstinence. Furthermore, changes in self-regulation and self-efficacy were largely independent. There are few studies in the field of educational psychology that have incorporated the presage variable of personal self-regulation. However, we find a few studies that confirm its importance in the educational context, including studies from de la Fuente et al. [5], where they seek to establish the relations between personal self-regulation and perception of maladaptive school behaviors in secondary students; and from de la Fuente and Cardelle-Elawar [40], who establish the relationships between self-regulation and coping strategies in university students.

In the former study [5], a total of 888 students from compulsory secondary education participated. The questionnaire used to assess personal self-regulation was the *Self Regulation Questionnaire*, SRQ [28], in its Spanish version, *CAR* [43]. The study showed that levels of total personal self-regulation modulate adolescents' perception of the school's social climate. Results from inferential analyses (ANOVAs) showed that the degree of personal self-regulation is interdependent with the perception of maladaptive or interpersonal problems at school. Specifically, low and high levels of total self-regulation, respectively, were accompanied by the perceived greater or lesser occurrence of maladjusted behaviors in the environment. We can thereby affirm that high self-regulation capacity is beneficial for personal and professional development, especially in preventing health-risk behaviors in adolescents, such as tobacco and alcohol use [40].

A total of 77 students from the University of Almería (Spain) participated in the second study [40]. Results revealed a statistically significant relationship between the study variables of personal self-regulation and coping strategies.

In order to assess personal self-regulation, the Spanish version of the *Self-Regulation Questionnaire* was used [28], and the *Coping with Stress Questionnaire* [44–46] was used to measure coping strategies. The results indicate that different levels of personal self-regulation determine the types of coping strategies. During a stressful situation, students with high levels of personal self-regulation manifest problem-focused coping strategies, while students with low levels of personal self-regulation have a more emotion-focused coping style.

9.4 Self-regulated Learning as a Meta-Cognitive and Meta-Motivational Process Variable of Students

The concept of self-regulated learning is emerging more from day to day, due to its great importance in the teaching-learning process. Interest in this construct began to appear in the mid-1980s, in answer to a big question: "how can students become the masters of their own learning process?" When we analyze this variable, we must not overlook its mediating role between students' mental ability and the acquisition of academic skills, such as reading or mastery of mathematics. Specifically, this construct refers to a self-directing process in the students, transforming their mental ability into academic skills.

Self-regulation is thus considered a proactive activity where the student takes the lead in helping himself, as well as developing learning strategies. For the definition of this variable, we must bear in mind the active role of students in the learning process, the feedback given to them during this process, and the role of motivation [47].

Researchers who study this variable suggest that students self-regulate when they take an active role, at the metacognitive, motivational and behavioral levels, in their teaching-learning process [48]. All the definitions that are given to self-regulated learning include these three properties, which allow students to be aware of their own learning process and of the importance of improving their academic performance. But these are not the only components in the definition of this construct, we also find what are known as feedback loops during learning [39, 48–50].

This refers to a cyclical process by which students direct the effectiveness of their learning methods or strategies to respond to feedback, with non-visible changes in self-perception as well as visible changes in behavior. The concept of self-regulated learning is a description of how and why students choose to use a self-regulated process in particular, a strategy or a response.

The vast majority of researchers are in agreement that motivation has a role in prompting these results. Many authors have shown interest in studying this variable [2, 7, 9, 51–56] in different educational contexts, mainly: Secondary Education [3] and University [9, 52, 56–59].

These studies have taken into account different variables such as: performance and academic success [9]; implementation of training programs in self-regulated learning [52]; motivation [60]; regulatory teaching [61]; attribution styles [62]; critical thinking [63]; acquisition of self-regulation competencies [58]; effects of self-assessment scripts in self-regulation [64]; action control and dispositional hope [56]; metacognitive knowledge [57]; and learning approaches [65].

Studies from Spain on self-regulated learning in higher education came later than for other stages of education, finally appearing at the end of the last century. The Spanish studies are characterized by use of theoretical models and methodologies created in other countries and originating from the sociocognitive perspective. A very important aspect when studying self-regulated learning is to know and identify the differences between competent/expert learners and beginners. After an exhaustive review of the different publications, Torrano and González [4] summarized six characteristics that distinguish self-regulating students from others (Fig. 9.3).

Specifically, the authors identify self-regulating learners as students who use learning strategies, have personal initiative, and are aware that academic success depends mainly on their involvement, on their engagement in the competencies exhibited, and on their perseverance on task [19, 50, 66, 67]. Research studies call attention to a gap between the self-regulation observed in university students and what is actually required in Higher Education [68]. However, Cabanach et al. [69] obtained a greater percentage for students that were high in self-regulated learning; they explain these results based on the composition of the higher self-regulation group, "almost half of them are enrolled in 3rd, 4th or 5th year" of their degree program [69]. It should also be noted that the groups were established without attention to the students' academic performance. It is possible that including this variable would further limit the conditions for belonging to the high self-regulation group and would ensure more real percentages.

Several researchers have taken action to address the need for improved self-regulated learning and for regulation in teaching in Higher Education. de la Fuente et al. used the DEDEPRO model to build two online tools for this purpose. Effects from an intervention with a sample of 728 students showed significant improvement in perceptions of the teaching-learning process, in components of both self-regulated learning and of regulation of teaching.

1) They know how to plan, monitor and direct their mental processes toward achieving personal goals (*metacognition*).

2) They know and can use a number of *cognitive strategies* (repetition, organization and elaboration) that help them attend to, transform, organize, elaborate and recover the information.

3) They show a set of *motivational beliefs* and *adaptive emotions*, such as a high sense of academic self-efficacy, the adoption of learning goals, development of positive emotions toward the tasks (e.g., joy, satisfaction, enthusiasm), as well as the ability to control and modify them, adjusting them to task requirements and the specific learning situations.

4) They *plan and monitor* the time and effort they are going to use on their tasks, and know how to create and structure favorable learning environments, such as finding a suitable place to study and seeking academic help from teachers and classmates when they have difficulties.

5) If the context allows, they show greater attempts to *participate in the control and regulation* of academic tasks, classroom climate and structure (e.g., how they will be evaluated, task requirements, the design of class assignments, organization of work groups).

6) They are able to implement a number of *volitional strategies*, oriented toward avoiding external and internal distractions, in order to maintain their concentration, effort and motivation while carrying out academic tasks.

Fig. 9.3 Characteristics of expert self-regulators [4, p. 3]

The study of learning as an active process of the learner has challenged traditional educational practices. Different theoretical perspectives and models have been investigating this emerging type of learning for decades, giving it different names: autonomous learning, self-directed learning, independent learning or self-learning. During these decades, educational psychologists have put emphasis on both the cognitive sphere (cognitive processes and strategies and metacognitive mechanisms) and the affective-motivational sphere of learners. Research and educational practices have tried to get closer to self-regulated learning and develop ways to encourage it, and so respond to one of the basic pillars of education: learning to learn [69].

Self-regulated learning is very much connected with motivation. These two constructs can be confused, but we find differences between the two. Namely, students may be motivated even when they are not able to make a personal choice, or because they perceive the need for something, or, they may even be motivated implicitly or unconsciously.

However, self-regulation requires a certain degree of choice or intentional selection of strategies or behaviors, which are planned in order to help achieve a goal. Motivational theories focus on how motivation may increase or decrease, as a function of personal and contextual factors, but they seldom look at how to intentionally monitor or regulate one's own motivation. By contrast, self-regulation models often deal with aspects related to how the individual may control his or her motivation, cognition and behavior [70]. Zimmerman [48] takes into account the relationship between motivation and self-regulation, defining self-regulated learning as the process by which students activate and sustain cognitions, behaviors and effects that are systematically oriented toward achievement of goals.

9.4.1 Dimensions of Self-regulation Learning

Zimmerman [22] developed a conceptual framework to address what self-regulation consists of, proposing six dimensions. Each dimension of self-regulated learning requires an action (task conditions) that will result in certain attributes and processes that favor or do not favor self-regulation.

This framework can be characterized according to six key questions, along with the corresponding self-regulation processes. An essential element of self-regulation is that students have some possible choice at least in some aspect or perhaps in more than one. This means that, inasmuch as not all aspects of the task are externally controlled, we may speak of self-regulation. When everything is controlled, it can be said that the behavior is externally controlled, or regulated by others. This type of situation occurs when teachers leave no margin for students in considering the why, how, when, what, where and with whom to complete the task.

Possibilities for self-regulation vary from low to high, depending on how many choices the learner has. For this reason, it is preferable to speak of self-regulation in terms of degree rather than in absolute terms (i.e., that one self-regulates or does not self-regulate).

9.4.2 Self-regulated Learning as a Socio-Cognitive Process

Sociocognitive theory considers that self-regulation contains three processes: self-observation, self-assessment and self-reactions [18, 19, 22]. *Self-observation* (monitoring) refers to deliberate attention given to aspects of one's own behavior. It is usually accompanied by a record of the frequency, intensity or quality of the behavior. Self-observation is essential for determining progress on an activity. Without it, selective memory of successes and failures would be at risk, because our beliefs about the results of an activity do not faithfully reflect what was actually obtained. The personal log can provide very good results in students with difficulties studying, since a log of their activity will tell them whether they actually take advantage of the time or if they use more than half the study time in non-academic tasks. It can also reinforce motivation, because students can realize what they are doing, and can react to this knowledge by modifying their behavior. However, the latter also requires self-assessment and self-reaction. We will therefore specify what these other two processes consist of.

First, when we speak of *self-assessment,* we refer to a comparison of our actual level of execution with the goal that we wish to reach. It depends on the type of self-assessment standards used, on the properties of the goal, on the importance of goal achievement and on attributions. The goal properties (specificity, proximity and difficulty) affect self-regulation and motivation. These properties increase the progress comparisons, so that students may maintain or modify their self-regulating strategies depending on their progress assessment. Self-assessments may also reflect the importance of achieving the goal. When persons are unconcerned with how they carry out the tasks, they may not assess its execution or increase their effort to try and improve [19]. People evaluate the progress of their learning when they are pursuing goals that they value.

Finally, *self-reaction* is defined as the behavioral, cognitive and affective response to the self-assessments. These self-corrections have the ability to motivate and to increase one's self-efficacy, stemming from the belief that one is making acceptable progress and from the anticipated satisfaction of achieving one's goal.

9.4.3 Cyclical Nature of Self-regulation Learning

Sociocognitive theory emphasizes the interaction of personal, behavioral and environmental factors [19, 50]. These factors normally change during learning and must be monitored, hence self-regulation is considered to be a cyclical process. Such monitoring leads to changes in the student's strategies, cognition, affect and behavior. This cyclical nature is stated in Zimmerman's three-phase self-regulation model [22]:

1. *Forethought phase*: A prior phase that precedes execution and refers to processes that prepare the scenario for action, giving thought to processes that occur during learning and that affect attention and action. During this initial phase, there are two different areas: task analysis processes and self-motivation beliefs. Task analysis involves a learner's efforts to break down a learning task into its key components. Students' task analyses influence their goal setting and planning.
2. *Performance control phase*: Two major classes of self-regulation processes are postulated during this phase: self-control and self-observation. The first of these processes refers to the actual use of different strategies to guide learning, such as task, cognitive, and behavioral strategies. The second process refers to specific methods to track one's performance; metacognitive monitoring deals with informal mental tracking of one's performance phase processes

and outcomes, whereas self-recording indicates creating formal records of the learning process and/or outcomes.

3. *Self-reflection phase*: This phase takes place after execution; students respond to the efforts they have made, where greater effort compensates for fewer self-regulation processes throughout the different phases. Students come to learning situations with different goals and different levels of self-efficacy for attaining them. While monitoring execution, they implement learning strategies, which then affect motivation and learning. Two types of processes occur during the self-reflection phase: self-judgments and self-reaction. Self-judgments refer to self-evaluations of the effectiveness of one's learning performance and causal attributions regarding one's outcomes. Learners' self-judgments are linked to two key forms of self-reactions: self-satisfaction and adaptive inferences. Self-satisfaction reactions refer to perceptions of satisfaction or dissatisfaction, and associated affect, with regard to one's performance. These emotions can range from elation to depression. A closely associated type of self-reaction involves adaptive or defensive inferences, which refer to conclusions about whether and how a learner needs to alter his or her approach during subsequent efforts to learn. These self-reactions influence forethought processes for further solution efforts, thus completing the self-regulatory cycle [47].

The cognitive, metacognitive, and motivational processes that underlie each of the phases are depicted in Fig. 9.4. This process makes clear that in order to carry on effective self-regulation, there must be goals and motivation [18, 19, 48, 70]. Students must regulate both their actions and their underlying cognitions with respect to their achievement, behavior, intentions and affect (including responses to stress, one focus of the present investigation). In order to attain this effective self-regulation, students must develop a sense of self-efficacy for self-regulating their learning and for properly executing the task. Processes of self-evaluating one's capacities and progress in acquiring skills are crucially important, for this reason students should self-evaluate regularly. In this way they draw attention to their improvements in execution, increasing their self-efficacy and sustaining self-regulation by promoting the learning of skills that are involved in the task [9].

Another model to keep in mind when studying self-regulated learning is the Pintrich model [70]. In this model, self-regulated learning is conceived as an active, constructive process, where learners set goals that guide their learning, direct, regulate and control their cognition, motivation and behavior—as well as contextual characteristics—toward the attainment of their goals.

The Pintrich [70] and Zimmerman [50] models have similarities: both are social cognitive models of motivation and cognition, for the purpose of constructing an integrated model of academic learning. One difference with respect to the Zimmerman [50] model is Pintrich's [70] characterization of the phases as nonsequential and recurring; the different phases, processes and components may be simultaneous and interactive. This model has become a powerful heuristic for conceptualizing and understanding self-regulated learning [10].

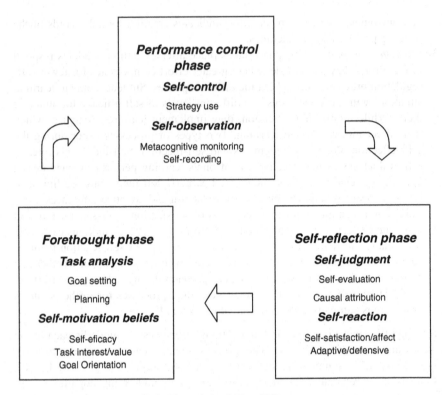

Fig. 9.4 Phases and processes of self-regulation [47, p. 402]

9.5 Strategies for Coping with Stress as a Meta-Affective Variable of Learning and Buffer of Academic Stress

We find ourselves at a very complex time socially, when the word stress plays a leading role in everyday life, and has become a familiar concept. It is an increasingly important phenomenon in modern society, and practically all population groups are experiencing increased stress. When a person's capacities do not match the demands placed on him or her, there is dissatisfaction and feelings of stress.

Despite being one of the most common and familiar life experiences, the term itself is difficult to precisely define. Many authors try to do so, Long [71] and Lazarus and Folkman [68] are among them. Long [71] defines it as the relationship between a person and his/her environment, where the environment is looked on as something that exceeds one's capacities and resources and puts one's wellbeing in danger. Stress is considered to be a physical and psychological reaction to a perceived or actual demand for change. The demand itself is called a stressor and the steps people take to resolve or avoid the stressor are referred to as coping. Lazarus and Folkman [68] suggest that psychological stress is "the result of a particular relationship between the individual and the environment, where the latter is assessed as threatening or surpassing one's resources, and endangering one's well-being" (p. 19). They focus on the relationship between the person and his or her environment, and an appraisal that the latter is too demanding, surpasses one's resources and is harmful to one's well-being. Therefore, it is the person's appraisal of the situation that determines a perception of stress. It must be kept in mind that individual differences play a very important role in stress processes. Whether or not the stress response is triggered depends primarily on aspects of perception. Lazarus and Folkman [68] determined that the physiological activation is triggered by one's assessment of the situation (primary assessment) and of one's ability to address it (secondary assessment). This will be dealt with later under Lazarus and Folkman's Transactional Model [68].

Human beings may experience stress from three main sources: the surroundings (noise, crowds, rigid schedules, etc.); one's own body or physiological urges, often reacting to threats in the environment that are in themselves stressful, producing observable physical changes (dilated pupils, heightened sight and hearing, tensed muscles, blood pumping more quickly to the brain in order to increase oxygen and favor mental processes, increased cardiac and respiratory frequency, etc.); and/or one's own thoughts, since the way that we interpret and label our experiences and the way we see the future can provoke stress.

We are aware of the importance of how we face stressful situations over the course of our lifetime. For this reason, *coping strategies* have been included as a process variable in the research. We hope to come to a better understanding of this concept. For this purpose, we define the concept of coping and we inquire into coping strategies. For delimitation of concepts, we mention recent researchers and the most important models in the field: *Transactional theory,* by Lazarus and Folkman [68] and the *Multiaxial Model of Coping,* proposed and studied by Dunahoo et al. [66]. We describe two of the most frequently used instruments for assessing coping strategies: the *Coping Strategies Scale* [44] and the *Coping Estimation Inventory* [72].

9.5.1 Concept of Coping

When we speak about coping we refer to cognitive and behavioral efforts to manage stress. However, most healthcare psychologists who study stress and coping would define coping broadly to include thought and behaviors that occur in response to stressful experience, whether the person is handling the situation well or poorly [73].

The concept of stress has been studied at length, and there are many authors who examine and seek to define it. These efforts have produced a variety of definitions that we present below. Schuler [74] defines coping as a "process of analysis and evaluation to decide how to protect oneself again adverse effects of any stressor and its associated negative outcomes yet to take advantage of its positive outcomes" (p. 351). Holroyd and Lazarus [75] define coping as "cognitive and behavioral efforts to master, reduce, or tolerate the internal and/or external demands that are created by the stressful transaction" (p. 843). Lazarus [76] defines coping as "cognitive and behavioral efforts to manage specific external or internal demands (and conflict between them) that are appraised as taxing or exceeding the resources of a person". Coping is regarded as a dynamic process that changes over time in response to objective demands and subjective appraisals of the situation (p. 648).

There are a variety of coping strategies that have been proposed by researchers in order to understand the discrepancies in how individuals act when dealing with stressful situations. We proceed now to discuss different coping strategies and the theories that study them [68, 77].

9.5.2 Coping Strategies

There are diverse definitions of *strategies for coping with stress,* but in general terms, we can say that this concept refers to behavioral and cognitive efforts that a person makes in order to deal with stress. In other words, these are strategies that one turns to in order to deal with either the external or internal demands that generate stress, as well as with the psychology discomfort that usually accompanies them [46].

Coping strategies have been studied in different age ranges: children [78]; adolescents [1, 79–81]; youth-adults [82]; and during the aging process [83]. They have mainly been investigated in healthcare contexts, since the way that the individual faces stress can act as an important mediator between stressful situations and health [84, 85]. For this reason, we find most of the studies in the area of Clinical Psychology. Different studies have established a relationship between coping strategies and other variables such as: anxiety [85, 86]; control of emotions or emotional intelligence [87]; sensitivity to pain [63]; professional stress [88]; chronic illness [89]; aging [83]; psychological well-being in students [82], and so on.

Coping strategies in the context of Educational Psychology are more related to academic stress and specifically to one of its main stressors, tests [90].

We consider it of vital importance to inquire into coping strategies, since all university students must face the external stressor of tests, as well as others. We must also keep in mind that university students are a very specific population, as are the ways that they deal with stress.

Hence the importance of introducing this variable in the present research study, as mentioned above. Fewer studies have been carried out in this field, but relationships have been found between coping strategies and academic performance [91] and student gender [92]. In addition, students' levels of stress have been studied in conjunction with the coping strategies they use [93].

Cohen et al. [91], in their study on academic performance and coping strategies, found that greater use of problem-focused and avoidance strategies predict better performance, confirming results from other previous studies [94, 95].

de la Fuente et al. [92] used a sample of 273 students from 2nd and 4th year Psychology at the University of Almería, in order to study the relationship between gender and coping strategies. Strategies were measured using the Spanish version of the Coping with Stress Questionnaire by Lazarus and Folkman [44, 46, 68] and yielded no general gender differences, although the girls made more use of problem-focused coping strategies than the boys.

These results are in the same direction as those of previous studies [82]. Ticona et al. [93] studied the level of stress and coping strategies present in nursing students. A total of 234 students from first to fourth year participated in the study, which used the Coping Estimation Inventory (COPE) by Carver, Scheier and Weintraub [72]. In this case, males were found to have a greater tendency toward managing emotions, and a lesser tendency toward managing the problem. First-year students presented the highest stress levels.

As we can observe, not many studies have been carried out in the Educational Psychology context, there are many unanswered questions, and the present study seeks to address a few of these.

After this empirical review, we now approach the distinction between *coping styles* and *coping responses* [94]. The former refers to the predisposition of one's personality to use different coping strategies depending on the situational context and the moment in time, in other words, it emphasizes stable ways of coping in different situations. The latter is understood as the particular thoughts and behaviors that are realized in response to stressful situations, and may change over time.

Fernández-Abascal [95] describes these responses as concrete processes that are used in each context and can be highly changeable depending on the triggering conditions. There is evidence of different patterns or styles of coping, but it is also evident that the specific situational factors play a role of utmost importance in coping reactions [94]. Based on the transactional model [68], there has been substantial consensus in classifying a large variety of possible coping strategies [96], [92] using the following categories:

1. Problem-focused strategies, directed toward solving the problem in order to eliminate stress.
2. Emotion-focused strategies, aimed at regulating, reducing or eliminating the emotional stress relative to a stressful situation.
3. And in some cases, we can find a category of *avoidance-focused coping strategies*. This refers to the use of evasive strategies that seek to avoid the stressful situation. These strategies are often included under the emotion-focused strategies [92, 97].

9.5.3 Transactional Model

Lazarus and Folkman [68] developed the *Transactional Model,* or *cognitive-mediational approach* [46], which focuses its attention on the concept of appraisal in order to address stress and coping.

Cognitive appraisal is considered to be a universal mental process, by which the significance of what is occurring is being constantly assessed and related to one's well-being and to the available resources for responding to the situation. Therefore, it is not the stressful agent itself that defines stress, but a particular person's perception makes of a stressful situation.

Lazarus and Folkman [68] distinguish three types of appraisal: *primary*, in which the person assesses the meaning of what is taking place, and the result is what determines whether the situation is considered unimportant or stressful; *secondary*, referring to the assessment of one's own resources for dealing with the situation, implying a cognitive search for available coping options and a prognosis of whether each option will be successful or not in dealing with the stressor, and; *reappraisal*, involving feedback processes that are developed during the person's interaction with external or internal demands and bring about corrections to previous appraisals during the coping process itself, and so refers to the change made to a previous appraisal, based on new information received from the environment.

This model allows us to conceptually delimit coping to comprise constantly changing cognitive and behavioral processes that are developed in order to manage specific external and/or internal demands that are perceived as excessive or surpassing the individual's resources.

It is characterized by a set of responses that come into play in order to reduce the adverse qualities of a stressful situation, as an attempt to manage stressors. Brannon and Feist [98] underscores three aspects to be considered with respect to coping: (1) it is a process that changes depending on whether the subject has experienced successful results when dealing with the stressful situation; (2) it is not only an automatic or physiological response, but is also learned by experience; (3) it requires an effort to manage the situation and reestablish homeostasis or adapt to the situation.

In order to define coping, three concepts are key: (1) it is not necessarily a behavior that has been completely executed; the attempt or effort to carry it out may also be considered coping; (2) the effort may not necessarily be expressed in visible behaviors, it may also be cognitions; (3) the cognitive appraisal of the situation as challenging or threatening is a prerequisite to making attempts to cope.

The Transactional Model also includes the context in its definition of coping, that is, coping is seen as a process inserted within a context. Another important contribution from this model is the idea that strategies should not be judged as adaptive or maladaptive; the question is rather, for whom and under what circumstances a particular way of coping has adaptive consequences, instead of an indiscriminate categorization of adaptive vs. maladaptive strategies.

Navarro [99] determined that coping depends on a person's internal or external capacities for facing the demands of the potentially stressful event or situation, called coping resources.

These play an important role within the coping process, because they can influence the choice of coping strategies to be used. They can be differentiated as: (1) Physical and biological: including environmental elements and a person's organic resources such as climate, diet, the house where he or she lives, immune problems, etc.; (2) Psychological or psychosocial: encompassing everything from intellectual capacity to level of dependence or autonomy, beliefs, values and problem-solving skills; (3) Social resources: ranging from social skills to social support.

The concept of coping strategies has been studied at length in the field of mental health and psychopathology, principally in relation to stress, emotions and problem solving. In relation to stress, we can define them as a set of resources and efforts, both cognitive and behavioral, which are directed toward solving a problem, reducing or eliminating the emotional response or modifying the initial appraisal of the situation [68]. Whether one strategy or another is used will depend on the situation itself, the cognitive appraisal and perceived control, emotions and/ or physiological activation.

But there is a tendency to generalize their use and this is what we call coping style, that is, characteristic, relatively stable ways that people use to face stressful situations. Lazarus and Folkman [68] consider one distinction to be extremely important: the difference between coping that is directed toward handling or altering the problem, and coping aimed at regulating the emotional response that the problem brings about.

The first is referred to as *problem-focused coping* and the second as *emotion-focused coping* [66]. In general, the former is more likely to appear when the harmful or stressful conditions are appraised as subject to change. Emotion-focused strategies are more likely to appear when the appraisal indicates that nothing can be done to modify the threatening conditions of the environment. The two types of strategies are specified and analyzed in more detail below [68]:

1. *Emotion-focused* ways of coping: The literature mentions a large number of such ways of coping, but we can divide them into two large groups: (a) Cognitive processes dedicated to decreasing the degree of emotional discomfort, including strategies such as avoidance, minimization, distancing oneself, selective attention, positive comparisons and finding positive value in negative events; (b) Cognitive strategies that are directed toward increasing the degree of emotional discomfort; some persons need to feel really bad before they can come to feel better; in order to find comfort they need to first experience intense discomfort, from which they can then move on to some kind of self-punishment. In other cases, they deliberately increase their degree of emotional discomfort in order to push themselves to action, such as when athletes challenge themselves in order to compete.

2. *Problem-focused* ways of coping: These strategies are similar to those used for solving the problem; they are directed at the definition of the problem, the search for alternative solutions, consideration of these alternatives based on cost and benefit, and the selection and application of alternative(s). An objective is also involved, an analytical process directed mainly at the environment. However, these ways of coping also include strategies internal to the person. We can therefore speak of two main groups of problem-focused strategies: those that refer to the environment and seek to modify environmental pressures, obstacles, resources, procedures, etc.; and those that refer to the subject, including strategies dedicated to motivational or cognitive changes, changing

one's level of aspirations, reducing involvement of the ego, seeking different channels for gratification, developing new behavior patterns, or learning new resources and procedures.

Different factors make up these two broad dimensions: the quantity of factors and their names have evolved over time and through the different investigations [45, 68, 76].

9.6 Initial Assessment

9.6.1 Prediction Between Personal Self-regulation, Self-regulated Learning and Coping Strategies

Based on SEM analysis, a consistent structural linear model appeared [Chi-square = 58.842, degrees of freedom = 9, $p < 0.001$], showing relationships between the factors that make up personal self-regulation (goals, perseverance, decision process, and learning errors), and self-regulated learning and coping strategies (emotion- and problem-focused strategies), as it is shown in Fig. 9.5. The indices reveal this model's adequacy (NFI = 0.965; RFI = 0.902; IFI = 0.970; TLI = 0.907; CFI = 0.970, and RMSEA = 0.06), offering evidence that goals, perseverance, and learning from mistakes are predictors of self-regulated learning (SRL), and SRL is predictive of the combined use of emotion- and problem-focused coping strategies.

9.6.2 Interdependence Between Personal Self-regulation, Self-regulated Learning and Coping Strategies: Transactional Model

MANOVAs were carried out on a sample of university students in order to establish any interdependence relationships, with the result that different levels of *personal self-regulation* (low-medium-high) were accompanied by corresponding levels

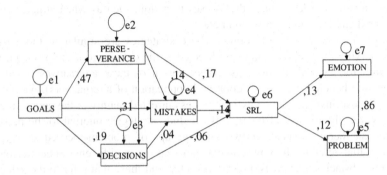

Fig. 9.5 Linear relations between components of personal regulation, self-regulated learning and coping strategies

Table 9.1 Interdependence between personal self-regulation and self-regulated learning

Self-regulated learning	Levels of personal self-regulation			Post
	1. Low	2. Medium	3. High	
	(n = 115)	(n = 179)	(n = 63)	
Planned learning	3.72 (0.48)	3.82 (0.70)	4.12 (0.55)	3 > 1, 2***
Meaningful learning	3.70 (0.64)	3.72 (0.72)	4.00 (0.62)	3 > 1, 2,*
Study techniques	3.95 (0.71)	4.05 (0.73)	4.36 (0.57)	3 > 1, 2,*

* $p < 0.05$; ** $p < 0.01$; *** $p < 0.001$

Table 9.2 Interdependence between levels of personal self-regulation and coping strategies

Coping strategies	Levels of personal self-regulation			Post
	1. Low	2. Medium	3. High	
	(n = 115)	(n = 179)	(n = 63)	
Help-seeking	3.72 (0.68)	3.81 (0.70)	4.12 (0.55)	3 > 1, 2***
Reduce anxiety/avoidance	3.70 (0.64)	3.72 (0.72)	4.00 (0.62)	3 > 2,*
Emot. venting/isolation	4.36 (0.73)	4.05 (0.64)	4.05 (0.57)	1 > 3,*

of self-regulated learning $F(6,582) = 3.03$ (Pillai test), $p < 0.01$, $eta^2 = 0.30$, see Table 9.1. Similarly, levels of personal self-regulation showed a significant main effect on some coping strategies, $F(6,582) = 3.03$ (Pillai test), $p < 0.01$, $eta^2 = 0.30$, see Table 9.2.

9.7 Applications: e-Assessment and e-Intervention Based on ICTs

Based on the well-established conceptual and empirical relationship, it is possible to carry out experiences for assisting university students through self-assessment and self-improvement, with a view to promoting proper levels of personal self-regulation, self-regulated learning, and better use of strategies for coping with academic stress. With this purpose in mind, an online utility has been created for e-assessment and self-help, under the name *e-Coping: Stress Management Tool for University* [100]. University students can log on and carry out a self-assessment of these variables through different standardized inventories. The utility provides immediate improvement feedback as a means of assisting the university students. Several examples of feedback are shown below in Tables 9.3, 9.4 and 9.5. The tables show a sequence of messages addressed to the user, followed by different kinds of feedback. These responses are given when students receive a low score on the variables discussed in this chapter.

Table 9.3 Feedback for students who are low in personal self-regulation [100]

Your score indicates that you should give some thought to this personal characteristic. Be aware you can train yourself and improve in this area. Recent research suggests that this personal characteristic contributes decisively to your accomplishments, whether personal, social or academic. Self-regulation is not a single behavior, but a combination of different groups of behaviors. We suggest that you analyze each of the aspects that make up self-regulation. This way you can improve on those specific aspects where you are most lacking

Feedback: Planning of goals. If you are low in planning, you should set out seriously to improve this aspect in your academic and personal life. Recent research has shown the importance of planning tasks well, since this is a predictor of optimal task execution afterward. This dimension refers to the ability to analyze the task you are about to do, and plan specific objectives for it, before performing the task itself

You can improve your planning by taking different steps:

• Analyze and break up the task into parts, defining their sequence

• List your reasons or motivations for doing it well

• Assign a number to represent how much you value this task, then decide how you can increase that value

• Establish specific objectives (goals) and the time needed for their execution

Feedback: Perseverance. If you are low in perseverance, it will be worth your while to work on this area and improve. Perseverance is defined as the ability to control yourself with willpower and keep up your motivation when executing a task. This is the skill that helps us to not give up on tasks when they require effort and sacrifice. Research has shown that perseverance is an essential characteristic of self-regulated people. It involves keeping up one's motivation and effort in order to carry out the task at hand. The following activities can help you to improve in perseverance:

• Keep up your motivation by cheering yourself on: "it's worth it to do a good job"

• While working, reinforce yourself by giving yourself positive messages: "very good!!!"

• Whenever you meet the timeframes or objectives you have set for yourself

• Don't give up on the task, even if it is difficult or complicated. Fight for your accomplishments

• Observe yourself while you work (keep a log of your behavior, if necessary)

Feedback: Decision making. If decision making is where you have a low score, resolve to improve in these behaviors. Decision making processes are fundamental while you are executing a task, in order to do a good job. These processes help you constantly evaluate and monitor your execution of a task. This way you can detect what is right, correct what is wrong, and ensure that you will meet the objectives you have set. Not making proper decisions while executing a task can mean mistakes or inadequate execution. The following activities can help you to improve:

• Observe how you go about performing the task

• Evaluate yourself and reflect on your progress: notice what you are getting right or wrong

• Make decisions about what you are doing well and what you ought to change

• Learn to give yourself instructions while you are executing a task

Feedback: Learning from mistakes. Perhaps learning from mistakes is where you have a low score. This set of behaviors refers to the ability to reflect after an experience or a completed task, to notice what you got right or wrong and learn from that. Learning from mistakes is very important and is characteristic of thoughtful persons. This skill increasing your likelihood of learning from experience and not making the same mistakes again

The following activities can help you improve this skill:

• Take pleasure in looking back over your behavior and analyzing it

• Take time each day to evaluate what you are getting right or wrong in your daily life

• After an experience, make a list of your accomplishments and mistakes

• Write up a resolution for improving your behavior in the future. Specific objectives for personal improvement are best

Table 9.4 Feedback for students who are low in self-regulated learning [100]

Your score should prompt you to reflect on *your perception of the teaching-learning process*. However, be aware that this score is specific to the teaching-learning process that you have just evaluated. It may vary according to the subject you are evaluating. This score can also be improved. Recent research suggests that academic achievement depends on contextual variables (the *teaching process*) and personal variables of how you go about learning (the *learning process*). On this questionnaire, the scores show: (1) your perception of how the teaching process is going; (2) your perception of how the learning process is going; (3) your satisfaction with the learning process; (4) your perception about having an achievement focus in learning

Perception of the teaching-learning process does not refer to a single behavior, but to a set of perceptions about the way your teacher teaches and the way you as a student are learning. Thus, this psychological construct refers to the specific way your teacher teaches this particular subject, plus how you yourself go about learning in this class subject

Perception of the teaching process refers to the extent that you feel your teacher's teaching activity is helpful and encourages proper learning of this subject. In other words, if he or she uses activities that prepare you for learning, if a specific syllabus is established and followed through, so that students can organize themselves, if the teaching method contributes toward students' self-regulation of the proposed learning activities, if the assessment system helps students identify the strong and weak points of their learning. In short, this represents your level of satisfaction with how your teacher teaches. As you can imagine, this information is important for *teachers*: if they know the overall view of their students (average ratings given by the class group, never by an individual), they can make improvements in how they teach and how students learn. In addition, your score can also help you as a *student* to become aware of your perception of your teacher. Research shows, for example, that a negative perception of the teaching produces demotivation toward learning in the student

Perception of a self-regulated learning process refers to your view of how you learn. Basically, this can help you understand how you usually go about learning, that is, whether you prepare yourself for learning, whether you use self-regulation strategies while learning and whether you use learning strategies. You can become aware of which aspects you use more and which you use less. Research has shown that a high level of self-regulation in the learning process is associated with a high level of performance, and the contrary also holds true

Satisfaction with the learning process refers to your perceived satisfaction with the way you have been learning, with the results of your learning, and with your performance in this subject. This aspect reflects your general satisfaction with how the subject is going and with your own learning behavior in that subject

Research has shown that students who receive poorer grades have less satisfaction with their learning process, and the contrary also holds true

Perception of an *achievement-focused learning process* refers to the extent that you perform learning behaviors for the exclusive purpose of achievement (grades), and not as much to learn well. This is called achievement-oriented learning. With this information you can make changes and improvements in the teaching-learning process

Feedback: Perception of the teaching process. If you have a low score in *perception of the teaching process*, this means that you are not satisfied with it. This may be due to different factors of how the teacher teaches. While this variable is not under your direct control, it can help you make suggestions to the teacher for making improvements, if he or she is open to that. However, it may also be that you have a tendency to look negatively on this type of teaching process, because of your own learning history

This negative perception may have led to a rather unsuccessful academic record in the recent past, because of the interference of stressful emotions that it causes. Perhaps you should learn to cope with this situation differently:

(continued)

Table 9.4 (continued)

• Analyze what aspects of the teaching process you notice that you perceive as negative, and the emotion that they produce (tension, nervousness, the urge to leave, anxiety, etc.)
• If the teacher gives you the opportunity, make constructive suggestions for improvement
• If you are unable to make any change in the teaching process, use self-talk to block out negative emotions
• Write down your self-instructions, put them in order and rate each one from 1 to 10, then begin to use them consistently
• Evaluate whether your level of stress has diminished. Be careful about avoidance strategies that may be harmful to you. Not all strategies are equally adequate
• If you cannot meet your objectives, seek professional help
Feedback: Perception of a self-regulated learning process. If you score low in *perception of a self-regulated learning process*, this means that you make little use of the learning behaviors you have just evaluated in yourself. These learning behaviors are important because they help you to learn well and to obtain good academic outcomes. Research has suggested that practicing self-regulated learning is essential to building a good knowledge base, being effective and properly managing the teacher's demands for learning
Therefore, you should take note that your learning process is less than adequate, and this can be a stress factor in itself. It is also likely that you have had only low to moderate levels of success in your recent learning history. It is appropriate for you to think about improving your learning process as a way to help you manage high-level learning situations, typical of university studies
The following activities can help you improve your learning process:
• Recall and analyze the learning behaviors that you use the least (activities to prepare you for learning, strategies for learning and study, self-assessment, study techniques, etc.)
• Analyze why you do not usually make good use of these behaviors. It is probably because you have not understood the importance of these types of learning behaviors
• Make a list of the learning behaviors that you are going to improve, and use self-talk to carry them out
• After trying these behaviors, put them in order and score them from 1 to 10 in how important and effective they are for you
• Check whether your stress level has diminished. For each person there are certain learning strategies that work best. Find which ones are best for you
• If you do not meet your objectives, seek professional help
Feedback: Satisfaction with the learning process. If you score low on *satisfaction with the learning process,* this means that you are not satisfied with how you are learning or with your achievement in this subject. The problem with lack of satisfaction is that it leads to demotivation in future teaching-learning processes related to the same teacher or the same subject matter. Research has suggested that being satisfied is a *positive emotion* that appears either during or after completing a task, and provides motivation for the next steps in learning. Therefore, you should be aware that this *lack of satisfaction with learning* (or lack of positive emotionality) can be a stress factor in itself. It is also likely that you have had only low to moderate levels of success in your recent learning history. It is appropriate for you to think about improving your satisfaction with the learning process, as you work to improve your manner of learning and your level of achievement, while the teacher is improving his or her manner of teaching. However, since the latter is not under your control, you should focus on satisfaction with the learning process, which is up to you

(continued)

Table 9.4 (continued)

The following activities can help you improve your learning process:
• Recall and analyze the aspects of satisfaction where you score the lowest (way of learning, meaningful learning, usefulness of what has been learned, achievements gained, enjoyment of learning, etc.)
• Analyze why you are dissatisfied with these aspects. You have probably not realized how important these aspects are for your learning process
• Make a list of the behaviors that are likely to help you improve your satisfaction with learning (way of learning, meaningful learning, usefulness of what has been learned, achievements gained, enjoyment of learning, etc.), and use self-talk to help you carry them out
• After trying these behaviors, put them in order and rate them from 1 to 10 in how important and effective they are for you
• Check whether your stress level has diminished. For each person there are certain learning strategies that work best. Find which ones are best for you
• If you do not meet your objectives, seek professional help
Feedback: Perception of an achievement focus in learning. If you are low in *perception of an achievement focus in learning*, this means that you seldom use learning behaviors aimed only at obtaining good academic outcomes (grades). Such learning behaviors, which you have just evaluated in yourself, are important in that they can trigger unproductive stress. Research has suggested that staying away from *exclusively achievement-focused learning* can help students have better *self-regulated learning*, and not be primarily worried about grades when learning. For this reason, your low score on this factor is adequate in helping you to experience less stress while learning
An appropriate response would be to try to maintain this level, while you seek to increase your self-regulated learning (D2)
This will help you enjoy the learning process, without the pressure of grades as your priority goal in learning
The following steps can help you take this approach:
• Recall and analyze in what situations you may adopt achievement-focused behaviors (thinking about grades while you are studying, evaluating yourself based on the test situation, starting off with a grade target from the beginning, etc.)
• Analyze why you adopt an achievement focus in these cases. You probably have not realized that these behaviors can produce unnecessary stress
• Make a list of the self-regulated learning behaviors that you are going to improve, and use self-talk to help you carry them out
• After trying these behaviors, put them in order and score them from 1 to 10 in how important and effective they are for you
• Check whether your stress level continues to be low. For each person there are certain learning strategies that work best. Find which ones are best for you
• If you do not meet your objectives, seek professional help

Table 9.5 Feedback for students who are low in coping strategies [100]

Your score should prompt you to reflect on how you cope with learning. Be aware that you can work on this and change. You should aim to increase certain coping behaviors. Recent research suggests that coping strategies contribute to academic achievement. Coping strategies refer to behaviors that people practice in order to manage and face stressful situations, in this case, academic stress. Coping strategies are not a single behavior, but a group of different types of behaviors. This psychological construct refers to how a person usually faces stressful situations. There are two main types of coping:

Emotion-focused coping refers to behaviors used for managing negative emotions produced by the stressful situation (anxiety, tension, irritability, etc.). This type of behavior, in turn, can take different shapes, such as *fantasy distraction, help-seeking for taking action, religious support, reducing anxiety and avoidance, preparing oneself for the worst, emotional venting and isolation* and *resignation*. These types of coping strategies help the person to minimize negative emotions; however, they do not help to solve the problem itself

Problem-focused coping is used to solve the problem that created the stress, or to minimize it if it cannot be solved. Different types of behaviors are included here: *seeking help and counsel, actions directed at the cause, self-instructions, positive reassessment and firmness, seeking support in others,* and *seeking alternative reinforcement*. Problem-focused coping helps solve the problem, but it does not help manage negative emotionality, at least not directly

Depending on the situation and the person, the two types of strategies can be used together. We suggest that you analyze the types of behaviors that you use, in order to make improvements. You can begin by working to improve the specific aspects where you are weakest

Feedback: Emotion-focused coping. If you are low in *emotion-focused coping*, be aware that you may have problems in managing negative emotions. Your low score can mean different things: (1) you make more use of problem-focused strategies; (2) you are not accustomed to managing your emotions; (3) you have little stress and do not need to manage negative emotions

This personal characteristic may have led to a rather unsuccessful academic record in the recent past, due to interference from stress-related emotions. You can improve by taking different steps:

• Analyze whether you have negative, stress-related emotions in academic situations (tension, nervousness, the urge to leave, anxiety, etc.)

• Analyze what you usually do and why you do not practice managing your emotions. It is probably because you have not realized the importance of these types of emotion-focused coping behaviors

• Make a list of emotion-focused coping strategies and use self-talk to start using them

• After trying these behaviors, put them in order and rate each one on a scale of 1–10, and begin to use them in priority

• Check whether your stress level has diminished. Be careful with emotion-focused strategies that may be harmful to your health. Not all strategies are equally adequate

• If you do not meet your objectives, seek professional help

Feedback: Problem-focused coping. If you are low in *problem-focused strategies,* you should know that remaining inactive in the face of problems does not help them be solved. It is appropriate to think about increasing your use of these types of strategies, which are adaptive and will help you manage situations that cause stress in the academic context

The following activities can help you increase your use of problem-focused strategies:

• Analyze the problems that cause stress in your academic context (work overload, excessive demands, pressures for grades, tight schedules, sustained effort, etc.)

• Analyze what you usually do to face these situations and why you do little to manage problems. It is probably because you have not realized the importance of these types of problem-focused coping behaviors

• Make a list of problem-focused coping strategies and use self-talk to start using them

(continued)

Table 9.5 (continued)

• After trying these behaviors, put them in order and rate each one on a scale of 1–10, and begin to use them in priority
• Check whether your stress level has diminished. For each person there are certain learning strategies that work best. Find which ones are best for you
• If you do not meet your objectives, seek professional help

9.8 Conclusions

Recent research has found linear associations and non-linear interdependence relationships between self-regulation (as a personal presage variable), self-regulated learning (as a meta-cognitive, process variable) and coping strategies (as a meta-motivational, meta-affective process variable) in university students experiencing academic stress [16]. However, these results should be confirmed with new studies that offer further consistency in establishing: (1) The importance of *personal self-regulation*, as an individual variable that determines the degree of *cognitive self-regulation* during the process of university learning. (2) The relationship between *personal self-regulation* and the type and quantity of *coping strategies*, where prior evidence has shown a significant positive relationship with problem-centered strategies, and a significant negative relationship with emotion-centered strategies. (3) The relationship between *self-regulated learning* and *coping strategies;* consistently with this evidence, some of the results found here show a stronger relationship with problem-centered strategies.

Zimmerman and Labuhn [47] have proposed the following directions for future work with regard to self-regulated learning. First, a clear difference should be established between self-regulated learning and self-regulation in performance, especially in adverse situations. The second problem would be to define the relationship between automated and meta-cognitive processes when learning, especially in expert individuals; in other words, how relatively automated cognitive processes relate to meta-cognitive processes needs to be clarified. Third, the dichotomy between objective and subjective measurements of self-regulated learning needs to be resolved, through an increase in online, real-time assessment processes.

In addition to these measurements, the role of meta-motivational and meta-affective variables (personal self-regulation) should be incorporated into the study of meta-cognitive processes. The present chapter seeks to address this relationship.

In addition, this chapter has shown how it is possible to work with university students using online tools for self-assessment and self-improvement in these psychological variables. This improvement refers to students explicitly improving their meta-motivational and meta-affective processes, as part of their meta-cognition.

Acknowledgments R&D Project ref. EDU2011-24805 (2012–2014). MICINN (Spain) and FEDER Fund (Europe). Motivational-affective strategies for personal self-regulation and coping with stress, in the teaching-learning process at university. http://www.estres.investigacion-psicopedagogica.com/english/seccion.php?idseccion=1.

References

1. de la Fuente, J., Zapata, L., Martínez-Vicente, J.M., Cardelle-Elawar, M., Sander, P., Justicia, F., Pichardo, M.C., García-Berbén, A.C.: Regulatory teaching and self-regulated learning in college students: confirmatory validation study of the IATLP scale. Electron. J. Res. Educ. Psychol. **10**(2), 839–866 (2012)
2. Rosário, P., Lourenço, A., Paiva, M.O., Núñez, J.C., González-Piedra, J., Valle, A.: Autoeficacia y utilidad percibida como condiciones necesarias para un aprendizaje académico autorregulado. (Self-efficacy and perceived utility as necessary conditions for self-regulated academic learning.) Anales de Psicología **28**(1), 37–44 (2012)
3. Elvira-Valdés, M.A., Pujol, L.: Autorregulación y rendimiento académico en la transición secundaria–universidad. (Self-regulation and academic performance in the secondary-to-university transition.) Revista Latinoamericana de Ciencias Sociales, Niñez y Juventud **10**(1), 367–378 (2012)
4. Torrano, F., González, M.C.: Self-regulated learning: current and future directions. Electron. J. Res. Educ. Psychol. **2**(1), 1–34 (2004)
5. de la Fuente, J., Peralta, F.J., Sánchez, M.D.: Autorregulación personal y la percepción de los comportamientos escolares desadaptativos. (Personal self-regulation and the perception of disadaptive school behaviors.). Psicothema **21**(4), 548–554 (2009)
6. Elliot, A.J., Dweck, C.S.: Handbook of Competence and Motivation. Guilford University Press, New York (2007)
7. Entwistle, N.J., Peterson, E.R.: Conceptions of learning and knowledge in higher education: relationships with study behavior and influences of learning environments. Int. J. Educ. Res. **41**(3), 516–623 (2004)
8. Pekrun, R., Goetz, T., Frenzel, A.C., Barchfeld, P., Perry, P.R.: Measuring emotions in students' learning and performance: The Achievement Emotional Questionarie (AER). Contemp. Educ. Psychol. **36**, 36–48 (2011)
9. Salmerón, H., Gutiérrez, C., Salmerón, P., Rodríguez, S.: Metas de logro, estrategias de regulación y rendimiento académico en diferentes estudios universitarios. (Achievement goals, regulation strategies and academic performance in different university studies.) Revista de Investigación Educativa **29**(2), 467–486 (2011)
10. de la Fuente, J., Justicia, F.: The DEDEPRO model for regulating teaching and learning: recent advances. Electron. J. Res. Educ. Psychol. **13**(5), 535–564 (2007)
11. Buff, A., Reuser, K., Rakoczy, K., Pauli, C.: Activation positive affective experiences in the classroom: "Nice to have" or something more? Learn. Instr. **21**, 452–466 (2011)
12. de la Fuente, J., Justicia, F., Casanova, P., Trianes, M.V.: Perception about construction of academic and professional competences in psychologists. Electron. J. Res. Educ. Psychol. **3**(1), 33–34 (2004)
13. DurlaK, J.A., Weissberg, R.P., Dymnicki, A.B., Taylor, R.D., Schellinger, K.B.: The impact of enhancing students´ social and emotional learning: a meta-analysis of school-based universal intervention. Child Dev. **82**, 405–432 (2011)
14. Biggs, J.: Calidad del aprendizaje universitario. (Quality of University Learning). Narcea, Madrid (2005)
15. de la Fuente, J.: Implications for the DEDEPRO model for interactive analysis of the teaching-learning process in higher education. In: Teixeira, R. (ed.) Higher Education in a State of Crisis, pp. 205–222. Nova Science Publisher, New York (2011)
16. Zapata, L.: Personal Self-Regulation, Learning Approach and Coping Strategies in university teaching learning process with stress. Doctoral Dissertation unpublisher. University of Almería, Almería (2013)
17. Vohs, K.D., Baumeister, R.F.: Handbook of Self-Regulation: Research. Theory and Applications. The Guilford Publications Inc., New York (2011)
18. Kanfer, F.J.: Implications of a self-regulation model therapy for treatment of addictive behaviours. In: Miller, W.R., Heather, N. (eds.) Treating Addictive Behaviors: Process of Change, pp. 121–138. Plenum Press, New York (1986)

19. Bandura, A.: Social cognitive theory: a genetic perspective. Am. Rev. Psychol. **52**, 1–26 (2001)
20. Miller, W.R., Brown, J.M.: Self-regulation as a conceptual basis for the prevention and treatment of addictive behaviours. In: Heather, N., Miller, W.R., Greely, J. (eds.) Self-control and the addictive behaviours, pp. 3–79. Maxwell Macmillan, Sydney (1991)
21. Brown, J.M.: Self-regulation and the addictive behaviours. In: Miller, W.R., Heather, N. (eds.) Treating Addictive Behaviors, 2nd edn, pp. 61–73. Plenum Press, New York (1998)
22. Zimmerman, B.J.: Developing self-fulfilling cycles of academic regulation: An analysis of exemplary instructional model. In: Schunk, D.H., Zimmerman, B.J. (eds.) Self-Regulated Learning: From Teaching to Self-Reflective Practice, pp. 1–19. Guilford, New York (1998)
23. Hoyle, R.H.: Handbook of Personality and Self-Regulations. Wiley, Oxford (2010)
24. de la Fuente, J., Trianes, M.V., Peralta, F.J., Sánchez, M.D.: Autorregulación personal y comportamientos de inadaptación sociopersonal y escolar. (Personal self-regulation and disadaptive socio-personal and school behaviors). I Congreso Internacional de Violencia Escolar (Bullying). University of Almería, Almería (2007)
25. de la Fuente, J., Berbén, A.B.G., Martínez, J.M.: La autorregulación personal de los estudiantes universitarios. (Personal self-regulation of university students.) In: Benítez, J.L., Berbén, A.B.G., Justicia, F., de la Fuente, J. (eds.) La Universidad ante el Reto del Espacio Europeo de Educación Superior: Investigaciones Recientes, pp. 68–88, EOS, Madrid (2006)
26. Muraven, M., Collins, R.L., Morsheimer, E.T., Shiffman, S., Paty, J.A.: The morning after: limit violations and the self-regulation of alcohol consumption. Psychol. Addict. Behav. **19**(3), 253–262 (2005)
27. Muraven, M., Collins, R.L., Nienhaus, K.: Self-control and alcohol restraint: an initial application of the self-control strength model. Psychol. Addict. Behav. **16**, 113–120 (2002)
28. Brown, J.M., Miller, W.R., Lawendowski, L.A. The self-regulation questionnaire. In: VandeCreek, L, Jackson, T.L., (eds.) Innovations in clinical practice: A sourcebook, vol. 17, pp. 281–292. Professional Resource Press/Professional Resource Exchange, Sarasota (1999)
29. Pichardo, M.C., Justicia, F., García-Berbén, A.B., de la Fuente, J., Martínez-Vicente, J.M.: Psychometric analysis of the self-regulation questionnarie (SRQ) at spanish universities. Span. J. Psychol. **17**, e62, 1–8 (2014)
30. Carey, K.B., Neal, D.J., Collins, S.E.: A psychometric analysis of self-regulation questionnaire. Addict. Behav. **29**, 253–260 (2004)
31. Neal, D.J., Carey, K.B.: A follow-up psychometric analysis of the self-regulation questionnaire. Psychol. Addict. Behav. **19**(4), 414–422 (2005)
32. Okun, M.A., Levy, R., Karoly, P., Ruehlman, L.: Dispositional happiness and college student GPA: unpacking a null relation. J. Res. Pers. **43**, 711–715 (2009)
33. Kogan, S.M., Bordy, G.H.: Linking parenting and informal mentor processes to depressive symptoms among rural african american young adult men. Cult. Divers. Ethn. Minor. Psychol. **16**(3), 299–306 (2009)
34. Creed, P.A., Fallon, T., Hood, M.: The relationship between career adaptability, person and situation variables, and career concerns in young adults. J. Vocat. Behav. **74**, 219–229 (2009)
35. Boekaerts, M., Maes, S., Karoly, P.: Self-regulation across domains of applied psychology: Is there an emerging consensus? Appl. Psychol. Int. Rev. **54**(2), 149–154 (2005)
36. Karoly, P., Boekaerts, M., Maes, S.: Toward consensus in the psychology of self-regulation: how far have we come? How far do we have yet to travel? Appl. Psychol. Int. Rev. **54**(2), 300–311 (2005)
37. Vancouver, J.B., Scherbaum, C.A.: Do we self-regulate actions or perceptions? A test of two computational models. Comput. Math. Organ. Theor. **14**, 1–20 (2008)
38. Latham, G.P., Locke, E.A.: New developments in and directions for goal-setting research. Eur. Psychol. **12**(4), 290–300 (2007)
39. Winne, P.H.: Comments on motivation in real-life, dynamic and interactive learning environments. Eur. Psychol. **9**(4), 257–264 (2004)
40. de la Fuente, J., Cardelle-Elawar, M: Personal self-regulation and coping style in university students. In: L.B., Nichelson, R.A. (eds.) Psychology of Individual Differences, pp. 171–182, Nova Science Publishers Inc., New York (2011)

41. Tangney, J.P., Baumeitser, R.F., Boone, A.L.: High self-control predicts good judgment, less pathology, better grades, and interpersonal success. J. Pers. **72**, 271–322 (2004)
42. Ferrari, J.R., Stevens, E.B., Jason, L.A.: The role of self-regulation in abstinence mainte-nance: effects of communal living on self-regulation. J. Groups Addict. Recovery **4**, 32–41 (2009)
43. de la Fuente, J.: Cuestionario de Autorregulación personal (CAR). [Personal self-regulation questionnaire (SRQ)]. Unpublished manuscript (2003)
44. Chorot, P., Sandín, B.: Escalas de Estrategias de Coping. (Scales of Coping strategies). UNED, Madrid (1987)
45. Chorot, P., Sandín, B.: Escalas de Estrategias de Coping revisado (EEC-R). (Scales of revised Coping strategies). UNED, Madrid (1993)
46. Sandín, B., Chorot, P.: Cuestionario de Afrontamiento del Estrés (CAE): desarrollo y vali-dación preliminar. (Questionnaire on Coping with Stress: Preliminary development and vali-dation.) Revista de Psicopatología y Psicología Clínica, **8**(1), 39–54 (2003)
47. Zimmerman, B.J., Labuhn, A.S.: Self-regulation of learning: process approaches to personal development. In: Harris, K.R., Graham, S., Urdan, T. (eds.) The Educational Psychology Handbook, Volume 1: Theories, Constructs, and Critical Issues, pp. 399–425. American Psychological Association, Washington (2012)
48. Zimmerman, B.J.: Models of self-regulated learning and academic achievement. In: Zimmerman, B.J., Schunk, D.H. (eds.) Self-Regulated Learning and Academic Achievement: Theory, Research and Practice, pp. 1–25. Springer, New York (1989)
49. Carver, C.S., Scheier, M.F.: On the structure of behavioral self-regulation. In: Boekaerts, M., Pintrich, P.R., Zeidner, M. (eds.) Handbook of Self-Regulation, pp. 41–84. Academic Press, New York (2000)
50. Zimmerman, B.J.: Becoming a self-regulated learner: an overview. Theory Pract. **41**(2), 64–70 (2002)
51. Abar, B., Loken, E.: Self-regulated learning and self-directed study in a pre-college sample. Learn. Individ. Differ. **20**, 25–29 (2010)
52. Cerezo, R., Núñez, J.C., Rosário, P., Valle, A., Rodríguez, S., Bernardo, A.B.: New media for the promotion of self-regulated learning in higher education. Psicothema **22**(2), 306–315 (2010)
53. Cleary, T.J.: Emergence of self-regulated learning microanalysis: Historical overview, essen-tial features, and implications for research and practice. In: Zimmerman, B.J., Schunk, D.H. (eds.) Handbook of Self-Regulation of Learning and Performance, pp. 329–345. Routledge, New York (2011)
54. de la Fuente, J., Lozano, A.: Design of the SEAI self-regulation assessment for young chil-dren and ethical considerations of psychological testing. In: Dettori, G., Persico, D. (eds.) Fostering Self-Regulated Learning through ICT, pp. 39–53. IGI Global, Hershey (2011)
55. Järvelä, S., Hadwin, A.F.: New frontiers: regulating learning in CSCL. Educ. Psychol. **48**, 25–39 (2013)
56. Papantoniou, G., Moraitou, D., Katsadima, E., Dinou, M.: Control de la acción y dis-posición a la esperanza: Un estudio de su incidencia en la autorregulación del aprendizaje. (Action control and dispositional hope: a study of their effect on self-regulated learning.). Electron. J. Res. Educ. Psychol. **8**(1), 5–32 (2010)
57. Escorcia, D.: Conocimientos metacognitivos y autorregulación: una lectura cualitativa del funcionamiento de los estudiantes universitarios en la producción de textos. (Metacognitive Knowledge and self-regulation: A qualitative lecture of students functioning in writing). Avances en Psicología Latinoamericana, **28**(2), 265–277 (2010)
58. Mauri, T., Colomina, R., Martínez, C., Rieradevall, M.: La adquisición de las competencias de autorregulación. Análisis de su concepción y aprendizaje en diferentes estudios universitarios. (Acquisition of self-regulation competencies. Analysis of their conception and learning in dif-ferent university studies.) Revista d' Innovacio' i Recerca en Educació, **2**, 33–60 (2009)
59. Mullen, C.A.: Facilitating self-regulated learning using mentoring approaches with doc-toral students. In: Zimmerman, B.J., Schunk, D.H. (eds.) Handbook of Self-Regulation of Learning and Performance, pp. 137–152. Routledge, New York (2011)

60. Gaeta, M.T., Tervel, M.P., Orejudo, S.: Aspectos motivacionales, volitivos y metacognitivos del aprendizaje autorregulado. (Aspects of motivational, volitional and Metacognitive of self-regulated learning). Electron. J. Res. Educ. Psychol. **10**(1), 73–94 (2012)
61. de la Fuente, J., Martínez-Vicente, J.M., Peralta, F.J., García-Berbén, A.B.: Percepción del proceso de enseñanza-aprendizaje y rendimiento académico, en diferentes contextos instruccionales de la Educación Superior. (Perception of the teaching-learning process and academic performance in different instructional contexts of Higher Education.). Psicothema **22**(4), 806–812 (2010)
62. Tavakolizadeh, J., Qavam, S.E.: Effect of teaching of self-regulated learning strategies on attribution styles in students. Electron. J. Res. Educ. Psychol. **9**(3), 1087–1102 (2011)
63. Phan, H.P.: Critical thinking as a self-regulatory process component in teaching and learning. Psicothema **22**(2), 284–292 (2010)
64. Alonso-Tapia, J., Panadero, E.: Effects of self-assessment scripts on self-regulation and learning. Infancia y Aprendizaje **33**(3), 385–397 (2010)
65. de la Fuente, J. (Coord.): Motivational-affective strategies for personal self-regulation and coping with stress, in the teaching-learning process at university. R & D Project ref. EDU2011-24805 (2012–2014). MICINN (Spain) and FEDER Found (Europe). (2012–2014) http://www.estres.investigacion-psicopedagogica.com/english/seccion.php?idseccion=1
66. Dunahoo, C.L., Hobfoll, S.E., Monnier, J., Hulsizer, M.R., Johnson, R.: There's more than rugged individualism in coping. Part 1: Even the lone ranger had tonto. Anxiety Stress Coping **11**, 137–165 (1998)
67. Núñez, J., Solano, P., González-Piedra, J., Rosário, P.: Evaluación de los procesos de autorregulación mediante autoinformes. (Assessing self-regulation processes through self-reports.). Psicothema **18**(3), 353–358 (2006)
68. Lazarus, R.S., Folkman, S.: Estrés y Procesos Cognitivos. (Stress and Cognitive Process). Ediciones Martínez Roca, S.A., Barcelona (1986)
69. Cabanach, R.G., Valle, A., Suárez, J.M., Fernández, A.: Diferencias en los componentes cognitivo y afectivo-motivacional entre distintos niveles de aprendizaje autorregulado en estudiantes universitarios. (Differences in cognitive and afectivo-motivacional components between different levels of learning self-regulated in university students). Bordón, **52**(4), 537–552 (2000)
70. Pintrich, P.R.: The role of goal orientation in self-regulated learning. In: Boekaerts, M., Pintrich, P.R., Zeidner, M. (eds.) Handbook of Self-Regulation, pp. 451–502. Academic Press, San Diego (2000)
71. Long, B.: Stress management for school personnel: stress-innoculation. Training and exercise. Psychol. Sch. **25**, 314–324 (1998)
72. Carver, C.S., Scheider, M.F., Weintraub, J.K.: Assessing coping strategies: a theoretically based approach. J. Pers. Soc. Psychol. **56**, 267–283 (1989)
73. DeLongis, A., Holtzman, S., Puterman, E., Lam, M.: Dyadic coping: support from the spouse in times of stress. In: Davila, J., Sullivan, K. (eds.) Social Support Processes in Intimate Relationships, pp. 151–174. Oxford Press, New York (2010)
74. Schuler, R.: Integrative transactional process model of coping with stress in organizations. In: T.A.B.R.S. Beehr (ed.) Human Stress and Cognition in Organizations: An integrated Perspective, pp. 211–224. Wiley, New York (1985)
75. Holroyd, K.A., Lazarus, R.S.: Stress, coping, and somatic adaptation. In: Goldberger, L., Breznitz, S. (eds.) Handbook of Stress: Theoretical and Clinical Aspect, pp. 124–140. Free Press, New York (1982)
76. Lazarus, R.S.: Emotion and Adaptation. Oxford University Press, New York (1991)
77. Hobfoll, S.E., Schröeder, K.E.E.: Distinguishing between passive and active prosocial coping: Bridging inner-city women's mental health and AIDS risk behavior. J. Soc. Person. Relat. **18**, 201–217 (2001)
78. Hampa, P., Petermann, F.: Age and gender effects on coping in children and adolescents. J. Youth Adolesc. **34**(2), 73–83 (2005)

79. Ader, E., Erktin, E.: Coping as self-regulation of anxiety: A model for math achievement in high stakes tests. Cogn. Brain Behav. Interdiscip. J. **14**(4), 311–332 (2010)
80. de la Fuente, J., Zapata, L., Martínez-Vicente J.M., Sander, P., Cardelle-Elawar, M.: Relationship between personal self-regulation, resilience and strategies for coping with stress at university. In: CIEAE 2013, Institute of Education, July, 15–18, Lisboa (2013)
81. Kirchner, T., Forns, M., Amador, J.A., Muñoz, D.: Stability and consistence of coping in adolescence: a longitudinal study. Psicothema **22**(3), 382–388 (2010)
82. El-Ghoroury, N.H., Galper, D.I., Sawaqdeh, A., Bufka, L.F.: Stress, coping and barriers to wellness among psychology graduate students. Training Educ. Prof. Psychol. **6**(2), 122–134 (2012)
83. Krzemien, D., Monchietti, A., Sebastian, U.: Afrontamiento activo y adaptación al envejecimiento en mujeres de ciudad de Mar de Plata: Una revisión de la estrategia de autodistracción. (Activing coping and adapting to aging in women from the city of Mar de Plata: a review of the self-distraction strategy.) Interdisciplinaria **22**(2), 183–210 (2005)
84. Boujut, E.: Development and validation of an exploratory measure to assess student coping: The student coping scale. Coll. Student J. **47**(1), 12–23 (2013)
85. Gantiva, C.A., Viveros, A.L., Dávila, A.M., Salgado, M. J.: Estrategias de afrontamiento en personas con ansiedad. (Coping strategies in persons with anxiety.) Psychologia: Avances de la Disciplina **4**(1), 63–70 (2010)
86. Castellanos, M.T., Guarnizo, C.A., Salamanca, Y.: Relación entre niveles de ansiedad y estrategias de afrontamiento en practicantes de psicología de una universidad colombiana. (Relationship between levels of anxiety and coping strategies in a Colombian university psychology practitioners). Int. J. Psychol. Res. **4**(1), 50–57 (2011)
87. Gaviria, A.M., Vinaccia, S., Riveros, M.F., Quiceno, J.M.: Calidad de vida relacionada con la salud, afrontamiento del estrés y emociones negativas en pacientes con cáncer en tratamiento quimioterapéutico. (Quality of life related to health, coping with stress and negative emotions in cancer patients undergoing chemotherapy treatment.). Psicología desde el Caribe. Universidad del Norte **20**, 50–75 (2007)
88. Marqués, A., Lima, M.L., Lopes da Silva, A.: Fuentes de estrés, burnout y estrategias de coping en profesores portugueses (Stress sources, burnout and coping among portuguese teachers). Revista de Psicología del Trabajo y de las Organizaciones **21**(1–2), 125–143 (2005)
89. Revenson, T.A., DeLongis, A.: Couples coping with chronic illness. In: Folkman, S. (ed.) Oxford Handbook of Coping and Health, pp. 101–123. Oxford Press, New York (2010)
90. Piemontesi, S.E., Heredia, D.E.: Afrontamiento ante exámenes: Desarrollo de los principales modelos teóricos para su definición y medición. (Coping with exams: development of the main theoretical models for its definition and measurement.) Anales de Psicología **25**(1), 102–111 (2009)
91. Cohen, M., Ben-Zur, H., Rosenfeld, M.J.: Sense of coherence, coping strategies, and test anxiety as predictor of test performance among college students. Int. J. Stress Manage. **15**(3), 289–303 (2008)
92. de la Fuente, J., Cardelle-Elawar, M., Martínez-Vicente, J.M., Zapata, L., Peralta, F.J.: Gender as a determining factor in the coping strategies and resilience of university students. In: Haumann, R., Zimmer, G. (eds.) Handbook of Academic Performance, pp. 205–217. Nova Science Publishers Inc, New York (2013)
93. Ticona, S.B., Paucar, G., Llerena, G.: Stress level and coping strategies in nursing students at the UNSA Faculty. Revista Electrónica Cuatrimestral de Enfermería **9**(2), 1–8 (2010)
94. Endler, N.S., Kantor, L., Parker, D.A.: State-trait coping, state-trait anxiety and academic performance. Person. Individ. Differ. **16**, 663–670 (1994)
95. Fernández-Abascal, E.G.: Estilos y estrategias de afrontamiento (Styles and coping strategies.) In: Fernández-Abascal, E.G., Palmero, F., Chóliz, M., Martínez, F. (eds.) Cuaderno de Prácticas de Motivación y Emoción (Practice Notebook Motivation and Emotion.) Pirámide, Madrid (1997)

96. Zeidner, M.: Adaptive coping with test situations: a review of literature. Educ. Psychol. **30**, 123–133 (1995)
97. Berjot, S., Gilt, N.: Stress and coping with discrimination and stigmatization. Front. Psychol. **2**, 1–13 (2011)
98. Brannon, L., Feist, J.: Comprensión del estrés y la enfermedad (Understanding stress and illness.). In: Brannon, L., Feist, J. (eds.) Psicología de la Salud, pp. 177–211. Paraninfo Thomson Learning, Madrid (2001)
99. Navarro, M.: Acercamiento al estrés en una muestra de estudiantes de medicina. (Versión electrónica) Revista del Instituto Superior de Ciencias Médicas de la Habana, **15**, 25–32 (2000)
100. de la Fuente, J.: e-Coping Stress Management Tool for University™. Almería (Spain): University of Almería. Land Registry AL-288635 (2014) (http://www.estres.investigacion-psicopedagogica.com/english/seccion.php?idseccion=7)

Part IV
Approaches

Chapter 10
What Makes Metacognition as Socially Shared in Mathematical Problem Solving?

Tarja-Riitta Hurme, Sanna Järvelä, Kaarina Merenluoto and Pekka Salonen

Abstract Mechanisms making metacognition as differentiator for successful collaborative problem solving has become an important topic for self-regulated and collaborative learning research. Recent empirical research on socially shared metacognition has examined the role of metacognition in technology supported learning situations. However, detailed research on operationalizing metacognition in collaborative learning remains scarce. Advancing understanding regarding socially shared metacognition in learning, the mechanisms and processes that advance the use of metacognition beyond the individual level to the group level should be defined more precisely. In this article, the aim is to contribute to the ongoing discussion and advance empirical understanding what makes metacognition as socially shared when pre-service primary teachers triads' (N = 18) collaborative mathematical problem solving is supported by an asynchronous and text-based WorkMates (WM) learning environment.

Keywords Computer-supported collaborative learning · Mathematics · Metacognition · Socially shared metacognition · Problem solving

T.-R. Hurme (✉) · K. Merenluoto · P. Salonen
Department of Teacher Education, Centre for Learning Research,
University of Turku, 20014 Turku, Finland
e-mail: tarja-riitta.hurme@utu.fi

K. Merenluoto
e-mail: kaarina.merenluoto@utu.fi

P. Salonen
e-mail: pekka.salonen@utu.fi

T.-R. Hurme · S. Järvelä
Department of Educational Sciences and Teacher Education, Learning and Educational
Technology Research Unit, Faculty of Education, University of Oulu,
P.O. Box 2000, 90014 Oulu, Finland
sanna.jarvela@oulu.fi

© Springer International Publishing Switzerland 2015
A. Peña-Ayala (ed.), *Metacognition: Fundaments, Applications, and Trends*,
Intelligent Systems Reference Library 76, DOI 10.1007/978-3-319-11062-2_10

Abbreviations

ANOVA   Analysis of variance
WM   WorkMates
SAGA   Students' appraisals of group assessment

10.1 Introduction

Metacognition has been recognized as an important part of individual's learning [1] and mathematical problem solving [2]. According to Flavell [3] and Brown [4], metacognition refers to knowledge of cognition, for example, "I know I have solved this kind of problem before", and regulation of cognition, for example, "I have to check if my result meets the task requirements". Recently researchers in the field of self-regulated learning have been interested in examining the role of metacognition in collaborative learning situations where metacognition could be seen as a socially shared phenomena [5, 6].

Most of the empirical research has focused on describing and operationalizing socially shared metacognition with various conceptualizations [7]. The findings suggest that in a pair's computer-supported problem solving, metacognition can be seen as a shared process and as a facilitator for peer thinking [8]. In addition, group and team learning studies have shown that effective knowledge construction requires group members to collectively monitor and control joint problem solving as regarding to the task requirements [9].

In all, research has recognized socially shared metacognition as an important component of collaborative problem solving; however, detailed description of what makes metacognition as socially shared metacognition remains still scarce and more is needed to understand about the mechanisms and processes which produce meta-cognitive processes in a group level. The aim of this study is to contribute to the empirical understanding what makes metacognition as socially shared when pre-service primary teachers triads' (N = 18) collaborative mathematical problem solving is supported by an asynchronous and text-based WM learning environment.

By extending the above mentioned ideas and using the qualitative content analysis of the computer notes, it is demonstrated how metacognition becomes socially shared in collaborative mathematical problem solving. For the analysis, a conditional definition for socially shared metacognition was proposed drawing on the recent literature of shared regulation [10].

10.2 Metacognition in Computer-Supported Collaborative Mathematical Problem Solving

Earlier research on metacognition in computer-supported collaborative mathematical problem solving has shown that group members influence one another's subsequent thinking by using agreement, disagreement and correct evaluations to develop a new idea [11]. Chiu and Kuo [12] found that during online discussions accurate evaluations and questions increased the quality of joint problem solving. These studies show that learners use metacognitions to advance their group's problem solving in a computer-supported collaborative learning context [11, 12]. In what follows, we use data-driven examples to show how metacognition plays an important role for computer-supported collaborative problem solving in mathematics.

In our study, the participants were 45 native Finnish pre-service primary teachers having their first university level course for teaching mathematics at primary school level. One half of them worked with an asynchronous and text based learning environment, and the other half of the participants worked without computers.

Participants formed 15 triads (comprised by one male and two females) based on the principal component analysis of the data from two self-report questionnaires on metacognition in mathematical problem solving [13] and group working skills, SAGA [14]. The questionnaire data were checked for non-normalized response patterns and the equality of the groups was confirmed by using the analysis of variance (ANOVA). After the equality confirmation, six triads (A, B, C, D, E and F) were randomly chosen to use an asynchronous and text-based WM learning environment in collaborative problem solving.

In this article, this specific problem solving situation is focused on because there is not much research on how the regulation of group problem solving processes get carried out across media where the lack of facial and gestural information makes the processes of problem solving different than those in face-to-face collaboration [15]. In addition, we had a parallel aim to develop detailed qualitative methods to examine how metacognition emerge and become shared in a computer supported collaborative problem solving process. To characterize the triads, the items in the metacognition and mathematics problem solving, and SAGA questionnaire data were analyzed using the principal component analysis where the most powerful tendencies were liking or not liking mathematics, assessing oneself as a good or a poor problem solver, and assessing oneself as a fast or a slow problem solver. In the SAGA data, the most powerful tendency was, respectively, liking or not liking to work in groups. The group characteristics in mathematics are described in Table 10.1.

According to the metacognition and mathematics problem solving questionnaire results, the members of the groups C and B like mathematics but they feel they need time to solve mathematical problems (Table 10.1). Groups D and A consist of participants who do not like mathematics and think they are not good in mathematics. The members in groups E and F like mathematics and they assess themselves as quick problem solvers. However, the items measuring metacognition did not form a powerful tendency. It is possible that the questionnaire items should have been more closely connected a certain type of mathematical problem such as an arithmetic word problem or a calculus task. Since we did not do that, it could be that the items were too general making the participants unaware of what kind of mathematical problem they should think when filling in the questionnaire.

It is also possible that the participants are not aware of their metacognitive processes during problem solving. In addition, they had had some lectures about

Table 10.1 The group profiles in mathematical problem solving

Groups	Mathematical profile
A, D	Group members do not like mathematics and they think are not good in mathematics
E, F	Group members like mathematics and they assess themselves as quick problem solvers
C, B	Group members like mathematics but they feel they need time to solve mathematical problems

educational sciences and mathematical didactics which made them fill in the questionnaire in way that they thought would be appreciated by the researchers. Further, in the SAGA questionnaire data, almost all of the participants reported that they liked to work in groups. This may be due to group working skills assessed in the entrance exam for pre-service teacher candidates. In order to pass the entrance exam, a candidate's positive attitude towards group work is appreciated. Therefore, the participants could be considered as an exclusive group.

10.2.1 Mathematical Tasks and Data Analysis

For this study, we selected a total of 16 various mathematical problems of different difficulty levels within the primary and secondary school curriculum. In this article we use two different problems: Task 1 is an algorithmic problem similar to river-crossing puzzles where information needed for the solution is provided in a task description and it is independent of task-specific knowledge [16, 17]. The second part of Task 2 is the challenge, since the relationship between two different sets of grades had to be calculated in percentage terms.

Task 1. The Dark Stairs. Matt, Grandmother, Little Sister and Dad are standing upstairs in the dark and they need to go downstairs. However, the stairs are rather narrow and in a state of dilapidation; therefore, they can take only two people's weight at a time. The stairs will collapse in 18 min. The family has only one flashlight and it is impossible to use the stairs without using the flashlight.

The sinuous flight of stairs is so long that it is impossible to throw the flashlight upstairs from downstairs. The members of the family are aware of the amount of time that is required for them to reach downstairs. Because Grandmother is in poor health, it takes 7 min for her to go down the stairs. Little Sister and Dad take 5 min and 3 min to cover the flight of stairs, respectively. Matt runs down the stairs in 2 min. Is it possible for everybody to get downstairs in 18 min? [18].

Task 2. A Strict Lecturer. A retired lecturer, who had a reputation of being a strict teacher, had given the following marks to his students during his long and self-denying career: 26,172 Ds; 11,583 Cs; 4,884 Bs; and 955 As. How many marks altogether did the strict lecturer give during his long and self-denying career to his students? In percentage terms, how many more D's as compared to A's had the strict lecturer given? In this task, you are not permitted to use a calculator. You can use approximate values to calculate the percentage [19].

For the two tasks reported here, the discussion forum data consists of the pre-service teachers' 599 posted computer notes in the WM learning environment. The focus of the analysis is on knowledge and thinking made visible by writing the computer notes without making generalizations into the thinking behind them.

In the data analysis, a process-oriented approach [20] was used to characterize socially shared metacognition. The unit of analysis was one computer note, and for reliability, the rater independence coefficient, Cohen's Kappa [21], was calculated.

After qualitative content analysis, the groups' problem solving processes were visualized as a problem solving process graph, as a function of time. The computer notes were differentiated on the basis of whether they were characterized as social, cognitive and metacognitive comments: Social comments were computer notes that were related to the perceptions of the task or the group's work. Computer notes that identified perceptions of the task included experienced task difficulty (e.g., "This is too difficult"), estimating task success or failure (e.g., "I think we can't solve this"), making suggestions on working procedure (e.g., "Should we think of this task ourselves first, shouldn't we?").

Perceptions of group's work comment were related to how to start and finalize the joint problem solving (e.g., "Have you read the problem already?" and "Let's move on another task"), and comments including humor, agreement, disagreement or engagement (e.g., "Let's try again, have strength!").

In characterizing mathematical problem-solving computer notes, a computer note could be classified into one or two of the following categories: analysis, exploration, implementation, or verification (Table 10.2). The classification relays strongly on Schoenfeld's [2] studies of mathematical problem solving process.

Identifying the process of socially shared metacognition requires recognizing a group-level metacognitive computer note [22] with the following three criteria: (1) the computer note should be identified as metacognitive, (2) it should be related to and focused on the earlier or ongoing discussion, and (3) the computer note

Table 10.2 Examples of computer notes from mathematical problem solving perspective

Problem-solving phase	Description	Example
Analysis	Fragmenting a problem into smaller parts, reporting earlier experiences of similar problems, clarifying task requirements	Does the problem sound like that someone has to run up and down the stairs many times…with the flashlight, doesn't it?
Exploration	Suggesting concrete ways to solve or questioning how to solve the problem	For Grandmother it takes 7 min. Who should go with her? P.S: A question raised in the task requirements "It is impossible to use the stairs without LIGHT….?!
Implementation	Reporting a result or the outcome of a calculation	My suggestion would be 2740.52 %
Verification	Evaluating the ongoing problem solving process or the received result	But it takes 12 min + then dad and the kids. It's not working

Table 10.3 The characteristics of metacognition in collaborative problem solving

Socially shared metacognition	Metacognitive monitoring made visible at group level but not used to progress joint problem solving	An individual's attempt to navigate the collaborative problem solving by metacognitive monitoring and control
A group-level metacognitive computer note is acknowledged and adapted by the group: the group converges on a new idea in the group's problem-solving process	A group-level metacognitive computer note was acknowledged with words such as "aha, now I understand". The new idea was not converged in the group's problem solving	A group-level metacognitive computer note was contributed to the discussion—no reply to the group level metacognitive computer note

should have an intention to interrupt, change, or promote the progression of the joint problem-solving process.

The process of socially shared metacognition begins when a group-level metacognitive computer note is acknowledged and adapted by the group. In other words, the group converges on a new idea in the problem-solving process, thereby fueling a shift in the group's collective approach. Different categories of metacognitive monitoring in collaborative problem solving are presented in Table 10.3.

In the following, groups with similar profiles are compared in cases of successful and unsuccessful problem solving and for metacognitive regulation with and without acknowledgement. The interpretations concerning the excerpts from the discussion forum data are presented with the characteristics of each computer note in brackets. The joint problem solving process is also presented as a process-oriented graph.

10.3 Success in Collaborative Problem Solving Requires Socially Shared Metacognition

10.3.1 Group's Intention to Solve a Problem Quickly and Reach a Right Solution Damages Metacognition

The results in our study show that in general the proportion of metacognitive messages in the total number of written computer notes was quite low. In Task 1, which was a brain teaser task, there were only 10 out of 270 computer notes characterized as metacognitive (3.7 %, M = 1.7, SD = 2.3).

In Task 2, there were altogether 329 messages of which 11 notes were characterized as metacognitive (3.5 %, M = 2, SD = 2.2). Figure 10.1 summarizes the proportions of metacognitive messages characterized as socially shared metacognition, metacognitive monitoring made visible at group level but not used to progress joint problem solving, an individual's attempt to navigate the collaborative problem solving by using metacognitive monitoring and control.

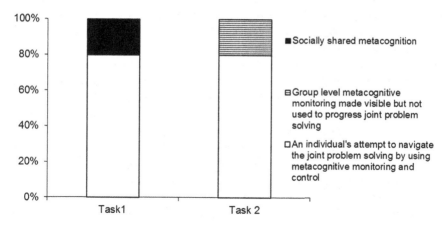

Fig. 10.1 Distribution of metacognitive computer notes in discussion forum data

The majority of the metacognitive computer notes indicated an individual's attempts to navigate the group's joint problem solving at a metacognitive level (Fig. 10.1). Among the metacognitive computer notes socially shared metacognition was recognized only in two cases. Both of them happened in group A's problem solving during Task 1. In groups B, C, and E, participants tried to navigate the group's problem solving without advancing problem solving at group level. In group B there were six metacognitive messages, and in groups E and C there was only one metacognitive computer note. Groups D and F did not have any metacognitive messages. There is a similar trend in Task 2. Group A had one metacognitive message, and group B had six messages but the group members did not acknowledge and use to advance the joint problem solving. Group C had three metacognitive messages of which one was acknowledged by a group member. Group F did not have any metacognitive messages.

Since the amount of metacognitive computer notes was low, a more detailed examination of the quality of the collaborative problem solving process is needed. Next, the results of the cognitive and social level content analysis in Task 1 are presented in the Fig. 10.2. The results show (Fig. 10.2) that the most of the posted computer notes were related to group processes how to begin and end the joint problem solving (varying 30–50 % of a group's total contribution) in Task 1. The large amount of social level computer notes is understandable because the group members were allowed to communicate only via WM learning environment.

It also indicates the participants' willingness to work as a group and make all the group members aware of how to proceed in collaboration. At the beginning of mathematical problem solving, the focused on searching for a solution. The groups did not use their time to think and analyze the problem much and they almost ignored the verification part of the joint problem solving process. The lack of analysis and verification could explain the low amount of metacognitive computer notes. Two groups did not post any metacognitive computer notes and in the other four groups'

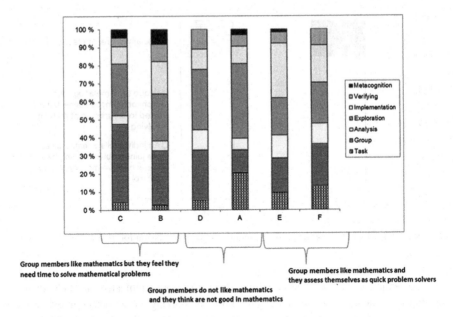

Fig. 10.2 Groups use most of their time to searching for a solution in Task 1

metacognitive messages varied from 2 to 10 % of a group's total contribution. Only group A yielded to the process of socially shared metacognition.

Task 2 was more demanding conceptually and findings has shown percent as a concept to be more challenging for students when percent is seen as a descriptor of the relationship of one set to another, or of the relative amount by which the sets differ from each other. This differs from the everyday idea of percent being a part of a whole varying between 0 and 100 % [23]. The results of the qualitative content analysis of the computer notes in Task 2 are presented in Fig. 10.3.

In the Task 2, four groups (B, D, A, E) have few metacognitive computer notes (Fig. 10.3). The challenge of the percentage concept explains the low amount of the metacognitive computer notes in Task 2. Those group members who were knowledgeable to navigate the group's joint problem solving shared their metacognitions to the others. However, the other group members were unable to apply the presented idea in the joint problem-solving process. They did not have enough conceptual understanding of percentage being more than 100 %.

As a result, this hindered the process of shared metacognition to emerge in groups' problem solving. Another reason for the lack of socially shared metacognition and metacognitive computer notes is that the analysis phase of mathematical problem solving was almost missing. It could be that the trivial part of the problem triggered the groups just to use the given numbers to perform calculations. The groups' joint problem solving processes were mainly focused on searching for an

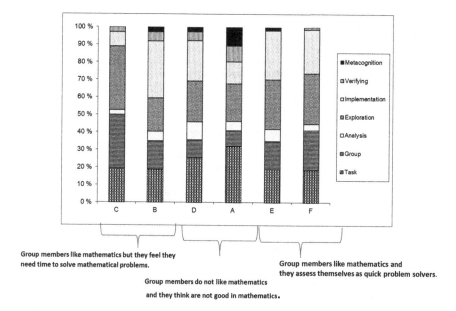

Fig. 10.3 The groups explored and implemented ideas for a solution in Task 2

explanation and implementing the provided solution effort with limited efforts to verify the problem. The lack of these two phases in mathematical problem solving damages metacognition. The example is presented below.

In our example, members in group D have reported that they do not like mathematics and they think are not good in mathematics. Group D's collaborative problem solving process is not at all successful:

- Leena (0:00): Somebody should run after going down with the one of the others because of the flashlight, right?
- Sirkka (0:00): That came up to my mind too. Is Matthew fast enough?
- Leena (0:01): I hope so.
- Leena (0:01): Because Matthew is the fastest, he runs.
- Janne: (0:01): OK. So it would be reasonable to make little Matthew to run and carry the flashlight up, and come down with one person.
- Janne (0:01): Yeah.
- Sirkka (0:02): Is it possible to show the light in halfway in stairs?
- Janne (0:02): I'm calculating on paper.

In group D, Leena and Janne present their individual thoughts, and Sirkka tries to create collaboration among the group members by asking clarification questions. The sparse interaction could indicate that Leena and Janne tried to solve the task individually without engagement in the joint processes (Fig. 10.4).

Fig. 10.4 Individual
thinking is made visible
in group D

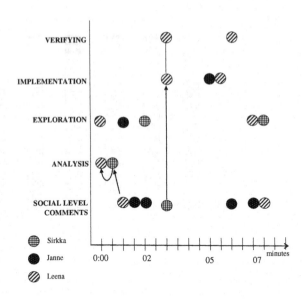

10.3.2 Metacognitive Messages Are Converged into Individuals' Own Thinking or Ignored in Discussion

The collaborative processes of groups B (Leevi (male), Hilkka (female), Leena (female)) and C (Jiri (male), Nelli (female), Maija (female)) are first discussed more detailed. These groups reported they like mathematics but they feel they need time to solve mathematical problems. In the example, the groups are working with the Task 2. In both groups, the first steps of problem solving were not discursive in nature. Further, the group members understanding of percentage was too miscellaneous.

The common denominator for these two groups is that the male participant clearly has more conceptual knowledge on how to calculate the percentage and they propose a solution process directly without providing rationale for their thinking (Fig. 10.5). Leevi (in group B) is inviting the others to contribute to the task by asking them to confirm his thinking although this did not happen.

The other members in the group do not even request Leevi to explain his thinking. Hilkka provides her alternative solution effort for elaboration and Leena support her. It could be that the lack of conceptual knowledge without metacognitive skills did not allow Hilkka and Leena to utilize Leevi's metacognitive message [24].

In group C, Nelli was only able to wonder what was happening, and Maija did not even get a chance to participate in the discussion before Jiri announced his suggestion for a result. In the following, we provide a detailed example how Nelli (group C) uses other group members' knowledge to improve her individual thinking.

In addition, we show how two group members in group B ignore their peer's attempts to regulate the ongoing problem solving at the metacognitive level. In both cases, there is no space for the process of socially shared metacognition.

Group C			Group B		
Jiri (0:02)	The percentage is 26172:599, let's calculate	(exploration)	Leevi (0:04)	26172:599x100? Would it go like this?	(exploration)
Nelli (0:03)	Oh, like that?	(social level)	Hilkka (0:05)	No, it wouldn't. It is 599/256172, isn't it and then it is multiplied by 100?	(exploration)
Jiri (0:05)	My suggestion would be 2740,52%	(implementation)	Leena (0:05)	Yes, let's calculate both grades' proportion of the whole and then compare our results	(exploration)
			Hilkka (0:05)	Let's calculate and then we can compare [received results] again. Okey?	(social level)
			Leevi (0:06)	No, because there are hundreds of percentages more D's.	(verification and metacognitive monitoring)

Fig. 10.5 The more knowledgeable peers suggest a solution to the Task 2

Group C			Group B		
Maija (0:09)	Are we ready?	(social level)	Leevi (0:13)	Hilkka was calculating the amount of D's of all grades, but is it useful?	(verification and metacognitive regulation)
Jiri (0:10)	If we all agree	(social level)	Leevi (0:15)	I don't think it goes like that. Imagine, if we would have 20% rabbits and 40% bunnies of all animals. It does not mean that there is 20% but 100% more bunnies	(verification)
Maija (0:12)	Yes we are, apparently	(social level)			
Nelli (0:11)	We could be but I need to think these percentage calculations cross my heart so I know exactly why, what and how to solve it	(social level)			
Maija (0:12)	The percentages are easy but the probability causes trouble	(social level)	Hilkka (0:15)	But we should calculate both grades [Ds and As] as percentage separately, and then compare, shouldn't we?	(exploration)
Jiri (0:12)	So, if it is asked how many percentages number 1 is from a hundred, the answer is 1%. But if it is asked other way round, so the division is 100:1=100, and then it should be multiplied with 100 which makes 10000%.	(exploration and metacognitive monitoring)	Leena (0:16)	Obviously, but now I don't even know how to do divisions. Hilkka, what result did you get with Ds?	(social level)
Maija (0:13)	It took me a while to understand [Jiri's explanation] but that's the way it is. Just enough zeros to be at stake.	(social level)	Hilkka (0:16)	But we don't know what is the amount of rabbits and bunnies of all animals in percentage. So, you bunny idea is not working.	(exploration)
Maija (0:14)	Did you understand, Nelli? Should we explain it to you some other way?	(social level)			
Nelli (0:16)	I wasn't thinking about that. But if it is askes how much more in percentages, so we should calculate 26172-955 and then divide the result, shouldn't we?	(social level)			
Jiri (0:18)	I think we did the right way, but…	(social level)			

Fig. 10.6 The emergence of socially shared metacognition is prevented

In group C, Nelli insists that the others should explain the procedure of the task more, and Jiri takes the role of an expert and explains the procedure in detail (Fig. 10.6). Nelli acknowledges Jiri's explanation, and she tries to analyse what the task was about. Nelli is now using her peer's knowledge to support her own thinking. Jiri's last reply reveals that he is satisfied when a common solution is reached and he does not take Nelli's thinking into account.

In group C, Jiri's metacognitive regulation message (at 0:12) has a long lasting effect on Maija's thinking and it takes 13 min time for her to understand the procedure (Fig. 10.7). Although interaction makes it possible for Nelli to advance her thinking, it is unbeneficial for the group's collaboration preventing socially shared metacognition to emerge [25].

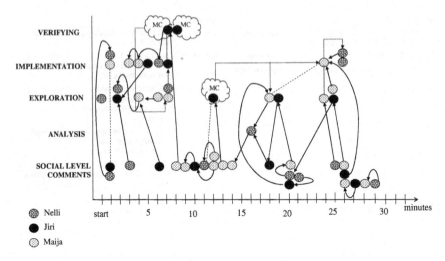

Fig. 10.7 The process of collaborative mathematical problem solving in group C

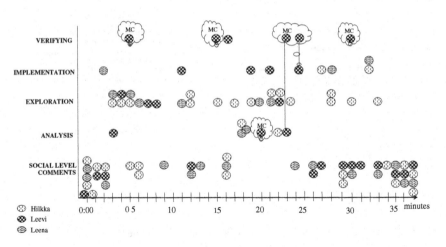

Fig. 10.8 The process of collaborative mathematical problem solving in group B

Socially shared metacognition is also hindered in group B's problem solving. Leevi contributes many metacognitive regulation messages but only one of them is acknowledged by Hilkka (Figs. 10.6 and 10.8). Leevi tries to change Hilkka and Leena's way of thinking by proposing questions and by using analogies.

Although Hilkka takes Leevi's metacognitive regulation message into account, she rejects the correct proposal [26], which could be due to the lack of conceptual knowledge, and metacognitive skills [24].

Thus, the process of socially shared metacognition is blocked by ignoring more knowledgeable peer's metacognitive computer notes.

In the process-oriented graph (Fig. 10.8) it can be seen that the most of the computer notes are explorative or social in nature. Leena and Hilkka's comments included only few attempts to analyse the task or use a strategy in order to solve a problem.

It seems like Leena and Hilkka were following only one solution path, that is calculate both grades in percentage separately. Leevi notives this and tries to regulate the group's work metacognitively by providing explanations and questions to forward the group's thinking.

At the end, the group has only social level discussion. The results of the qualitative analysis of the computer notes and the process-oriented graph of group problem solving taken together it can be argued, that acknowledgement alone is not a sufficient prerequisite for socially metacognition to emerge.

10.3.3   Socially Shared Metacognition

There has been research concerning socially shared metacognition in the fields of social psychology [27] and learning research [5, 8, 28], suggesting that socially shared metacognition is monitoring and regulation of cognitive processes on the interindividual level. The evidence of the phenomena under study is scarce, and more knowledge is needed to understand what socially shared metacognition is, and why it is important for the collaborative problem solving process. For this purpose we focus on socially shared metacognition in group A's problem solving process in the Task 1.

After 10 min, group A enters a crossroads situation [29], which now provides them an opportunity to jointly regulate their joint problem solving where socially shared metacognition could be involved (Fig. 10.9). When 10 min have passed, Alina states that they should stop the problem solving process, she writes, "I think that they cannot make it with any arrangements." Anna (at 0:12) encourages a joint

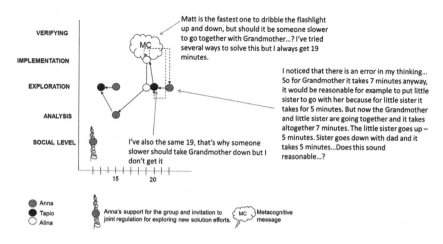

Fig. 10.9   The process of socially shared metacognition in group B

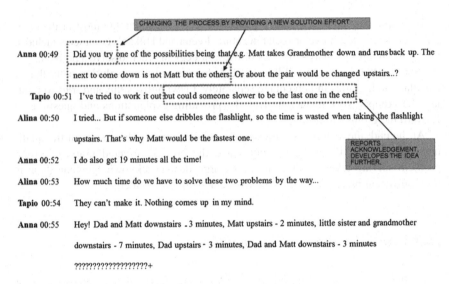

Fig. 10.10 Successful collaboration leads to a right solution

group reflection by providing strong support to the others by typing "Let's try again, have strength!". This leads to continue the process and explore the new suggestions for possible solution methods, and to reanalyze the given problem as follows:

Alina's (at 0:19) message illustrates metacognitive regulation because she brings up a new way of thinking. The others in the group acknowledge Alina's idea and take advantage of her suggestion and try to develop the presented idea further. Tapio agrees and he appreciates Alina's idea, but he still does not quite understand how the prerequisites of the task are fitting together. Anna is monitoring her own thinking and then she adopts Alina's idea and tries to develop it further and asks the others to make comments about it. This kind of regulation of the group's cognitive activity could be considered as a sign of socially shared metacognition. It could also indicate the trust and interdependence among the group members considered to be a crucial aspect of shared processes [10]. Finally, group A reached the solution (Fig. 10.10).

Working almost 50 min, Anna provides a solution effort to the others (Fig. 10.10). In her metacognitive message, Anna combines the ideas presented earlier in joint problem solving and she proposes an essential component needed for solution. Tapio acknowledges Anna's suggestion, and he brings up another essential feature for solution. Also, Alina reports that she had tried to solve the problem as Anna suggested. In this situation, we consider socially shared metacognition becoming visible, although Tapio's suggestion is not clearly formulated. But it makes it possible for the group to combine the essential features of a problem and thus receive the solution.

This groups' working shows that the process of socially shared metacognition is a complex and extraordinary phenomenon but it is a differentiator, making problem solving successful at the group level.

10.4 Discussion

In this study, by comparing cases of group's successful and unsuccessful mathematical problem solving in a computer-supported collaborative learning context, it was established that socially shared metacognition is a differentiator making problem solving successful in groups. In the group where socially shared metacognition emerged, a group member regulated the ongoing problem solving and the other group members acknowledged metacognitive message and developed it further.

As shown in this study, for shared metacognition, it is not essential that all group members regulate the ongoing problem solving simultaneously. In the process of socially shared metacognition the focus is on the mutual construction of understanding where the participants intentionally engage themselves in reciprocal interaction, and where the regulatory processes and products are shared [10]. Our study contributes to the ongoing discussion regarding socially shared metacognition and the individual cognitive activity and metacognitive regulation in collaborative learning [6, 8, 10, 28, 30]. The results are also in line with group and team learning studies showing that collective monitoring and controlling of task requirements are needed for effective knowledge construction [9, 31].

By contrasting cases of acknowledged and unacknowledged metacognitive messages, we showed that acknowledgement is not a sufficient prerequisite for metacognition to become shared. If the metacognitive message was acknowledged but due to lack of conceptual knowledge and metacognitive skills it was not built upon, the process of socially shared metacognition was impeded.

It is possible that, without reciprocal attempt to build upon a metacognitive message, the group could be left with a false sense of understanding [32]. Students' own admissions of whether they do really understand the procedure by saying "aha, that way" or "I get it" may not always be accurate [33].

However, in the problem solving discussion, the groups did not reach the metacognitive level very often. The results of this study are also consistent with findings suggesting that a technology-based learning environment itself does not guarantee high-level collaboration [15] where socially shared metacognition can be embedded.

From the mathematical problem solving point of view, the results are consistent with Schoenfeld's [2] findings. The groups mostly ignored the analysis phase, and they were not verifying the ongoing problem solving process where metacognition plays an important role. Further, due to a lack of content knowledge the correct suggestions where ignored [26] and poor metacognitive decisions were made [24]. This is in line with Malmberg, Järvelä and Kirschner's study showing that the task types effects on students' allocation of learning strategy use and the lack of content knowledge misleads the groups to focus on irrelevant aspects of the problem solving process [34].

The concept of socially shared metacognition is not thoroughly studied in the field of educational psychology. To contribute to the understanding of what socially shared metacognition is, we proposed a tight operational definition for socially shared metacognition since literature provides only a little evidence of shared processes that can not be reduced to the individual [10]. In our case, we found the operational definition useful and worth developing further in our future studies. In addition, the term of socially shared metacognition is challenging because it addresses the complex relation between cognition and metacognition at the group level. Thus, further studies are needed to understand the intertwined role of individual and shared metacognition in collaborative problem solving, that is, how individual's metacognitive processes are linked to group level metacognitive processes.

In this study, discussion forum data was used to examine how the groups regulated their ongoing problem solving. Computer notes were considered to be thinking made visible although no generalizations into the thinking behind them were made.

This is a limitation in studies where participants' written computer notes or other documents are analysed. In this study, this means that metacognition is needed when individuals read each other's computer notes, and when they formulate a reply to the others' messages.

These metacognitive processes do not become visible in their written computer notes, and multiple data gathering methods, like on-line questionnaires, are needed to gain this information. However, the participants wrote their messages as precisely as possible by themselves, in order to mediate their thinking during the process to their peers. In interpreting these results, a focus on group level characteristics and utilization of the process-oriented graph of groups' problem solving offer a strategy for examining the interplay between social, cognitive and metacognitive dimensions in joint mathematical problem solving. In this process, socially shared metacognition seems to be a complex and extraordinary phenomenon but it is important for successful problem solving at group level.

The results of this study can be utilized in teaching and developing metacognition not only in collaborative learning situation but also in mathematics teaching in general. In this study, it was shown that the lack of analysis and verification phases in mathematical problem solving was the most harmful for metacognition. When these phases are missing, a learner does not use his existing knowledge to implement and select appropriate strategies for the task. Further, a learner does not learn to monitor his own problem solving process and to check whether the received result meets the task requirements.

Computer-supported collaborative problem solving could be used as an instructional method in mathematics classrooms, for example in form of homework or an exploratory task before instruction. In future, a learning environment allowing the usage of mathematical symbols and graphs but also requiring a learner to explain one's thinking to others would benefit the development of metacognition in mathematics learning.

Acknowledgments This study was funded in part by Academy of Finland/ADDRESS (127640) Adaptive Motivation Regulation in Individual and Socially Shared Learning Situations, and by the Grant of Finnish Cultural Foundation (first author).

References

1. Veenman, M.V.J., Van Hout-Wolters, B., Afflerbach, P.: Metacognition and learning: conceptual and methodological considerations. Metacognition Learn. **1**(1), 3–14 (2006)
2. Schoenfeld, A.H.: Mathematical Problem Solving. Academic Press, Orlando (1985)
3. Flavell, J.: Metacognition and cognitive monitoring: a new area of cognitive-developmental inquiry. Am. Psychol. **34**(10), 906–911 (1979)
4. Brown, A.L.: Metacognition, executive control, self-regulation and other mysterious mechanisms. In: Weinert, F., Kluwe, R. (eds.) Metacognition, Motivation and Understanding, pp. 65–115. Lawrence Erlbaum, Hillsdale, NJ (1987)
5. Hadwin, A.F., Järvelä, S., Miller, M.: Self-regulated, co-regulated, and socially shared regulation of learning. In: Zimmerman, B.J., Schunk, D.H. (eds.) Handbook of Self-Regulation of Learning and Performance, pp. 65–84. Routledge, New York (2011)
6. Järvelä, S., Hadwin, A.F.: New frontiers: Regulating learning in CSCL. Educ. Psychol. **48**(1), 25–39 (2013)
7. Hadwin, A.F., Oshige, M.: Self-regulation, co-regulation, and socially shared regulation: Exploring perspectives of social in self-regulated learning theory. Teachers Coll. Rec. **113**(2), 240–264 (2011)
8. Iiskala, T., Vauras, M., Lehtinen, E., Salonen, P.: Socially-shared metacognition within primary school pupil dyad's collaborative processes. Learn. Instr. **21**(3), 379–393 (2011)
9. DiDonato, N.C.: Effective self- and co-regulation in collaborative learning groups: An analysis of how students regulate problem solving of authentic interdisciplinary tasks. Instr. Sci. **41**(1), 25–47 (2013)
10. Winne, P.H., Hadwin, A.F., Perry, N.E.: Metacognition and computer-supported collaborative learning. In: Hmelo-Silver, C., Chinn, C., Chan, C.K.K., O'Donnell, A. (eds.) The International Handbook of Collaborative Learning, pp. 462–478. Taylor & Francis, New York (2013)
11. Moss, J., Beatty, R.: Knowledge building in mathematics: Supporting collaborative learning in pattern problems. Int. J. Comput. Support. Collaborative Learn. **1**(4), 413–440 (2006)
12. Chiu, M.M., Kuo, S.W.: From metacognition to social metacognition: Similarities, differences, and learning. J. Educ. Res. **3**(4), 1–19 (2009)
13. Howard, B.C., McGee, S., Shia, R., Hong, N.S.: Metacognitive self-regulation and problem solving. Expanding the theory base through factor analysis. Paper presented at the Annual Meeting of the American Educational Research Association, New Orleans, LA (2000)
14. Volet, S.E.: A Manual for the Use of Students' Appraisals of Group Assessment (SAGA). School of Education, Murdoch University, Murdoch, WA (1998)
15. Järvelä, S., Häkkinen, P.: Web-based cases in teaching and learning—the quality of discussions and a stage of perspective taking in asynchronous communication. Interact. Learn. Environ. **10**(1), 1–22 (2002)
16. Greeno, J.G.: Hobbits and orcs: acquisition of a sequential concept. Cogn. Psychol. **6**(2), 270–292 (1974)
17. Miyake, N.: Making internal process external for constructive collaboration. In: Proceedings of Humanizing the Information Age, 2nd International Conference on Cognitive Technology, pp. 119–123. IEEE Computer Society, Aizu-Wakamatsu City (1997)
18. Björklund, J., Lehto, S., Pasanen, S., Viljanen, M.: Sukkia ja muuta matematiikkaa: Monipuolisuutta matematiikkaan [Socks and Mathematics: Versatility in Mathematics Learning, in Finnish] MFKA, Helsinki (2002)
19. Oster, G.: Problem Book: Mieletön matikka. [Problem Book: Creative Problems in Mathematics. In Finnish.] WSOY, Helsinki (2002)
20. Järvelä, S., Järvenoja, H., Näykki, P.: Analyzing regulation of motivation as an individual and social process—a situated approach. In: Vauras, M., Volet, S. (eds.) Interpersonal Regulation of Learning and Motivation: Methodological Advances, pp. 170–187 EARLI series: New Perspectives on Learning and Instruction, Routledge (2013)

21. Cohen, J.: A coefficient of agreement for nominal scales. Educ. Psychol. Measur. **20**, 37–46 (1960)
22. Artzt, A.F., Armour-Thomas, E.: Development of cognitive-metacognitive framework for protocol analysis of mathematical problem solving in small groups. Cognition Instr. **9**(2), 137–175 (1992)
23. Parker, M., Leinhardt, G.: Percent: a privileged proportion. Review of Educational Research. **65**(4), 421–481 (1995)
24. Goos, M., Gailbraith, P., Renshaw, P.: Socially mediated metacognition: creating collaborative zones of proximal development in small group problem solving. Educ. Stud. Math. **49**(2), 193–223 (2002)
25. Kieran, K.: The mathematical discourse of 13-year-old partnered problem solving and its relation to the mathematics that emerge. Educ. Stud. Math. **46**(1–3), 187–228 (2001)
26. Barron, B.: When smart groups fail. J. Learn. Sci. **12**(3), 307–359 (2003)
27. Tindale, R., Kameda, T.: Social sharedness' as an unifying theme for information processing in groups. Group Processes Intergroup Relat. **3**(2), 123–140 (2000)
28. Khosa, D. K., Volet, S.: Productive group engagement in cognitive activity and metacognitive regulation during collaborative learning: can it explain differences in students' conceptual understanding? Metacognition and Learning, Springer, Heidelberg. (2014). doi:10.1007/s11409-014-9117-z
29. Dewey, J.: How we think. D.C. Heath & Company, Boston (1910)
30. Saab, N., Van Joolingen, W.R., Van Hout-Wolters, B.: Support of the collaborative inquiry learning process: influence of support on task and team regulation. Metacognition Learn. **7**(1), 7–23 (2012)
31. Fransen, J., Kirschner, P.A., Erkens, G.: Mediating team effectiveness in the context of collaborative learning: The importance of team and task awareness. Comput. Hum. Behav. **27**(3), 1103–1113 (2011)
32. Webb, N.M., Nemer, K.M., Ing, M.: Small-group reflections: parallels between teacher discourse and student behavior in peer-directed groups. J. Learn. Sci. **15**(1), 63–119 (2006)
33. Shavelson, R.J., Webb, N.M., Stasz, C., McArthur, D.: Teaching mathematical problem solving: Insights from teachers and tutors. In: Charles, R., Silver, E. (eds.) Teaching and Assessing Mathematical Problem-Solving: A Research Agenda, pp. 203–231. Lawrence Erlbaum, Hillsdale, NJ (1988)
34. Malmberg, J., Järvelä, S., Kirschner, P.A.: Elementary school students' strategic learning: Does task-type matter? Metacognition and learning, (2013). doi:10.1007/s11409-013-9108-5

Chapter 11
Enhancing the Metacognitive Skill of Novice Programmers Through Collaborative Learning

Margaret Bernard and Eshwar Bachu

Abstract Computer Supported Collaborative Learning (CSCL) aims to improve education by combining collaborative learning with modern information and communication technology. The opportunity exists to develop successful CSCL applications due to the increase in popularity of social networking and online gaming among students. In this chapter, we present an approach for promoting metacognition in computer programming using collaboration and computer games. We show that CSCL can improve the students' metacognitive skills and the use of games motivates and engages students in the learning process. Together, they enhance the qualities of a successful problem solver and low problem solving skill has been identified as the main challenge faced by novice computer programmers.

Keywords CSCL · Programming · Metacognition · Problem solving · COPS · Multiplayer game · Collaborative learning

Abbreviations

CL   Collaborative learning
COPS   Collaborative online problem solving
CSEC   Caribbean secondary education certificate
CSCL   Computer supported collaborative learning
CXC   Caribbean examinations council
DIV   Division

M. Bernard (✉) · E. Bachu
Department of Computing and Information Technology, The University
of the West Indies, Circular Road, St. Augustine, Trinidad and Tobago
e-mail: margaret.bernard@sta.uwi.edu

E. Bachu
e-mail: eshwar.bachu@sta.uwi.edu

© Springer International Publishing Switzerland 2015
A. Peña-Ayala (ed.), *Metacognition: Fundaments, Applications, and Trends*,
Intelligent Systems Reference Library 76, DOI 10.1007/978-3-319-11062-2_11

ICT   Information and communications technology
IT   Information technology
MOD   Modulo

11.1 Introduction

Teaching and learning of computer programming is a major challenge worldwide. A significant contributor to that challenge is the low problem solving ability of students. Students may understand the individual programming building blocks such as an 'if' statement or a 'while' loop but have difficulty in knowing how and when to use them.

Metacognition is a complex concept that relates to the higher order thinking that enables students to understand, analyze and control their own thought processes. This skill is particularly important in developing good programmers. Problem solving certainly involves cognition but more is required: students must constantly reflect on their strategies for problem solving and critically appraise their approach, thereby improving their connections between concepts. They need to build their knowledge of how and when to use particular strategies for problem solving.

Collaborative Learning has proved to be useful in many disciplines. In Computer Programming, the most commonly used strategy that involves collaboration is pair-programming; here students work in pairs, encouraging and correcting each other. A relatively new area of research is Computer Supported Collaborative Learning, in which the computer system is directly supporting the collaboration.

Bachu and Bernard [1] have shown that CSCL can increase the benefits of collaboration by enhancing the metacognitive abilities of students in the problem solving stage of programming. They present a framework for the development of a CSCL environment that incorporates a number of characteristics: the environment must promote positive interdependence where each member of the group becomes personally responsible for the group's success.

It should promote argumentative discussion where each member of the group must be aware of the need to make the best decision and encouraged to discuss and defend their reasoning for a given action, and it should promote equal participation where all members of the collaborative group take full responsibility for their learning and learn through the experiences of the other members of the group. Such systems aim directly at enhancing the metacognitive skill of students so that they know what they know and know how and when to apply basic concepts in a constructive manner to produce an algorithm that is the solution to a problem.

In this chapter, in Sect. 11.2 we first present a detailed description of the problem. We give an overview of some of the research work that has been done on analyzing and addressing the challenge of poor programming, particularly of novice programmers. We put this section into a context that is the Caribbean region, where all countries have adopted a common exam. In Sect. 11.3 we discuss the issue of metacognition in programming.

We review the significant work of others in this area and develop and discuss three programming problems that illustrate the challenges students have with the problem solving phase of programming. Section 11.4 addresses an approach for promoting metacognition in programming using collaboration and computer games. Again we review the literature in this area of Collaborative Learning and Computer Supported Collaborative Learning, particularly as they relate to programming.

In Sect. 11.5, we present some important elements of a CSCL game environment that specifically targets enhancing metacognitive skills of the players. The game is a multi-player, turn-based game that encourages students to collaboratively build an algorithm that is the solution to a given problem. Section 11.6 presents the experimental results and findings from the use of the CSCL game. We conclude in Sect. 11.7 with some thoughts for future development and research.

11.2 Description of Problem

The teaching and learning of programming has posed a major challenge for educators worldwide for many decades; most students are able to learn basic programming skills but do not achieve any level of programming fluency [2]. In Trinidad and Tobago and the wider Caribbean region, students are introduced to computer programming at the secondary school level while preparing to sit the Caribbean Secondary Education Certificate (CSEC) Information Technology (IT) exam.

Prior to 2010, IT was offered in both General and Technical Proficiencies, but from 2010, all students were required to sit the General Proficiency exam. Table 11.1 presents the number of students who sat the General and Technical exams between 2005 and 2009.

The percentages in Table 11.1 indicate that only 4 % candidates sat the general exam prior to 2010. The main reason for the small number of general candidates was that the general exam placed greater emphasis on computer programming which the students had significant difficulties with. As a result, most schools allowed their students to sit the technical exam which contained very little

Table 11.1 CSEC IT candidate figures for June exam sittings [6]

Year	General	Percent (%)	Technical	Percent (%)	Total
2005	762	3.6	20,511	96.4	21,273
2006	898	3.8	22,446	96.2	23,344
2007	980	4.0	23,775	96.0	24,755
2008	1,210	4.4	26,064	95.6	27,274
2009	1,106	3.8	27,706	96.2	28,812

programming. From 2010, the lone general exam attempted to strike a balance in the programming requirements.

In 2010, CSEC IT had a pass rate of 84 and 79 % in 2011; while these are acceptable pass rates, a more in depth look at the examiners' reports produced for each exam sitting raises some major concerns in the areas of problem solving and programming in Pascal. The following are excerpts from these reports:

"Part (c) was poorly done as the majority of candidates could not write the correct algorithm for the given rule."; "Many candidates avoided this question. The use of arrays in programming challenged candidates" [3].

"The question was poorly done by the majority of candidates. Many candidates could not identify and correct the errors in the programming statements given in Part (a)"; "This question tested candidates' ability to use control structures and their knowledge of terms and concepts associated with programming. It was poorly done by the majority of candidates" [4].

"Part (c) was poorly done by the majority of candidates; they could not provide a correct arithmetic statement to calculate the final price"; "Part (a) was poorly done by the majority of candidates who appeared unfamiliar with the use of loops" [5].

"This question tested candidates' knowledge of concepts associated with problem solving and programming. Many candidates did not have a clear understanding of the concepts required to answer the various parts of the question" [7]. "Part (a) of the question was poorly done by the majority of candidates who were unable to identify the correct line numbers containing input, declaration and output statements. Part (b) was also poorly done. The majority of candidates could not identify the errors in the program segment and hence, could not provide the corrected codes. Candidates did badly on Part (c) as well. The majority of them seemed unfamiliar with the concepts of variable, data type and conditional statement" [8]. "The mean performance on this question was 3.75 out of a maximum of 15" [9].

These comments indicate that students experience significant difficulty with problem solving and programming. Beaubouef and Mason [10] also identified poor problem solving skill as a major contributing factor. There has been an abundance of research carried out to investigate the teaching and learning of programming; publications of this nature between 2005 and 2008 were analyzed by Sheard et al. [11] and suggestions which they offered as a way forward in programing education included the use of social networking and group work.

Problem solving requires reflecting on the solution, communicating the problem solution [12]; and the designing of a program to solve a particular task. Deek et al. [13] presented two challenges which students face when learning the task of program development: deficiencies in problem solving strategies and tactical knowledge; and ineffective pedagogy of programming instruction. They also presented a six step problem solving and program development model:

- Formulating the Problem: Preliminary Problem Description, Preliminary mental model, and Structured Problem Representation.
- Planning the Solution: Strategy discovery, goal decomposition, and data modelling.
- Designing the Solution: Organization and Refinement, Data/function specification, and Module Logic Specification.

- Translation: Implementation, Integration, and Diagnosis of Errors.
- Testing: Critical Analysis, Evaluation, and Revision.
- Delivery: Documentation, Presentation, and Dissemination.

The first three steps are those which present the toughest task for novices since it requires the problem solving ability which they lack. While problem solving remains their greatest challenge, there are other important skills which are necessary. Step four also requires novices to be able to comprehend and generate program code. Watson et al. [14] define programming comprehension as the ability to read and understand the outcomes of an existing piece of code and generation as the ability to create a piece of code that achieves certain outcomes. Achieving programming fluency would require developing students' problem solving and generation skills, however, this chapter focuses on increasing their problem solving ability. Recognizing these concerns, it becomes imperative to address the challenges which these students face. A repercussion of this is that students become disenchanted with computer programming at the secondary school level, and in the future, are hesitant to pursue to higher degrees in the fields Information Technology and Computer Science.

Their difficulties are also worrying since Sardone [15] recently highlighted that there is a need for producing college graduates who are considered to be fluent in information technology and programming lies at the core of information technology. Concerns about the high attrition rates in programming and computer science courses have also been raised [10, 16]. The authors of this chapter also have first-hand experience of the difficulties of introductory programming students.

11.3 Metacognition in Programming

Mayer [17] identified the importance of cognitive, metacognitive, and motivational skills in problem solving. Mayer argued that most students are able to solve routine problems (problems they are familiar with or have met before and know how to solve) but all three skills are vital for students to be able to solve non routine problems (problems they have failed to solve in the past or have never met). In the context of introductory programming, the basic cognitive skills can be broken down as: variables and constants; logical and comparison operators; selection statements; iterative statements; and arrays.

Mayer continued that mastering each of these component skill is not enough to promote non-routine problem solving, students need to know not only what to do, but also when to do it. He referred to this aspect of problem solving as metaskill or metacognition. Metacognition can take many forms; it includes knowledge about when and how to use particular strategies for learning or for problem solving [18].

Jonassen [19] describes the two main components of metacognition as knowledge of cognition and self-regulation. Knowledge of cognition requires knowledge of task requirements, knowledge of self (personal skills) and knowledge of learning styles. Self-regulation requires being able to monitor learning, plan and select strategies and evaluation of regulation. Consider the next examples of programming problems:

1. Write a program which prompts the user to enter an integer and returns the sum of the digits in the number. E.g., if the user enter 123, your program should return 6 (1 + 2 + 3).
2. Write a program which tests each number between 20 and 60 to determine if it is even and prints the even numbers.

The first step in solving the above problems is to formulate the problem, i.e., to understand what the task requires and this presents a significant challenge to students [20]. Consider problem 1, while this may seem simple to understand and an example was given; students have experienced difficulty with understanding what is required especially when they have not met similar problems before.

Although the problem specifies that a single integer is to be inputted, a common misconception is that each digit will be inputted separately or that only three-digit integers will be inputted based on the example. Problem two requires that students be able to iterate through the numbers between 20 and 60 and determine the even numbers which are to be displayed, some students recognize that they know what the even numbers are and simply output/display them individually.

These fallacies are a result of the students' lack of determining task requirements which has been identified as key component of metacognition. Formulating or defining the problem would also entail correctly identifying the sub tasks involved in solving the problem in its entirety.

After correctly understanding what the problem requires, a solution must be derived and tested. In cases where the problem was misunderstood, it is possible to produce solutions which give correct results sometimes.

For problem two, a program which outputs all even numbers individually by adding 2 such as 20, 22, 24, ..., 58, 60 would produce correct results each time the program is executed but it is still incorrect as it does not test the number to determine if it is even. Similarly a solution for problem one which expects only three-digit integers would be correct for those cases only. This highlights the importance of students being able to verify whether or not their solution is correct, however, developing this skill is difficult since the students' ability to evaluate their solution is linked to their knowledge of the task requirements. In practise, the verification of the solution is normally done by someone other than the developer.

In most cases, there are multiple correct solutions for a given problem and students are required to choose between them. At the introductory programming level, program efficiency is not mandatory but choosing the best solution should be encouraged. For problem one, the solution would entail performing a series of integer division (DIV) and modulo (MOD) of the inputted integer by ten (10). The modulo operation return the least significant digit and this would be added to a continuous sum while the integer division operation would remove the least significant digit from the integer until the integer is zero. The next pseudocode represents this solution:

```
Begin
  read n   /*let the inputted integer   be n*/
  sum = 0
  while n   <> 0
  sum = sum + (n MOD 10)
  n = n DIV 10
  end while
  print sum
End
```

Another possible solution would be to convert the inputted integer into a string (array of characters) and traversing through the array and summing all the digits. The following pseudocode represents this solution:

In the below pseudocode, TO_CHAR and TO_INT are pseudo-functions; students would be required to use the respective functions in the programming language they are using. There are also variations to the two solutions presented for problem one which are correct. Some students are able to recognize both solutions and would be required to choose between them; this relates to the component of metacognition regarding the planning and selecting of strategies.

```
Begin
  read n /*let the inputted integer be n*/
  char[]nums = TO_CHAR(n) ; /* TO_CHAR converts the inte-
ger to array of characters*/
  i = 0
  sum = 0
  while i < length(nums)
  sum= sum + TO_INT(nums[i]) /*TO_INT converts a charac-
ter to integer*/
  i = i + 1
  end while
  print sum
End
```

Before the students can select a strategy, they must ensure that they possess the necessary programming skills to implement the solution this relates to the knowledge of self (personal skills) component of metacognition.

To implement the first solution, students need to know about the usage and purpose the DIV and MOD operations and about repetition loops (while, for, etc.). For the second solution, knowledge of repetition loops, arrays and the conversion functions are necessary.

For the second problem, the solution requires the traversal of all the numbers between 20 and 60 using a loop and determining whether each number is even using the MOD operation (MOD 2) and checking if the result is equal to zero.

If the result is 0 (no remainder when divided by two), the number is even, this can be done by checking to see if the result of MOD 2 was not equal to 1 but this requires a further understanding that MOD 2 would only return one of two values (0, 1).

For both problems, although they are different, the MOD operator is useful. The use of the while loop was different in both problems. For problem one, the loop condition was 'while $n <> 0$', i.e., while n was not equal to zero and the loop variable 'n' was modified within the loop using the DIV operator.

For problem two, the loop condition was i < length(nums), i.e., while i was less than the length of the array nums and the loop variable i was modified within the loop by increasing its value by 1.

A key problem solving skill that students should possess is to be able to recognize problems they have previously done and be able to transfer their knowledge from solving other problems which utilized the same skills. This is often referred to as 'far transfer' of learning [21]. Consider the following problem:

3. The year is 2013, in a small company, the CEO appointed new managers of accounts, finance and sales. Accounts managers are appointed for 2-year terms, finance managers for 3-year terms and sales managers for 4-year terms. Write a program which determines the next year in which there will be new appointees to all three positions.

For problem three, students are required to determine the next year after 2013 in which the difference between that year and 2013 is exactly divisible by 2, 3 and 4. A repetition structure and the use of MOD operator are useful for this problem which suggests that students who would've done problems one and two should be able to solve problem three.

However, for problem three, good problem solving students would recognize that they do not need to check every successive year since the longest term is 4 years; they only need to check every 8 years after 2013 so the loop variable can be incremented by 4. Also, when incrementing by 4, the variable will always be exactly divisibly by 4 and all numbers which are exactly divisible by 4 will be exactly divisible by 2, so the only necessary check is to determine if the variable is exactly divisible for 3 as given in the following solution:

```
Begin
  int yrDiff=4
  while yrDiff MOD 3 <> 0
    yrDiff = yrDiff + 4
  end while
  newYr = 2013 + yrDiff
  print newYr
End
```

The ability of students to transfer knowledge from previous problems they have done to new problems represents their ability to regulate and their learning and develop their learning strategies. For all three problems, students also need to know about variables, arithmetic calculations, program input/output and program sequencing. Most of these can be easily taught with the exception of sequencing. Sequencing represents the order in which the various program components form the solution and is the most important part of the solution. In both given solutions for problem one, if the initialization of the sum variable was done within the while loop, incorrect sums would be calculated.

After determining whether they lack any of the necessary skills to implement the solution, it is the students' responsibility to learn/acquire these skills. For introductory programmers, this normally requires the intervention of the teacher since the students are normally unaware of their deficiencies at the introductory level, however, as they increase their problem solving skill, they will be able to better monitor and control their learning. Finally, there are certain operations where the order in which they are done can vary and a correct result is still obtained; for example, in the second solution for problem two, the order of the 'i' and 'sum' variables can be interchanged without affecting the main purpose of the program. We propose four dimensions of metacognitive skills that are required by programmers:

- Properly understand and correctly interpret the problem and what is required.
- Determine the steps required to solve the problem and know the correct sequence of the steps.
- Identify whether they possess the required skills for a particular solution, choose between different solutions and choose the best solution.
- Correctly verify whether their solution is correct.

The main challenge for educators and researchers lies in developing teaching methodologies which addresses the metacognitive and motivational aspects of problem solving. Collaborative learning has been identified as one such opportunity.

11.4 Promoting Metacognition in Programming Using Collaboration and Computer Games

11.4.1 Collaborative Learning

Ben-Ari [22] suggests that programming concepts need to be actively acquired by students and cannot be directly transferred from instructor to student. The application of constructivism to the teaching of programming is a possible solution [23–25]. Constructivist theory proposes that knowledge is actively constructed by students throughout the learning process and not absorbed from teachers or textbooks.

The theory also suggests that knowledge construction is recursive; therefore students will continuously build on what they already know and can also build on the experiences of other students and their teachers. Collaborative learning is an instruction method which utilizes constructivism.

Collaborative learning is an instruction method in which students work in groups towards a common academic goal [26]. Panitz [27] distinguished collaborative learning and cooperative learning as follows:

"Cooperative learning is defined by a set of processes which help people interact together in order to accomplish a specific goal or develop an end product which is usually content specific."

"Collaborative learning (CL) is a personal philosophy, not just a classroom technique. In all situations where people come together in groups, it suggests a way of dealing with people which respects and highlights individual group members' abilities and contributions."

However, the terms cooperation and collaboration have always been used interchangeably in research conducted in the areas of collaborative learning or cooperative learning since both are founded on constructivist learning theory. Sardone's comparison [15] of the traditional and constructivist learning environments showed that constructivist learning environments where active learning strategies are used, negate the influence of preferred learning styles. This suggests that collaborative learning approaches can meet the learning preferences of all students. Some of the major achievements of collaborative learning are:

- Motivation: Students are driven by a reward or goal structure and the only way they can attain their personal goal is if the entire group succeeds so they would encourage the other members of the group.
- Social Cohesion: Students may care about the other members of the group and would therefore help and encourage them.
- Development: Students will be exposed to the viewpoints and explanations of their group members and this will enhance their own cognitive processes.
- Cognitive Elaboration: Students will be required to explain their contributions and decisions and by having to provide to these explanations in a social context, it can help clarify and reinforce their own thought processes.

Alavi [28] as reported by Jara et al. [29] stated that collaborative learning methods tend to encourage the construction of knowledge, deeper understanding, and greater skill development since students are engaged in dynamic learning. Roger and Johnson [30] presented the following criteria for tasks which are deemed applicable to collaborative learning:

- The task is complex or conceptual.
- Problem solving is desired.
- Divergent thinking or creativity is desired.
- Mastery or retention is important.
- Quality of performance is expected
- Higher level reasoning strategies and critical thinking are needed.

Clearly all the listed criteria apply to computer programming. Acknowledging this, and the former stated benefits of collaborative learning (specifically, motivation and cognitive elaboration), collaborative learning appears to be a tool which can be successfully utilized for teaching programming to novices.

Additionally, Kelleher and Pausch [31] identified a lack of social context in programming and suggested that the use of group work can make the task of learning programming easier and fun. Collaborative learning is also a more realistic model of how software is developed in industry as opposed to the solitary programming which is normally used in introductory programming courses [32].

Furthermore, most software development companies utilize software development methodologies like extreme programming which incorporate team work as a fundamental component. In extreme programming, managers, customers and developers are all equal partners in a collaborative team.

11.4.2 Collaboration Enhances Programming Metacognition

Encouraging novice programmers to collaborate can help alleviate some of the challenges which they face while programming. Through structured collaboration, the students will be forced to engage in argumentative discussion where they are required to listen to the viewpoints and opinions of their peers and offer their own.

Referring to the programming process, when the students collaborate, they can avoid misinterpreting of the task requirements since they can correct each other's misconceptions. The collaboration also increases the chances of multiple solutions being developed for the same problem and through argumentative discussion, the best one can be chosen.

The implementation of the solution also becomes easier since it is more likely that the all skills required to implement the solution exists within the group than for an individual programmer. Finally, the evaluation of the solution also becomes more effective since it can be tested by different persons within the group.

A study which investigated the usefulness of collaboration for Java programming concluded that collaboration was deemed most important while conceptualizing, brainstorming, and formulating the problem and its requirements; also the more complex the problem, the greater the importance of the collaboration [33].

This result follows with research which suggested that the major cause of students' failure in introductory programming is the lack of basic problem solving skill. Most of the research on the use of collaboration to teach programming is in the pair programming pedagogy. Pair programming is described as:

A style of programming in which two programmers work side-by-side at one computer, continuously collaborating on the same design, algorithm, code, or test. One of the pair, called the driver, types at the computer or writes down a design. The other partner, called the navigator, has many jobs. One is to observe the work of the driver, looking for defects. The navigator also has a more objective point of view and is the strategic, long-range thinker. Together, the driver and the navigator continuously brainstorm a solution. Periodically, the programmers switch roles between the driver and the navigator [34].

Davidson [35] identified the key attributes of collaborative learning as:

- Common Task or Learning Activity.
- Small Group Learning.
- Cooperative Behavior.
- Positive Interdependence.
- Individual Accountability and Responsibility.

Preston [36] used the above framework to analyze the pair programming pedagogy and concluded that pair programming is a model for collaborative learning. DeClue [37] concluded that pair programming has positive motivational characteristics; produces higher quality programs with coherent design and code documentation; leads to decreased time to complete assignments; and improves understanding of the programming and software engineering processes.

Collaborative learning was mainly adopted in classroom based environments which required face to face interaction between learners, as is the case with pair programming. This approach has shown to be very useful to learners, but it needs to be extended and enhanced to make its benefits more accessible to teachers and students.

11.4.3 Computer Supported Collaborative Learning (CSCL)

Computer Supported Collaborative Learning (CSCL) has been identified as one of the most promising innovations to improve teaching and learning with the help of modern information and communication technology (ICT) [38]. CSCL aims to enhance learning by combining computer support and collaborative learning [39].

Originally, collaborative learning was mainly adopted and confined to the classroom based environments which required face to face interaction between students and lecturer; by utilizing technology, there is no longer the need for this physical interaction. Through CSCL, the opportunity also exists to extend the learning process from the classroom and make it easily available to students.

Newman et al. [40] claims that a clear link between critical thinking, social interaction, and deep learning has emerged. An abundance of social interaction takes place on the Internet and therefore it is possible for good CSCL systems to do just as well in promoting learning as conventional group work.

A study conducted by Pifarre and Cobos [41] found an increase in students' metacognitive skills after using a CSCL system and their result was similar to other findings which the authors reported as part of their literature review. This suggests that the use of CSCL systems enhance the development of metacognitive learning processes. Lee et al. [42] concluded that while engaged in group or community learning, students analyzed the community discourse and improved their own understanding.

Diziol et al. [43] stated that when students collaborated on conceptual-problem solving steps, they talked to each other and provided mutual explanations. This led to improved learning when compared to individual learning. Chen [44] also proposed that the use of collaborative tools such as discussion boards or emails can be exploited to stimulate student motivation and encourage problem solving. These findings all indicate successful applications of CSCL for problem solving.

Argumentation and how students can benefit from it has always been a main focus in the field of CSCL [45]. Lu et al. [46] suggested that argumentation tools and visualization can be designed in CSCL to facilitate students' problem solving by promoting collaboration and shared understanding. Collaboration Scripts have been identified as a solution for improving the quality of argumentation. A collaboration script is defined as:

"A set of instructions regarding to how the group members should interact, how they should collaborate and how they should solve the problem" [47]. Implementing CSCL scripts has resulted in improved learning outcomes [48, 49]. Scripts can also be useful in helping students structure their argumentative discourse [50].

11.5 Multiplayer Games to Support Programmers Problem Solving

11.5.1 Educational Multiplayer Games

Games have been known to create interest, cooperation, and competition for its players. These are all qualities which educators have strived to inspire in the classroom [51], therefore it makes sense to merge the motivation of games with learning.

Games provide a challenge and deliver rewards which encourage students to work harder and can be used to encourage learning. The long term effects of game playing are as follows [52]:

- Heightened concentration.
- Increased Intelligence.
- Batter hand-eye coordination.
- Increased stamina and determination.
- Better multi-tasking.
- Better awareness.

Doherty and Kumar [53] recognized the highly abstract nature of core programing concepts and suggested that games which are successful at teaching programming are those which cause the learner to develop and understand concepts from the content of the game as a consequence of its system and interface. Doherty and Kumar defined a game environment as one in which the concepts that emerge from interacting with it are created by the goal. Games can also help to alleviate some of the difficulties which students face while programming [54].

Recognizing the important benefits of collaboration in learning, researchers began exploring the possibility of building educational or collaborative multiplayer games. The possibility of using multiplayer games as educational tools has been explored using factors such as frequency of game play, gender, self-esteem, computer self-efficacy, and academic performance [55].

Their findings strengthened the possibility of multiplayer games becoming educational tools that can engage students and lead to accomplished learning outcomes. Li et al. [56] agreed that online multiplayer games can be used as educational tools if they are guided by an appropriate learning theory like constructivist theory.

It has been suggested that the principles of CSCL and problem solving can be applied to multiplayer games [57]. They put forward that the game mechanics should not only encourage but rather require players to engage in collaborative interactions in order to solve the problem; and that collaborative interactions should be enforced, rather than competitive ones.

11.5.2 Guidelines for Designing Multiplayer Games to Support Problem Solving

The following are guidelines for creating multiplayer games which aim to enhance the metacognitive skills of novice programmers. The game's design and characteristics should stimulate collaboration not only as part of the task, but as an integral part of the learning process. It should also promote collaborative rather than competitive interactions. Strategy games are preferred since it requires careful thinking and planning by the players to ensure success.

The two common types of strategy games are real-time strategy (RTS) and turn-based games. In turn-based games, each player is required to pay attention to the moves made by every other player, whereas RTS games allow players to play independently for portions of the game. In both cases, the game environment changes to reflect the actions performed by the players.

The game should engender or embody the educational content. A game which embodies educational content is one which has the educational content as a core part of its system; and a game which engenders educational contents causes the learner to develop and understand concepts from the content as a consequence of its system and interface [53]. The educational content should not be a simple addition to the game in the form of multiple choice or fill in the blank questions; instead, the game should be designed around the concepts being taught.

A game can motivate students to learn by rewarding learning, practice or mastery with in-game success [53]. This means that players cannot be successful at the game unless they understand the educational concepts which are being taught. A main challenge of many games is to earn rewards or get the highest score; students should not be allowed to attain these unless they are successful at the learning tasks in the game.

A game which requires the players to earn points to progress from one level to the next; or to unlock new features and get bragging rights can also be very useful in motivating students. The use of games can provide the motivation which is important to ensure that learning occurs and the students acquire the necessary skills.

However, it is important that the gameplay and aesthetics do not overwhelm the educational content. The learning tasks should always be the priority, the players should not be allowed to stray away from the required learning tasks. The game should implement a cooperative rather than a competitive reward structure. Players will work in groups to accomplish a learning task and all members of the group should be rewarded or penalized equally.

This ensures that the members of the group understand that their success lies in the success of the entire group and they can only succeed if the group as a whole succeeds; this is referred to as positive interdependence.

Positive interdependence also encourages the development and cognitive elaboration benefits of collaborative learning [58]. Members of the group would be more likely to help their group members and receive help in return; during this exchange they would be exposed to the viewpoints of other members and presented with the opportunity to express their thoughts, which in turn contributes to their development.

Promoting positive independence will also foster higher level critical thinking and reasoning strategies; and encourage a willingness to take on more difficult tasks [59]. Each member of the group becomes personally responsible for the team's success and would be encouraged to try harder because they know that their group members are dependent on them.

Argumentative Discussion is a key feature of the game design and it is related to positive interdependence. Each member of groups feels responsible for the team's success so they would make every effort to ensure they succeed; however, the game should ensure the group members are always aware of the need to make the best decisions throughout the entire game.

This can be enforced using time or other constraints. For example, the group can be required to accomplish an objective within a fixed time period, within a fixed number of moves, or to achieve an outcome which satisfies specified criteria.

Group members will be encouraged to compare and contrast their reasoning and decisions with their group members promoting higher quality decision making, creativity and problem solving.

This will also lead to greater productivity by the entire group since the members would strive to make the best possible decisions throughout the game.

Equal participation within a group is a vital area of concern since experiences with traditional group work have always shown cases where certain members of the group take full responsibility and the other members do not participate. The easiest approach for ensuring equal participation would be implementing a turn based approach. However, if complex tasks are being targeted, they can be divided into smaller tasks and each member can be assigned as the lead for a subtask.

Equal participation entails not only ensuring that each member works on their own task but that each member of the group has the opportunity to contribute to what the other members have done. Each activity or move made in the game should be a result of group collaboration.

In CSCL research, individual accountability refers to an individual evaluation after the collaborative process is completed. It suggests that in order to accurately evaluate the collaborative learning process, each group member must be individually assessed as they are responsible for their own learning.

However, in this context individual accountability is interpreted as encouraging each group member to explain their actions or moves to their group members to promote argumentative discussion. The game should create scenarios which require each member to explain their decisions to their group members. In return, the members must assess and contribute to what was done. This helps to ensure that the best possible group decisions are made throughout the game.

11.6 Implementation and Experimental Findings

11.6.1 Implementation

The main output of problem solving is an algorithm. An algorithm is the sequence of steps required to solve a problem. Flowcharts and pseudocode are two common program design tools used for the representation of algorithmic solutions. Pseudocode is a text based representation which consists of English-like statements.

It is designed to fill the gap between the informal (spoken or written) description of the programming task and the final program code [60]. Flowcharts are a visual representation of program flow using a combination of arrows and symbols to represent the actions and sequence of the program.

Collaborative Online Problem Solving (COPS) is a turn based strategy game in which groups of two, three or four players are required to collaboratively build a program flowchart for a given problem within a target number of moves.

COPS was developed using the guidelines presented in the previous section. There are two different games in COPS:

- SWAP: The group is shown a flowchart with pieces out of order and the players are required to swap pieces to correct the flowchart. This game is designed to be the easier level of COPS.
- JIGSAW: The group is required to build a flowchart similar to how they would construct a jigsaw puzzle. This game is designed to be the harder level of COPS.

Each member of group receives 10 points for each SWAP game and 20 points for each JIGSAW game which is completed within the target number of moves. For both games, the group is deducted one point for each extra move they make beyond the target number of moves. For games which the group quits, no points are awarded.

While each member of the group is awarded equally, a player is allowed to play with different groups and increase the number of points which they earn. This allows players who play regularly to score more points and stimulates a competitive atmosphere amongst players.

Each problem in COPS has an associated question type. The question types are categorized based on the basic skills which novice programmers should acquire. Each puzzle also has an associated difficulty level. In order to progress to more difficult puzzles, players are required to reach a minimum number of points.

The collaboration in COPS is enforced through a voting system. COPS is turn based and each time a player makes a move, the other group members are required to vote on whether they agree or disagree with the move; if the move receives a majority vote, the game accepts the move otherwise it is rejected. To avoid ties, the player making the move is given a higher weighted vote.

The voting system aims to encourage argumentative discussion, positive interdependence and individual accountability. The player who makes the move would be required to explain their move and decisions using the chat system to their group members to convince them to accept the move. Similarly, the other group members would be required to explain why they may not agree with the moves.

The target number of moves is also meant to encourage the group make the best possible move each time since they only earn maximum points if they solve the puzzle within the target number of moves. The enforced collaboration between the group members ensures that every member of the group is involved in each game move and makes them accountable for their decisions. The overall design of COPS also ensures that the individual players only succeed when the entire group succeeds.

COPS provides intelligent feedback to players through graphical and textual alterations to help the students visualize their problem solving. For each accepted move, COPS automatically generates the pseudocode equivalent of the flowchart regardless of whether the flowchart is correct or incorrect.

The pseudocode guide is useful since it can help the players when they become stuck and it also shows them the pseudocode for their flowchart which will be useful in implementing the solution to the problem. Within the SWAP game, the places in the flowchart which are in incorrect positions are highlighted to guide the learning process; un-highlighted pieces indicate to the players that the pieces are correct and they can focus on solving other parts of the puzzle. In the JIGSAW game, the parts of the flowchart which are correct are highlighted to offer the same guidance to players.

Referring back to the definition of metacognition in programming given in Section three, COPS aims to help students in many ways. All problems in COPS are done by a group and through the chat system provided; members can help each other clarify any misinterpretations with the problem description. In both the SWAP and JIGSAW games, the group is given the general components which form the solution but it is the responsibility of the group to determine the sequence of the pieces/components to build the solution. COPS also uses intelligent feedback to let the players know whether they solution is correct or incorrect and accepts multiple correct solutions for the same problem, so the players are allowed to consider varying solutions.

11.6.2 Experimental Findings

The primary target users of COPS are secondary school students between the ages of 13 and 17 who are learning problem solving and programming for the first time. However, COPS focuses on problem solving and is language independent so it can be used for introductory programming courses at any level or institution. A survey of secondary school students was done asking about their difficulties with problem solving. The responses were categorised based on the four dimensions of programming metacognition given at the end of section three. The findings indicated that:

- 52 % of respondents admitted difficulty in understanding and interpreting a problem and what is required.
- 19 % had difficulty with determining the steps required to solve the problem and knowing the correct sequence of the steps.
- 7 % had challenges choosing between different solutions and choosing the best solution.
- 12 % admitted having problems with the syntax of the programming language.
- 10 % responded that they don't know what their major difficulties were.

The first statistic reiterates the findings of previous research and supports the case for the use of collaboration for teaching programming. Two studies were done to investigate the usefulness of COPS for secondary school students across Trinidad and Tobago in learning problem solving and programming.

A control version (non-collaborative/single player) of COPS was built as compared against the collaborative (multiplayer) version of COPS while being used by introductory programming students who had never done programming before.

An ancova analysis of the pre and post test results from study one showed significantly (p-value: 0.002) better performance by the students who used the collaborative version (mean improvement 14.27 %) than those who used the controlled version (mean improvement of 11.58 %). For the second study, there was no control, but the collaborative version of COPS was used by students who had previously done programming A paired t-test of the pre and post test results showed a significant (p-value: 0.000) improvement by the participants by 21.56 %.

The results from both studies indicated that COPS was useful both as a learning and revision tool for novice programmers. A more detailed analysis of the collaboration amongst participant in both studies showed that the first time programmers from the first study benefitted more from collaborating with the same students more often. However, the participants from the second study who had done programming before benefitted more from collaborating with different students. These findings indicate that COPS can be used to successfully enhance the problem solving skill of novice programmers.

11.7 Conclusion

The chapter addresses the challenge of problem solving in computer programming and presents an approach for Collaborative Learning that enhances the metacognitive skill of novice programmers. Many students acquire basic coding skills but they are unable to utilize them in a meaningful way to solve non-routine problems and they are unable to verify whether their solution is correct. Improving their metacognitive ability would help students identify and understand what a problem requires and analyse and evaluate the different alternative solutions to the problem.

Collaborative learning, which is founded on constructivist learning theory, has been shown to help students improve their problem solving skill by promoting metacognition. In experiments conducted using a collaborative strategy game, students improved their metacognition skill.

They were better able to understand the requirements of the problem and, through visualizing the solution using flowcharts and pseudocode, improve in decomposing the problem into manageable chunks, all the while enjoying the interaction with other players. Students were captivated with the game and learning took place transparently. The chapter provides a comprehensive review of the state-of-the-art of research on Collaborative Learning and Metacognition.

We developed a framework for successful computer supported collaborative learning environments; the collaboration should encourage equal participation, argumentative discussion and positive interdependence.

We have developed a multiplayer strategy game which conforms to this framework and which improves the metacognition of each player. By enforcing equal participation, each player is motivated to understand the problem and analyse the logic in the programming solution; the argumentative discussion means that they must be able to defend their solution to the other players in the team and positive interdependence means that all players must learn if the team is to complete the game successfully and no player is left behind.

There are still several open areas for research and future work. One of these has to do with integration of CSCL environments into traditional classroom teaching on a large scale. Developing a truly blended approach is not a straightforward task. We are convinced that teachers and educators play an invaluable role in the success of CSCL games. The success of the games relies on its adoption and addition to the classroom. Secondly, future work could focus on the whole program development life cycle.

We have focused on the early phases of understanding the requirements of the problem and developing a solution. After solving a puzzle in COPS, the players/students would have an algorithmic solution (flowchart/pseudocode) for the problem but they are still required to write the program code afterwards. This has its own set of challenges. We also want to examine the relationships between different types of players collaborating in the group.

Do students prefer to play with other inexperienced players that they may know well or is there some benefit in playing with more advanced programmers? People are naturally drawn to people that they are comfortable with. We want to study how groups are formed online, the dynamics of the group, and the impact of different combinations of players with differing abilities.

References

1. Bachu, E., Bernard, M.: A computer supported collaborative learning (CSCL) model for educational multiplayer games. In: 11th International Conference on e-Learning, e-Business, Enterprise Information Systems, and e-Government. Las Vegas (2012)
2. Bachu, E., Bernard, M.: Enhancing computer programming fluency through game playing. Int. J. Comput. 1(3) (2011)

3. CXC. Information technology general proficiency examination May/June 2011. Report on Candidates' Work in the Secondary Education Certificate Examination. St. Michael, Barbados (2011)
4. CXC. Information technology general proficiency examination May/June 2010. Report on Candidates' Work in the Secondary Education Certificate Examination. St. Michael, Barbados (2010)
5. CXC. Information technology general proficiency examination May/June 2012. Report on Candidates' Work in the Secondary Education Certificate Examination. St. Michael, Barbados (2012)
6. CXC. Annual reports for year 2005–2009. St. Michael, Barbados
7. CXC. Information technology general proficiency examination January 2011. Report on Candidates' Work in the Secondary Education Certificate Examination. St. Michael, Barbados (2011)
8. CXC. Information technology general proficiency examination January 2010. Report on Candidates' Work in the Secondary Education Certificate Examination. St. Michael, Barbados (2010)
9. CXC. Information technology general proficiency examination January 2012. Report on Candidates' Work in the Secondary Education Certificate Examination. St. Michael, Barbados (2012)
10. Beaubouef, T., Mason, J.: Why the high attrition rate for computer science students: some thoughts and observations. ACM SIGCSE Bull. 37(2), 103–106 (2005)
11. Sheard, J., Simon, S., Hamilton, M., Jan L.: Analysis of research into the teaching and learning of programming. In: 5th International Computing Education Research Workshop, pp. 93–104. ACM, New York (2009)
12. Gomes, A., Mendes, A.J.: Learning to program-difficulties and solutions. In: International Conference on Engineering Education, vol. 2007. Coimbra, Portugal (2007)
13. Deek, F.P., McHugh, J.A., Turoff, M.: Problem solving and cognitive foundations for program development: an integrated model. In: Sixth International Conference on Computer Based Learning in Science (CBLIS), pp. 266–271. Nicosia, Cyprus (2003)
14. Watson, R., de Raadt, M., Toleman, M.: Teaching and assessing programming strategies explicitly. In: 11th Australasian Computing Education Conference (ACE 2009), Wellington, New Zealand (2009)
15. Sardone, N.B.: Developing information technology fluency in college students: an investigation of learner environments and learner characteristics. Inf. Technol. Educ. 10(1), 101–122 (2011)
16. Hundhausen, C.D., Farley, S.F., Brown, J.L.: Can direct manipulation lower the barriers to computer programming and promote transfer of training?: an experimental study. ACM Trans. Comput. Hum. Interact. 16(3), 1–40 (2009)
17. Mayer, R.E.: Cognitive, metacognitive, and motivational aspects of problem solving. Instr. Sci. 26(1), 49–63 (1998)
18. Metcalfe, J., Shimamura, A.: Metacognition: Knowing About Knowing. Bradford Books, Cambridge (1994)
19. Jonassen, D.: Learning to solve problems: A Handbook for Designing Problem-Solving Learning Environments. Taylor & Francis, United Kingdom (2011)
20. Bachu, E.: A Framework for Computer Supported Collaborative Learning (CSCL) Using Online Multiplayer Games. M.Phil., dissertation, The University of the West Indies, St. Augustine, Trinidad and Tobago (2013)
21. Viviene, C., Macaulay, C.: Transfer of Learning in Professional and Vocational Education. Psychology Press, United Kingdom (2000)
22. Ben-Ari, M.: Constructivism in computer science education. In: 29th SIGCSE Technical Symposium on Computer Science Education, pp. 257–261. ACM, New York (1998)
23. Gonzalez, G.: Constructivism in an introduction to programming course. J. Comput. Sci. Coll. 19(4), 299–305 (2004)

24. Boyer, N.R., Langevin, S., Gaspar, A.: Self direction and constructivism in programming education. In: 9th ACM SIGITE Conference on Information Technology Education, pp. 89–94. ACM, New York (2008)
25. Lui, A.K., Kwan, R., Poon, M., Cheung, Y.H.Y.: Saving weak programming students: applying constructivism in a first programming course. ACM SIGCSE Bull. 36(2), 72–76 (2004)
26. Gokhale, A.: Collaborative learning enhances critical thinking. J. Technol. Educ. 7(1), 56–65 (1995)
27. Panitz, T.: Collaborative versus cooperative learning: a comparison of the two concepts which will help us understand the underlying nature of interactive learning (1999)
28. Alavi, M.: Computer-mediated collaborative learning: an empirical evaluation. MIS Q. 18(2), 159–174 (1994)
29. Jara, C.A., Candelas, F.A., Torres, F., Dormido, S., Esquembre, F., Reinoso, O.: Real-time collaboration of virtual laboratories through the Internet. Comput. Educ. 52(1), 126–140 (2009)
30. Roger, T., Johnson, D.W.: An overview of cooperative learning. In: Thousand, J., Villa, A., Nervin, A. (eds.) Creativity and Collaborative Learning. Brookes Press, Baltimore (1994)
31. Kelleher, C., Pausch, R.: Lowering the barriers to programming: a taxonomy of programming environments and languages for novice programmers. ACM Comput. Surv. 37(2), 83–137 (2005)
32. Urness, T.: Assessment using peer evaluations, random pair assignment, and collaborative programing in CS1. J. Comput. Small Coll. 25(1), 87–93 (2009)
33. Bagley, C.A., Chou, C.C.: Collaboration and the importance for novices in learning Java computer programming. In: 12th Annual SIGCSE Conference on Innovation and Technology in Computer Science Education, pp. 211–215. ACM, New York (2007)
34. Williams, L.: Lessons learned from seven years of pair programming at North Carolina State University. SIGCSE Bull. 39(4), 79–83 (2007)
35. Davidson, N.: Cooperative and collaborative learning: an integrative perspective. In: Thousand, J., Villa, R., Nevin, A. (eds.) Creativity and Collaborative Learning: A Practical Guide to Empowering Students and Teachers, pp. 13–30. Paul H. Brookes Publishing Co., Baltimore, MD (1994)
36. Preston, D.: Pair programming as a model of collaborative learning: a review of the research. J. Comput. Sci. Coll. 20(4), 39–45 (2005)
37. DeClue, T.H.: Pair programming and pair trading: effects on learning and motivation in a CS2 course. J. Comput. Sci. Coll. 18(5), 49–56 (2003)
38. Ehtinen, E., Hakkarainen, K., Lipponen, L., Rahikainen, M., Muukkonen, H.: Computer supported collaborative learning: a review. The JHGI Giesbers Reports on Education (1999)
39. Stahl, G., Koschmann, T., Suthers, D.: CSCL: an historical perspective. In: Sawyer, K.R. (ed.) Cambridge Handbook of the Learning Sciences, vol. 5, pp. 409–426. Cambridge University Press, UK (2006)
40. Newman, D.R., Webb, B., Cochrane, C.: A content analysis method to measure critical thinking in face-to-face and computer supported group learning. Interpersonal Comput. Technol. 3(2), 56–77 (1995)
41. Pifarre, M., Cobos, R.: Promoting metacognitive skills through peer scaffolding in a CSCL environment. Int. J. Comput. Support. Collaborative Learn. 5(2), 237–253 (2010)
42. Lee, E.Y.C., Chan, C.K.K., Van-Aalst, J.: Students assessing their own collaborative knowledge building. Int. J. Comput. Support. Collaborative Learn. 1(1), 57–87 (2006)
43. Diziol, D., Rummel, N., Spada, H., McLaren, B.M.: Promoting learning in mathematics: script support for collaborative problem solving with the cognitive tutor algebra. In: 8th International Conference on Computer Supported Collaborative Learning, pp. 39–41. ISLS, USA (2007)
44. Chen, J.W.: Designing a web-based Van Hiele model for teaching and learning computer programming to promote collaborative learning. In: 5th IEEE International Conference on Advanced Learning Technologies, pp. 313–317. IEEE, NJ (2005)

45. Stegmann, K., Weinberger, A., Fischer, F.: Facilitating argumentative knowledge construction with computer-supported collaboration scripts. Int. J. Comput. Support. Collaborative Learn. 2(4), 421–447 (2007)
46. Lu, J., Lajorie, S.P., Wiseman, J.: Scaffolding problem-based learning with CSCL tools. Int. J. Comput. Support. Collaborative Learn. 5(3), 283–298 (2010)
47. O'Donnell, A.M., Dansereau, D.F.: Scripted cooperation in student dyads: a method for analyzing and enhancing academic learning and performance. In: Hertz-Lazarowitz, R., Miller, N. (eds.) Interaction in Cooperative Groups: The Theoretical Anatomy of Group Learning, pp. 120–141. Cambridge Universirty Press, UK (1992)
48. Rummel, N., Spada, H.: Can people learn computer-mediated collaboration by following a script? In: Fischer, Frank, Kollar, Ingo, Mandl, Heinz, Haake, JörgM (eds.) Scripting Computer-Supported Collaborative Learning, pp. 39–55. Springer, USA (2007)
49. Weinberger, A., Fischer, F., Mandl, H.: Fostering computer supported collaborative learning with cooperation scripts and scaffolds. In: 5th International Conference on Computer Supported Collaborative Learning, pp. 573–574. ISLS, USA (2002)
50. Bures, E.M., Abrami, P.C., Schmid, R.F.: Exploring whether students' use of labelling depends upon the type of activity. Int. J. Comput. Support. Collaborative Learn. 5(1), 103–116 (2010)
51. Prensky, M.: Digital game-based learning. Comput. Entertainment 1(1), 1–4 (2003)
52. Tsiatsos, T.A., Konstantinidis, A.: Utilizing multiplayer video game design principles to enhance the educational experience in 3D virtual computer supported collaborative learning environments. In: 12th IEEE International Conference on Advanced Learning Technologies, pp. 621–623, IEEE, NJ (2012)
53. Doherty, L., Kumar, V.: Teaching programming through games. In: International Workshop on Technology for Education, pp. 111–113. IEEE, Bangalore (2009)
54. Rajaravivarma, R.A.: Games-based approach for teaching the introductory programming course. SIGCSE Bull. 37(4), 98–102 (2005)
55. Paraskeva, F., Mysirlaki, S., Papagianni, A.: Multiplayer online games as educational tools: Facing new challenges in learning. Comput. Educ. 54(2), 498–505 (2010)
56. Li, Y., Tian, X., Gao, P.: Research on the application of MMO games in education. In: International Conference on Industrial Control and Electronics Engineering (ICICEE), pp. 535–538. IEEE, NJ (2012)
57. Voulgari, I., Komis, V.: Massively multi-user online games: the emergence of effective collaborative activities for learning. In: 2nd IEEE International Conference on Digital Game and Intelligent Toys Based Education (DIGITEL), pp. 132–134. IEEE, Banff, BC (2008)
58. Slavin, R.E.: Research on cooperative learning and achievement: what we know, what we need to know. Contemp. Educ. Psychol. 21(1), 43–69 (1996)
59. Johnson, D.W., Johnson, R.T., Karl Smith, K.: The state of cooperative learning in postsecondary and professional settings. Educ. Psychol. Rev. 19(1), 15–29 (2007)
60. Roy, G.G.: Designing and explaining programs with a literate pseudocode. ACM J. Educ. Resour. Comput. 6(1), 1–18 (2006)

Chapter 12
Designing a Metacognitive Approach to the Professional Development of Experienced Science Teachers

Osnat Eldar and Shirley Miedijensky

Abstract In this chapter we present a metacognitve approach to the professional development of science teachers. We designed two courses based on the Design Principles Database (DPD). Our aims were: to characterize the design principles of both courses, to expose expressions of metacognition among the teachers and examine the changes they designed and applied in their teaching units and teaching processes, and to characterize the resources which effected the development of the teachers' metacognitive knowledge. The participants were 21 teachers, 17 children, a laboratory assistant, and the two researchers. The data included: both courses' design and activities, the researchers' reflections, interviews, the teachers' teaching units, and observations of the children's physics lessons. The data was analyzed using qualitative methods. Our findings show that the courses we designed engaged teachers in constructing their own knowledge as well as collaborating as a group of learners. We identified six key resources which influenced the metacognitive development of the teachers: the courses in this study; other courses; the researchers' insight; the teachers' teaching experience; peer suggestions and children's reflections. The present study strongly encourages teachers to develop and design activities, as well as testing them in a supportive environment which can help them understand their own beliefs as well as promote their metacognitive knowledge.

Keywords Metacognitive knowledge · Professional development · Design principles · Peer learning · Reflection

O. Eldar and S. Miedijensky contributed equally to this work

O. Eldar (✉) · S. Miedijensky
Oranim Academic College of Education, Tivon, Israel
e-mail: eldar@oranim.ac.il

S. Miedijensky
e-mail: shirley_m@oranim.ac.il

© Springer International Publishing Switzerland 2015   299
A. Peña-Ayala (ed.), *Metacognition: Fundaments, Applications, and Trends*,
Intelligent Systems Reference Library 76, DOI 10.1007/978-3-319-11062-2_12

Abbreviations

AAAS   American association for the advancement of science
CDC   Collaborative diagnosis of conceptions
DPD   Design principles database
FD   Forum discussions
KI   Knowledge integration
ME   Metacognitive experiences
MK   Metacognitive knowledge
NCTM   National council for teachers of mathematics
NRC   National Research Council
PCK   Pedagogical content knowledge
PD   Professional development
PL   Professional learning

12.1 Introduction

This chapter presents a collaborative self-study of two teacher educators who recognize the importance of researching their own practice, leading to a better understanding of the complex nature of teaching and learning about teaching.

Loughran [1] stated that teacher educators need to be good teachers of teaching so that the complex nature of teaching and learning becomes more evident to their students of teaching. Yet being a good teacher of teaching requires much more than just being a good teacher, and this is where self-study comes in as an important force shaping teacher education practices. The authors belong to the Science Education community, and in this chapter we wish to elaborate the design and implementation of two courses which are part of a two-year M.Ed. program designed for experienced high school science teachers who are interested in their own personal and professional development and intend to continue teaching in school. In both courses, "Principles of developing teaching units in Physics" and "Excellence in Science and Mathematics", the teachers were asked to develop teaching units in science and mathematics.

In the first course the teachers were also given the opportunity to teach the units they had developed to children participating in an enrichment program at the college. Since both authors are interested in promoting metacognitive knowledge (MK), and the fact that according to Zohar and Barzilai [2], fewer than half of the studies (37.9 %) on metacognition in science education dealt with MK, we chose to examine our work through the metacognitive lens, checking our own understanding of this concept together with the professional development (PD) of the teachers we teach. We believe that MK and metacognitive experiences (ME) are essential for the development of good, established teachers.

12.2 Theoretical Background

12.2.1 Professional Development of In-service Science Teachers

The reform of science and mathematics education in recent decades has led to the design and implementation of various professional development (PD) programs for science teachers, supported by the major science and mathematics bodies (American Association for the Advancement of Science (AAAS) [3], National Council for Teachers of Mathematics (NCTM) [4–7]).

Teachers' PD is part of the understanding that teachers must be lifelong learners, and that teaching is a complex and demanding profession which should involve the development of teachers' skills and attitudes so they may improve their students' learning [8]. Constructivism serves as the philosophical foundation of this reform, which calls for a new way of thinking about science teaching and learning.

Since most high school physics teachers come from a very conservative, teacher-centered way of teaching, a good PD program must involve the changing of the teachers' beliefs and understanding of teaching and learning, in order to deal with the quality of science teaching which seems to remain largely confined to knowledge transfer and is threatened in many countries by a shortage of qualified teachers. Scholars have shown that teachers consider new teaching practices as practical when (a) efficient procedures are available to translate innovative ideas into concrete instruction; (b) the change proposed sufficiently fits their current practice and goals; and (c) implementation of the innovation requires limited investment and the expected benefits are substantial [8].

One major characteristic of effective professional development for Science teachers is the development of MK about teaching and learning science. The National Science Education Standards [6] outlined goals for the PD of both in-service and pre-service science teachers. These goals suggest that the preparation of science teachers should include constructivist experiences which allow learners to gain both content knowledge and pedagogical skills through developing their MK. These experiences should enhance the science teachers' ability to provide similar experiences in their own classrooms [9]. This ability is interwoven with metacognitive knowledge about science teaching and learning.

Choosing the characteristics and content may be the most important decision taken in planning a PD program. The research literature [8] emphasizes six core features of an effective PD programs which are used as an organizing frame: focus, active and inquiry-based learning, collaborative learning, duration and sustainability, coherence and school organizational conditions. Eylon et al. [10] developed an evidence-based professional development program for physics teachers. Focusing on knowledge integration, they emphasize the need for a long PD program centered on evidence-based materials which the teacher tests in class and can afterwards discuss in the PD environment.

Nilsson [11] discusses professional learning (PL), describing a group of teachers working together collaboratively with a researcher to identify important aspects of students' learning regarding a specific topic, attempting to improve their teaching in a systematic manner. This is a cyclical process in which teachers reflect on the necessary conditions for learning specific content and how to meet these conditions in the learning situation. The teachers explore their teaching in order to identify which features may be critical for their students' learning. It seems self-evident that teachers need to know the subject matter they are required to teach.

The reform papers indicate that teachers need to have a deep and complex understanding of science concepts, the ability to make connections among them, and apply them in explaining natural phenomena or real-world situations [3, 6]. Teachers must also have content-specific teaching strategies. Moreover, teachers should have meta-strategic knowledge; they need to recognize the role of prior knowledge, particularly students' misconceptions, in shaping student comprehension.

In general, when teachers are more comfortable with teaching a particular topic, they are more likely to allow student questioning and discussion, which are essential features of inquiry.

PD focusing on content providing teachers with opportunities for active learning and connected to daily life in school is more likely to produce enhanced knowledge and skills. Teachers require rich pedagogical knowledge in order to create highly interactive learning environments which address the needs of all students.

As student preconceptions are a central issue in Science teaching, teachers need to be able to identify their students' preconceptions and design instructional strategies accordingly. Another changing perspective on PD suggests that learning activities should preferably be situated and meaningful, implying that they should be embedded within the regular work context. Learning activities are more effective when they are characterized by clear connections with daily practice in which problems, questions and solutions are integrated.

12.2.2 Teachers' Metacognitive Knowledge

Metacognition is a key component of active learning. It is defined as cognition about cognition [12], and involves thinking about one's own cognitive processes [12, 13], as well as the ability to monitor, regulate and evaluate one's thinking. Flavell [12] suggests that metacognition consists of both Metacognitive Experiences (ME) or regulation, and Metacognitive Knowledge (MK). MK includes knowledge of strategies that might be used for different tasks, knowledge of the conditions under which these strategies might be used, knowledge of the extent to which the strategies are effective, and knowledge regarding persons (the self and others) [12].

Schraw and Moshman [14] categorize metacognitive knowledge into three kinds of metacognitive awareness: declarative knowledge (what), procedural knowledge (how), and conditional knowledge (why and when). ME involve monitoring and regulation of cognitive processes, which consist of planning and monitoring cognitive activities and checking their outcomes [13, 15]. If teachers wish to relate to the pragmatic aspect of metacognition development or enable its expression among children, they need sound knowledge of metacognition and pedagogical knowledge in the context of teaching metacognition [2]. Unfortunately, research indicates that metacognition is almost invisible to Science teachers. Ben-David and Orion [16] showed that Science teachers could not explain or provide detailed examples of metacognitive-level thinking. However, following a training program, the teachers' pedagogical thinking in the context of metacognition has improved.

Baumert et al. [17] found that compared to teachers' subject-matter knowledge, pedagogical content knowledge (PCK) was a more powerful predictor of instructional quality. In analyzing 90 lessons taught by 10 teachers, [18] provided evidence linking teachers' PCK to the mathematical quality of instruction: teachers with stronger PCK made fewer mathematical errors, responded more appropriately to students, and chose examples which helped students construct meaning of the targeted concepts and processes. Teachers with weaker PCK were not successful at selecting and sequencing examples, presenting and elaborating upon textbook definitions, and using representations [19]. Well-established PCK cannot be separated from teachers' MK which includes knowledge about curriculum materials, tasks and strategies.

12.2.3 Metacognition in the Present Study

In the present study we focus mainly on metacognitive knowledge, relating to the following components (see Fig. 12.1):

- Knowledge of people. How one examines and monitors his/her own thinking processes; how one examines others' ways of thinking; how one interacts and collaborates with his/her colleagues and praises or gives constructive feedback.
- Knowledge of the task. How one presents his/her understanding of the task: what is the task about, its rationale, structure and components, reasons and

Fig. 12.1 MK components in the present study

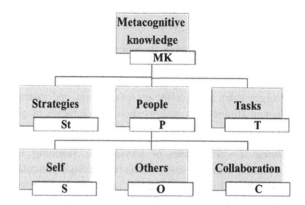

explanations regarding the task structure/activities, the importance of the task, and when to apply the task or its components.

- Knowledge of strategies. How one can employ strategies to improve performance with regard to the task.

12.3 The Study

12.3.1 The M.Ed. Program Structure

The fundamental assumption of this program [20] is the constructivist philosophy that individuals build their own knowledge by incorporating what they learn into what they already know [21, 22]. In translating this philosophy into practice in the development of the program, we addressed two main aspects: learning from active engagement and learning based on personal experience.

Radford (1998, p. 74) [23] claimed that "teachers are most likely to internalize science concepts and teaching methodologies when both their hands and minds are engaged in the process". The curriculum of the M.Ed. program aims at developing concepts and teaching methods which emphasize the development of scientific and pedagogical skills as well as disciplinary enrichment, along with learning and experiencing research methods in Science and Mathematics education.

The program is designed for teachers with a Bachelor's degree and a Teaching Certificate in Science or Mathematics in secondary schools, with at least three years' teaching experience, who are interested in their own personal and professional development and intend to continue to teach in school. The program consists of 24 credits earned over two years, and is divided into two main parts: one part for all the students, in which they are presented with a wide pedagogical basis encompassing terms and concepts shared by all science and mathematics teachers; and a second part, which covers three specializations: biology, mathematics and physics education, in which the students learn updated scientific and pedagogical content knowledge.

The curriculum offers advanced studies in different areas: teaching and learning, science and mathematics education, research on teaching and learning mathematics and science, and advanced scientific topics in the teacher's professional discipline. The two courses described below are part of the M.Ed. program.

Course 1—"Excellence in Science and Mathematics" The course focuses on gifted and talented students, revolving around math and science activities which enhance the development of higher order thinking skills as well as creativity. The teachers learn about effective curriculum and instructional models for gifted and talented students, as well as the characteristics of good teachers for the gifted, in addition to the issues of gender and giftedness with regard to science and math.

Course 2—"Principles of Developing Teaching Units in Physics" The course focuses on comprehension of the main design principles for effective Physics lessons, and is divided in two. In the first semester the teachers learn how to design a physics unit and how to evaluate students' comprehension. In the second semester the teachers are given the opportunity to teach these units to children participating in an enrichment program at the college. The implementation is documented and supervised by the researcher, and following each lesson the teachers and the researcher analyze the lesson and decide how to improve the next lesson's design. The teachers were asked to write a journal, documenting all the design decisions and their on-going reflections. The children who participated in the program were asked to reflect after each lesson on its design, suggesting how it may be improved. The reflections were documented in a forum which was part of a web-site that the teachers developed throughout the semester.

12.3.2 The Study Goals

The study goals are the following: (a) characterize the design principles of both courses as shown in the researchers' descriptions; (b) expose expressions of metacognition among the teachers and examine the changes they designed and applied in their teaching units and processes; (c) study the interactions among researchers, teachers, children in order to understand the teachers' metacognitive knowledge and skills.

12.3.3 Participants

The participants included 21 teachers from the M.Ed. program participating in the course "Excellence in Science and Mathematics". Four of them also participated in the course "Principles for developing teaching units in Physics". Additional participants were 17 children (6th grade) participating in an enrichment program for talented students in the college, a laboratory assistant and the two researchers who were the course advisors.

12.3.4 Data Collection

The data included: (a) both courses' design and activities; (b) researchers' reflections; (c) interviews with the teachers; (d) an in-depth interview with the laboratory assistant; (e) the teachers' teaching units and activities; (e) observations of the teacher's Physics lessons. The teachers' reflections and their FD references are analyzed based on the metacognitive knowledge components as defined in this study (Fig. 12.1).

12.4 Findings

12.4.1 The Courses' Design Principles

Four meta-design principles based on the Design Principles Database (DPD) were implemented in the courses. The DPD is a public mechanism enabling researchers and curriculum designers to share design knowledge and activities. Designers can publish, connect, discuss and review design ideas [24]. The design principles in each course are presented below:

Course 1: Excellence in Science and Mathematics

Principle 1: learning from and with peers This principle enables the teachers to benefit from their colleagues' ideas, introduces new perspectives and motivates them to interpret their ideas. As mentioned earlier, this course was on-line. It included seven forum discussions (FD) which usually lasted one to two weeks. This time period enabled the teachers to think, raise questions and discuss relevant issues with their colleagues and the course instructor.

The discussed subjects were related to the knowledge the teachers had acquired in each lesson. Each FD had its guiding instructions. These were also used as assessment criteria to evaluate each discussion.

For example, in the fourth FD, which focused on curriculum models for gifted and talented students, teachers were asked to critically examine the various models according to what they have learned and read in the course. They were then asked to provide feedback to peers and discuss their preferences. The course instructor gave feedback and clarification regarding the teachers' discussion when necessary.

In the last discussion, the teachers introduced a part of their teaching activity to their peers and asked them to participate or try to solve a problem and then provide formative feedback and suggestions for improvement.

Principle 2: making thinking visible This principle allows the teachers to integrate knowledge, reflect upon new information and engage in linking, distinguishing, or reconciling their own or their peers' ideas. In the course, the teachers met this principle on their own when they wrote their reflections regarding their activity and collaboratively when they participated in FD throughout the course.

Principle 3: make contents accessible The course included a variety of components: simulations, media resources, academic materials, on-line activities and FD. These components exposed the teachers to a rich learning environment and allowed them to connect ideas and reconsider their existing ideas. It scaffold their inquiry process and generated new connections within their science content knowledge and regarding their ways of thinking. Making the content accessible enabled the teachers to reflect on their reasoning. This reflection helped motivate them to develop and design science teaching activities, to gain meaningful learning experience.

Principle 4: Promoting autonomy so students can become lifelong learners This principle involves establishing a rich, comprehensive learning environment encouraging teachers to take responsibility for their learning and develops them as lifelong learners. In this course, the teachers were required to design their own activities, explore a variety of resources, and introduce their activities in the FD on the course web-site. This enabled them to relate to colleagues' questions and feedback.

The teachers' activities were open-ended and allowed for expression of creativity and higher order thinking skills. The teachers' reflections were used as a 'window' to their thinking processes as well as a mirror to their development as learners.

The teachers were asked to find the proper resources which can help them in constructing the tasks for the children, involving them in a reflective process eliciting their own thinking regarding teaching and learning, helping them to think critically about their teaching practice, and making effective decisions in real time in the classroom.

Course 2: Principles for developing teaching units in Physics

Principle 1: learning from and with peers This course was designed for supporting social interaction among the teachers, so that they could benefit from their peers', instructors' and students' ideas.

Learning from their peers includes planning together, observing their peers teaching in class, and collaborative reflection on the planning and implementing of the teaching units. Learning from their advisor, the course moderator, and the laboratory assistant includes all the talks (both formal and informal) which took place between the teachers and the course advisor.

The laboratory assistant introduced new laboratory equipment and gave the teachers an opportunity to check the equipment, making sure they know how to operate it and understand the physics it demonstrates. Learning from the children includes listening to the children's talks, conducting open discussions about the teaching units, and reading the children's reflections on the course forum.

During implementation of this project the teachers experienced a new way of interacting with students, asking them to reflect on the lesson activities and design. This kind of interaction was not used by them in school, and they did not have any prior knowledge of it.

Moreover, this kind of experience was unfamiliar to them even as learners. It might be a frightening experience, listening to the students, understanding what they enjoy, what the thought should be improved, and how.

The experienced teachers come from an environment where they are the authority. They do not ask their students those kinds of questions and the feedback they usually get originates from formal subject matter tests.

The children also indicated that this was a new experience for them; their teachers in school have never asked them those kinds of questions.

> In school my teachers aren't interested in me, they don't care if we enjoy anything or understand it, school is really boring, here we felt that you really appreciate our opinion, you were listing to us it was a delightful experience.

Principle 2: making thinking visible This meta-principle is divided into two types of ideas: (a) making thinking visible so teachers can learn about others' ideas and communicate their own ideas to others; (b) designing models or communicating complex concepts using visualization in order to make complex scientific phenomena visible.

The course design used this principle in two forms, the first by giving the teachers an opportunity to share their MK about the proper activities in the physics laboratory. Moreover, the teachers had to write a reflective journal about their experiences during the design process, so that they could explicitly understand the mental processes they were experiencing.

Because the teachers had to explain their decisions to their peers and instructors, it broadened their perspectives, promoting their knowledge integration process. Obtaining a combination of perspectives helped the teachers acquire more coherent knowledge about designing scientific tasks for children. The second way was leading the teachers towards use of various visualization tools and integrating them in the lesson (simulations, demonstrations, videos, etc.).

Principle 3: make contents accessible The implementation of this meta-principle was divided into two types. The first was connecting to the teachers' prior knowledge regarding science teaching by building on their own knowledge and experience.

During each cycle the course moderator asked the teachers what do they usually do in their classes and what kind of activities do they implement at school, giving them confidence in the task; then helping them to articulate the tacit knowledge they have about teaching through class discussions on those activities, connecting them to the theory and redesigning them to fit to the specified objectives.

The second type of implementation of this meta-principle was by directing the teacher to designing activities connected to the children's world, bringing examples from everyday life, helping the students to connect the scientific concepts to the day-to-day knowledge the already have. Thus the teachers could acquire the MS about building experiences which can help the students integrate their scientific knowledge.

Principle 4: Promoting autonomy so students can become lifelong learners Implementation of this meta-principle in the course was achieved through encouraging the teachers to design their own activities and find the proper resources that can help them construct the tasks for the children; involving them in a reflection-on-action process that will elicit their own thinking on teaching and learning, helping them to think critically about their teaching practice, and making effective decisions in real time in the classroom.

12.4.2 Expression of Teachers' Metacognition

In this section we present teachers' metacognitive knowledge as expressed by the teacher's reflections in both courses. It seems that the design of both courses promoted the MK of the teachers in relation to people, tasks and strategies. In the following section we focus on two case studies of experienced physics teachers who participated in both courses. Those stories demonstrate the teachers' PD throughout both courses and their MK development (see Table 12.1).

Khaled's story: This story was constructed from data collected from Khaled's reflections in both courses, from observations of the lessons he had implemented in the talented children's course, and from interviews. Khaled is an experienced high

Table 12.1 Teachers' metacognitive knowledge expressions

Metacognitive knowledge		Quote
Knowledge of people	Self	*I felt discomfort, I was not sure if I could build an activity that actually teaches in a constructivist way. This was my weakness.* So, I started to read about the Pythagoras sentence and examined the complexity of the quiz
		The task was very special to me. *I thought differently. I put myself in the place of the student and began to think how they understand the subject and enjoy learning it. It took me a while!*
		I started to look differently at students! I went through a change! Today, *I find myself thinking how to introduce a problem in a challenging way*
		While thinking how to develop my activity (fractal), *I realized that there is no reason to "hold" with subjects that I am familiar with and that I can give a place to my imagination!* But, another problem arose. I had a lot of ideas. All were good! I started to think which idea is the most suitable to gifted youth and to the process-product model. *I knew that as a tutor, my skills are not so good,* so I prepared very detailed inquiry instructions within the task
	Others	Nivin and Jakob supported my idea. They actually directed me! *Jakob analyzed the problem and its complexity* and Nivin's *clarifying questions* caused me to change part of the activity
	Collaboration	The exposure to peer feedback enabled me to understand my colleagues' thinking techniques and the assessment process they conducted. At first, I was afraid of the exposure, but the *discussion was conducted in a constructive way and with respect. I raised a question to Jacob and he explained his rationale. I felt that we contributed to one another;* they helped me understand that my task is good!
Knowledge about the task		My activity is suitable for gifted children. *It develops scientific thinking skills, creativity and requires concentration. It's based on the epistemology model as it focuses on the meaning of "structures" and on interdisciplinary learning because, the "golden ratio" appears in nature, in art, in architecture, and in math*
		The task structure was different at the start. Later on, I organized it in a way that *it was logical to go through two dimensions to 3D*
		The activity includes multidisciplinary aspects and represents that the whole is greater than the sum of its parts. I will give it at the end of the learning process, in the "Amirim" program
Knowledge about strategies		*I learned that inquiry skills are not so simple to apply. Teachers need to check at what stage the students are in the process of inquiry, to give them instructions and to stimulate their curiosity with a short but attractive activity or trigger.* One that can motivate them to do the task. In the summary stage (of the activity), I prepared reflection questions, but now *I know that I will enable students to choose which question they prefer to answer.* The ability to choose is very important to the gifted students

school physics teacher, teaching in an Arab school in northern Israel. During this project he was in his second year of the M.Ed. program, and he participated in both courses. He is part of a small group of physics teachers attending the M.Ed. program. Khaled had taken the first course, "Excellence in Science and Mathematics", online and developed a teaching unit about diving. Khaled claimed that diving is a very broad topic, incorporating different disciplines such as Physics, Biology, Chemistry and Mathematics, and he wanted to focus on decompression sickness.

This issue relates to the respiratory system, the vascular system, diving depth, and rate of increase of the water. All these combine Biology and Physics as well as Mathematics, and can be taught to talented children. Khaled underwent a profound change during the course. His thinking has changed and now he is more open, thinks of interdisciplinary subjects and includes authentic and interesting topics in his lessons.

By the end of the course, he wrote:

> The sea is a mystery. It gave me inspiration. I know the dangers associated with it and think that this is an important subject to teach. Now, I'm thinking about the other way,

what happens to the human body in heights? Is the process reversed? What are the dangers? I also think of skydiving as an interesting subject to teach and explore next time.

In the second course, "Principles of developing teaching units in Physics", Khaled planned to use some of the materials but could do only the beginning of the unit. He had many doubts concerning his ability to teach Hebrew-speaking talented students. He thought it would be hard for him to connect to the children, he was afraid they would not understand his language and even his jokes. During the first lesson the children were divided into four groups, each group headed by one of the teachers.

The group that worked with Khaled immediately connected with him and participated in the task he gave them. At the end of the lesson one of the children came to Khaled, asking him to be his school teacher. Khaled told the child that he teaches in an Arab speaking school, so it would be hard for him to learn. The child said that it does not matter, he wants to learn with him. This interaction contributed to Khaled's confidence regarding his ability to connect to the children, even from a different culture, and to be a significant mentor for them. Moreover, some children asked Khaled for his personal email and started to communicate with him, asking for his advice regarding problems in their regular school.

In one of the lessons we had a mother who came to observe. She was sitting at the end of the class, away from her son, near the researcher. In the middle of the lesson she told the researcher that Khaled is an extraordinary teacher, that he has a unique way to reach the children, that he interacts with each child in the correct way in order to encourage him/her. She also mentioned the interaction Khaled had with the children in the course forum, noting his gentle approach. At the end of the lesson she went to Khaled, shook his hand and told him how she appreciates the work he is doing with all the children, and particularly with her son.

Khaled pointed out in his reflections that while observing Michal he analyzed her interaction with the children, and could clearly understand the better way to teach them. He realized that the children enjoyed the experiments and wanted to do them by themselves, rather than being passive in the lesson. He noted that the children ask a lot of questions and are very interested in the scientific phenomena that Michal had introduced to them.

One of the main features in the course design was getting feedback from the children by asking their opinion about the lessons in an open discussion in the class and in the forum that was opened for them on the course website. Khaled wrote about it:

> The concept of getting feedback from the students was a new experience for me, I think it is missing in the school where I teach.
> The truth is that this resource helps teachers to assess and evaluate the teaching methods they use, and teachers can improv e themselves by listening to the students, especially by asking questions like: what was missing in the activity, what do you want to know, what kind of activity suits you? I am sure I'll use reflection of students in future classes.

The story of Michal: Michal is an experienced physics teacher working in a religious girls' school in northern Israel. She has six children and is juggling between her commitment to her home as a wife and mother, to the school where she is the Physics coordinator, and to her studies at the college. During this project she was in her third year of the M.Ed. program at the college. Throughout the development phase she constantly tried to connect to her practical knowledge, bringing examples from her prior classroom experience. In her school she teaches physics in addition to science to non-science students, believing that all children should learn some science in high-school, even if they major in Arts or other subjects.

Michal had some experience working with talented students at her school. In the course "Excellence in Science and Math" she developed an activity that included scientific ideas relating to radiation and the principle of energy conservation. The activity was developed for a regular class that included talented children as well. The activity goals were to identify students' misconceptions, reconstructing them in a more accurate manner. She designed an activity that combined other disciplines (e.g., photosynthesis in biology).

According to her, even when students declare they know there is something called The Energy Conservation Law, in everyday life they treat it as if the energy disappears. Misconceptions in science were an important subject for Michal. This was expressed in the activities she designed in the course "Principles for developing teaching units in Physics" as well.

Michal chose the physics of the rainbow, a subject she had done scientific research on and is part of her M.Ed. final project. From the beginning of the course, she tried to connect every piece of new teaching concept to her understanding of her own practical experience, for example while discussing the need to identify students' preconceptions, she immediately tried to connect it to her own project, discussing how to identify children's pre-conceptions of the rainbow. She discussed this issue in class with her peers and with the class moderator.

Doing this, she could get a better understanding of the concept, and have the ability to implement it properly in class. Michal tried it later with the children, testing her assumptions, discovering what really works. She implemented the CDC teaching strategy [25] introduced in class with the children. Because we applied a cyclical procedure, whereby after each lesson we analyzed the classroom practice, Michal could get a better understanding of the teaching strategy and how it is actually carried out in class.

Michal is a highly reflective teacher, extremely self-critical, but although she is very verbal, she does not tend to write down her reflections. She had a firm conception of what one should do in class. The experience with children that were not her usual students opened her to new insights and challenged her own beliefs about what and how one could teach.

While teaching the physics of the rainbow, she used a PowerPoint presentation about several physical phenomena that can be observed in the rainbow. It had a lot of information, and after the lesson she felt that the children had learned a lot. However, reading the children's reflections caused her to rethink the effectiveness of the presentation. One of the children wrote:

> We have learned a lot during this lesson, and it was very interesting, but we are a special group of children that are very interested in science, I think that presenting a PowerPoint

presentation during the whole lesson could be very hard to children that are not interested in science like us, I think you should do a more active lesson...

Another child wrote:

> You presented us with a presentation that included a number of phenomena that can be observed in the rainbow, I think you should have taken each phenomenon, teach it, explain it and afterwards go to the next phenomenon, but I enjoyed the lesson very much and learned a lot.

In her last lesson Michal demonstrated the creation of a rainbow with a device that was developed by Cohn [26]. Using this device, the students could see with their own eyes all the phenomena they had learned about in class. It was a very special experience for the students, for the teachers and for Michal herself. In the class discussion with the children after the demonstration, we asked them what was so special in the demonstration, and how they thought it contributed to their understanding the physics of the rainbow. One of the children said:

> During the lesson you told us about the different phenomena, I think I understood what you wanted to teach, but I had to believe you that those phenomena really can be seen in the rainbow, now I saw them with my own eyes, and that makes the difference for me, I could see that physics works...

Michal is now preparing to teach about the rainbow at her school. It will be interesting to see the changes she will make to the design of her activity. Michal, Khaled and the other Physics teachers participating in the course designed the course for talented children. The design process was a collaborative effort facilitated by the course advisors. Figure 12.2 presents the course environment, including the main components developed by the teachers.

The activities developed were both hands-on and minds-on, and were constructed according to the models taught in the course. The environment includes many modern technologies which can be incorporated in the lesson. The activities used the children's mobile phones as lab equipment, combined with more traditional equipment. Simulations were used as an important visualization tool which

Fig. 12.2 The course environment

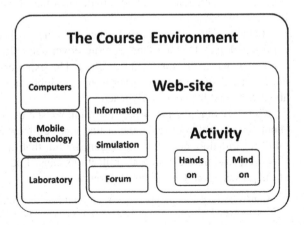

can help the students understand the physical concepts by making the complex scientific phenomena visible.

The course design includes construction of an internet site serving as a place where the children can find more information and participate in a FD sharing their reflections after each lesson. The internet site was open for parents to view what was happening in class.

12.4.3 Resources for the Teachers' Meta-Cognitive Development

Viewed comprehensively, we note that a number of key resources influenced the teachers' meta-cognitive development as it is shown in Fig. 12.3. We address each resource, providing an explanation and examples of the data collected in the study.

The courses in the present study: The courses stated in this chapter were developed based on the rationale that meaningful learning occurred when teachers experience significant learning while planning and constructing activities, as well as interacting with fellow learners, course instructors, professional experts and their students. The principles enabled a learning environment where teachers integrate knowledge, conduct analysis and synthesis, consider their ideas and those of their colleagues and promote their professional development as teachers and as lifelong learners. The teachers noted use of the various materials and media included in the course:

> As aids I used the articles studied in the course, the material so clearly worded in the partitions, the videos included in the course. In addition, studies in the course helped me

Fig. 12.3 Resources for teachers meta-cognitive development

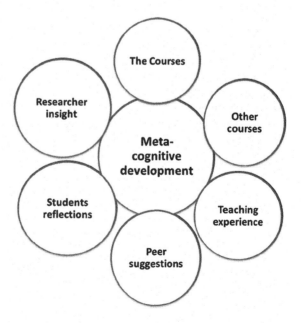

greatly with the planning stages. For instance, knowing what the conditions are for promoting investigative learning, I made sure to incorporate them in the planned activity (T).

Other courses: The data we collected show that teachers were also influenced by other courses studied, helping them design and build their own activities, as expressed by Neta and Khaled:

> The courses "Thinking about thinking" and "Critical thinking", greatly aided me in the steps required of me to make changes to the activity. For instance, I implemented what I learnt in your previous course, to make use of clarification questions, raise arguments, present examples and come up with ideas for future solutions (Neta).
>
> The two online courses I took with you greatly helped me in planning and provided me with new concepts that I was unaware of previously, through which I learned of the power of reflection which is a source of immense influence on planning and improving. Also, at the training seminar in which we dealt with non-conventional solutions of mathematical problems, we also used reflection to adjust and improve the content I chose. I must admit that I have a fear of choosing content, in having to choose carefully in order to interest the students and of course we mustn't forget the additional course in which we conveyed material to talented students and all this was connected to the course materials learnt (Khaled).

Teaching experience: An additional source is teaching experience. The teachers who participated in this study are experienced teachers, and indicated whether this assisted them in developing activities and/or influenced their meta-cognitive thought processes. Khaled, who is a very experienced teacher, noted that conducting thinking about thinking was new to him and he had not experienced it before:

> I must admit that the "thinking about thinking" process is critical to the field of teaching, a process I did not use at all in my teaching two years ago (before beginning studies in the faculty). Although this process throws you into an infinite spiral of thoughts, it can help form opinions and new approaches for the next plan.

Peers suggestions: Feedback collected from colleagues taking the two courses discussed in this chapter was stated as one of the main resources affecting the teachers' and their colleagues' thinking processes; particularly with knowledge regarding the task, in relation to the self as a learner and in relation to others. This was reflected in the changes made in planning the activities, in the manner in which teachers gave feedback to their colleagues and in the interaction during the lessons themselves (both with the students and as part of the online forum). D and M are relating these issues:

> Peers feedback is one of the most important stages of planning; the colleagues' feedback is a powerful tool for improvement. Their comments helped a great deal in making alterations. I recall N.'s reference made me look at the activity from a completely different standpoint; switch to a broader perspective, three dimensional. Overall, I understood that beyond technique, it is worthwhile to integrate other realms such as the presence of the phenomenon in nature and architecture. (D)
>
> From the colleagues' feedback and the sessions presented I learnt about their planning processes, the structure of the sessions and the nature of the questions raised. The entire issue of learning from colleagues is vital and you can essentially evaluate yourself and see where you stand. (M)

Children's reflection: The children reflections as well as the classroom discussions with the children about the course served as important resources for the teachers. Although all the teachers were experienced with many hours of classroom practice behind them, they did not have the habit of listening to student's views. This was a new experience for the children as well. Some of the children had very developed metacognitive awareness and could reflect very deeply about the strategies and tasks used in the lessons.

Many children expressed how much they enjoyed the course: "Physics is the best course we have". "We like the experiments and the hands-on activities." "This is much better than the Science lessons at school, Science at school is boring...."

Researchers' insight: Throughout the online course, face to face meetings were held with a number of students. These meetings were a continuation of discussions conducted in the forum during the course. In these meetings we discussed the processes they experienced during the different stages of developing the activities. Some came during the planning stages and others during more advanced stages. Students also related to that. For example, T. mentioned that following a discussion she understood that there were other areas available to her and she could expand the activity planned and include interdisciplinary aspects and future references. In the face-to-face course the design process was cyclical, conducted with the researcher, reflecting the researcher's insight at each step. Thus the teachers had the opportunity to discuss the design process with their peers and with the researcher.

12.5 Discussion

In this chapter we focused on the design and study of a metacognitive approach to the PD of high-school science teachers. We designed two courses based on meta-design principles known to provide a well-established environment enabling meaningful learning [27]. We provided the teachers with some new experiences that we think are essential for their PD. The first one is participating in a long term PD program for teachers that can promote their knowledge integration (KI) [28, 29] by connecting new theoretical concepts to their practical knowledge, using an evidence-based approach [10]. This was particularly evident in the teachers who participated in both courses.

Those teachers were exposed to a variety of content such as theories, models and examples of teaching resources including lab experiments and simulations. The teachers designed, made changes to their design, and experienced an iterative process of improving their activities.

This practice is supported by several researchers who argued that PD of a longer duration is more likely to contain the kinds of learning opportunities necessary for teachers to integrate new knowledge into practice [8, 30]. The second is providing an opportunity for the teachers to elicit their conceptions with regards to the activities they developed for the talented students, to discuss them with their peers and with their instructors, and to reflect on their own thinking. Our findings show that the courses we designed engaged teachers in constructing their own knowledge as well as collaborating as a group of learners.

The knowledge they acquired was not restricted to content knowledge, including the teachers' MK as well. These conditions of shared knowledge enabled the teachers to recognize the differences between engagement and awareness while guiding students to implement metacognitive strategies [31]. The third experience gives the teachers the opportunity to apply their teaching units in a supportive environment, leading to a successful experience in implementing their new and inspiring ideas.

This required close supervision by the courses advisors together with the laboratory assistant in order to prevent flaws leading to failures. Once having discussions with the teachers after each lesson, analyzing the stronger and weaker parts of their experience, led to a deeper understanding of the design process as well as of the main components of a task able to promote scientific understanding.

In this study we identified six key resources which influenced the metacognitive development of the teachers: (a) the courses in this study; (b) other M.Ed. courses; (c) the researchers' insight; (d) the teachers' teaching experience; (e) peer suggestions and (f) children's reflections. We found that each resource empowered the teachers, adding new layers to their metacognitive knowledge.

The interviews with the teachers and their reflections indicate that both courses, as well as other courses in the M.Ed. program, have impacted the teachers in designing their activities. In addition, it seems that the researcher's insight provided throughout the courses helped the teachers become aware of their own thinking processes and able to express them verbally. This awareness is significant, as we know that teachers are often unacquainted with the knowledge they possess. Moreover, in their regular practice they do not need to express it [25].

Teachers' teaching experience was also a major contributor to their metacognitive development. Teachers related to knowledge in physics, biology, chemistry and mathematics they already had, to strategies that they were familiar with and to diverse types of tasks (e.g., simulations, lab experiments, and riddles). The fifth resource was the teachers' peer feedback. It seems that this was one of the elements that affected the teacher's knowledge regarding the activities, and raised their self-awareness in relation to strategies they have adopted and the interactions they had with the students. It is worth noting that, in a regular class, experienced teachers teach on their own. Here, teachers had an opportunity to be present in a lesson taught by an experienced colleague, to watch and learn from him/her and even to experience co-teaching.

This experience along with the teachers' peer feedback facilitated the teachers' cognitive apprenticeship [32] and their MK development throughout the year. The last recourse we identified was the children's reflections.

Children were asked to write a reflection at the end of each lesson and refer to the knowledge they acquired, the teaching strategies that the teachers implemented, and the lab experiments. At the end of each lesson, the teachers rushed to read the children's feedback. This feedback regarding the teachers' presentations, quizzes, simulations and lab experiments helped to construct the teachers' MK. This was expressed in the teachers' reflections and in changes the teachers enacted in their lessons. One of the interesting references that the children raised in their reflections was the fact that the teachers actually listened to them, and wanted to know their opinions.

The children explicitly stated in their feedback that the teachers were really interested. This was something they had never experienced in their regular school where the teachers never asked them to express their opinions nor provide suggestions for improving the lesson/activity/teaching strategies. This ought to be considered in teaching all students, but in particular in teaching talented students. Research shows that talented students who experience inquiry learning in an environment enabling them to give and receive reflective feedback to and from different agents (e.g., peers, teachers, experts, tutors, parents) develop their reflective and inquiry skills [33].

To sum up, it seems that the courses we developed and implemented offered the teachers a supportive and constructive learning and teaching environment. They enhanced their knowledge of students' understanding of science and thinking processes. This is similar to Darling-Hammond et al. [34] who argued that "to understand deeply, teachers must learn about, see, and experience successful learning-centered and learner-centered teaching" (p. 598).

In conclusion, we agree with Zohar and Barzilai [2] who claim that teachers need to understand what metacognition means and practice metacognitive thinking with respect to classroom activities. They need to explicate MK and practice MS with respect to the activities that take place in the classroom. We concur with other researchers [2, 16, 19, 35] that teachers' metacognitive proficiency is crucial, and that metacognition should find its way into routine Science instruction. Our study strongly encourages teachers to develop and design activities and test them within a supportive environment, practice which can help them in understanding their own beliefs as well as promote their metacognitive knowledge.

We believe that the teachers' own understanding of the complex teaching mechanism has changed, and they will take this experience into account while designing activities for their students in the future. As Khaled stated:

> The process of getting feedback from the students is missing from the school where I teach, now I realize that this resource can help teachers to assess and evaluate their own teaching methods, and can give the teacher the opportunity to improve himself by asking the students questions like: what do you want to know; what was missing in the lesson; what kind of activity concerns you more; therefore, I am sure I will use students' reflection in the future in my classes.

References

1. Loughran, J.: Researching teaching about teaching: Self-study of teacher education practices. Studying Teach. Educ. 1(1), 5–16 (2005)
2. Zohar, A., Barzilai, S.: A review of research on metacognition in science education: current and future directions. Stud. Sci. Educ. 49(2), 121–169 (2013)

3. American Association for the Advancement of Science: Benchmarks for Science Literacy. Oxford University Press, Oxford (1993)
4. National Council of Teachers of Mathematics: Assessment Standards for School Mathematics. National Council of Teachers of Mathematics, Reston (1995)
5. National Council of Teachers of Mathematics: Principles and Standards for School Mathematics. National Council of Teachers of Mathematics, Reston, VA (2000)
6. National Research Council: National Science Education Standards. National Academy Press, Washington (1996)
7. National Research Council: Inquiry and The National Science Education Standards. National Academy Press, Washington (2000)
8. van Driel, J., Meirink, J., van Veen, K., Zwart, R.: Current trends and missing links in studies on teacher professional development in science education: a review of design features and quality of research. Stud. Sci. Educ. **48**(2), 129–160 (2012)
9. Haney, J., McArthur, J.: Four case studies of prospective science teachers beliefs concerning constructivist teaching practices. Sci. Educ. **86**(6), 783–802 (2002)
10. Eylon, B.S., Berger, H., Bagno, E.: An evidence-based continuous professional development program on knowledge integration in physics: a study of teachers' collective discourse. Int. J. Sci. Educ. **30**(5), 619–641 (2008)
11. Nilsson, P.: When teaching makes a difference: developing science teachers pedagogical content knowledge through learning study. Int. J. Sci. Educ. **36**(11), 1794 (2014)
12. Flavell, J.H.: Metacognition and cognitive monitoring: a new area of cognitive developmental inquiry. Am. Psychol. **34**(10), 906–911 (1979)
13. Brown, A.L.: Knowing When, Where, And How to Remember. A Problem of Metacognition. University of Illinois, Urbana-Champaign (1977)
14. Schraw, G., Moshman, D.: Metacognitive theories. Educ. Psychol. Rev. **7**(4), 351–371 (1995)
15. Schraw, G.: Promoting general metacognitive awareness. Instr. Sci. **26**(1–2), 113–125 (1998)
16. Ben-David, A., Orion, N.: Teachers voices on integrating metacognition into science education. Int. J. Sci. Educ. **35**(18), 3161–3193 (2013)
17. Baumert, J., Kunter, M., Blum, W., Brunner, M., Voss, T., Jordan, A., Tsai, Y.-M.: Teachers mathematical knowledge, cognitive activation in the classroom, and student progress. Am. Educ. Res. J. **47**(1), 133–180 (2010)
18. Hill, H.C., Blunk, M.L., Charalambous, C.Y., Lewis, J.M., Phelps, G.C., Sleep, L., Ball, D.L.: Mathematical knowledge for teaching and the mathematical quality of instruction: an exploratory study. Cogn. Instr. **26**(4), 430–511 (2008)
19. Charalambos, Y.C., Hill, H.C.: Teacher knowledge, curriculum materials, and quality of instruction: unpacking a complex relationship. J. Curriculum Stud. **44**, 443–466 (2012)
20. Trumper, R., Eldar, O.: The effect of a M.Ed. program in science education on teachers' professional development. A research report prepared for the Research and Evaluation Authority, Oranim, Academic College of Education, Kiryat Tiv'on (2014) (in Hebrew)
21. Matthews, M.R.: Constructivism and empiricism: an incomplete divorce. Res. Sci. Educ. **22**, 299–307 (1992)
22. Yager, R.: The constructivist learning model. Sci. Teach. **58**, 52–57 (1991)
23. Radford, D.L.: Transferring theory into practice: a model for professional development for science education reform. J. Res. Sci. Teach. **35**, 73–88 (1998)
24. Kali, Y., Ronen-Fuhrmann, T.: Teaching to design educational technologies. Int. J. Learn. Technol. **6**, 4–23 (2011)
25. Eldar, O., Eylon, B.S., Ronen, M.: A metacognitive teaching strategy for pre-service teachers: collaborative diagnosis of conceptual understanding in science. In: Zohar, A., Dori, Y.J. (eds.) Metacognition in Science education: Trends in Current Research, Contemporary Trends and Issues in Science Education, vol. 40, pp. 225–250. Springer, Netherlands (2012)
26. Cohn, A.: Thinking and Doing in Science: A Guide for Parents and Teachers. TOV MA'ASE, Mofet Institute, Tel-Aviv (2001) (in Hebrew)

27. Kali, Y., Levin-Peled, R., Dori, Y.J.: The role of design-principles in designing courses that promotes collaborative learning in higher-education. Comput. Hum. Behav. **25**(5), 1067–1078 (2009)
28. Linn, M.C., Eylon, B.S.: Science education: Integrating views of learning and instruction. In: Alexander, P.A., Winne, P.H. (eds.) Handbook of Education Psychology, pp. 511–544. Lawrence Erllbaum Associates, Mahwah (2006)
29. Linn, M.C., Eylon, B.S.: Science learning and instruction: taking advantage of technology to promote knowledge integration. Routledge/Taylor & Francis Group, New York (2011)
30. Brown, J.L.: Making the most of understanding by design. Association for Supervision and Curriculum Development, Washington, DC (2004)
31. Wilson, N.S., Bai, H.: The relationships and impact of teachers' metacognitive knowledge and pedagogical understandings of metacognition. Metacognition and Learning **5**(3), 269–288 (2010)
32. Yerushalmi, E., Eylon, B.S.: Supporting teachers who introduce curricular innovations into their classrooms: a problem-solving perspective. Phys. Rev. Spec. Top. Phys. Educ. Res. **9**(1), 10–121 (2013)
33. Miedijensky, S., Tal, T.: Embedded assessment in project-based science courses for the gifted: insights to in for teaching all students. Int. J. Sci. Educ. **31**(18), 2411–2435 (2009)
34. Darling-Hammond, L., Chung Wei, R., Andree, A., Richardson, N., Orphanos, S.: Professional learning in the learning profession: a status report on teacher development in the United States and abroad. National Staff Development Council, Dallas (2009)
35. Veenman, M.V.J.: Metacognition in science education: definitions, constituents, and their intricate relation with cognition. In: Zohar, A., Dori, Y.J. (eds.) Metacognition in Science Education: Trends in Current Reasearch, Contemporary Trends and Issues in Science Education, vol. 40, pp. 21–36. Springer, Dordrecht (2012)

Part V
Tools

Chapter 13
Modeling Metacognitive Activities in Medical Problem-Solving with BioWorld

Susanne P. Lajoie, Eric G. Poitras, Tenzin Doleck and Amanda Jarrell

Abstract Medical diagnostic reasoning is ill-defined and complex, requiring novice physicians to monitor and control their problem-solving efforts. Self-regulation is critical for effective medical problem-solving, helping individuals progress towards a correct diagnosis through a series of actions that informs subsequent ones. BioWorld is a computer-based learning environment designed to support novices in developing medical diagnostic reasoning as they receive feedback in the context of solving virtual cases. The system provides tools that scaffold learners in their requisite cognitive and metacognitive activities. Novices attain higher levels of competence as the system dynamically assesses their performance against expert solution paths. Dynamic assessment in this system relies on a novice-expert overlay and it is used to develop feedback when novices request help. When help-seeking occurs, help is provided by the tutoring module which applies a set of pre-defined rules based on the context of the learner's activity. The system also provides cumulative feedback by comparing the novice solution with an expert solution following completion of the case. This chapter covers the essential design guidelines of this scaffolding approach to metacognitive activities in problem-solving within the domain of medical education. Specifically, we review recent advances in modeling metacognition through online measures, including concurrent think-aloud protocols, video-screen captures, and log-file entries. Educational data mining techniques are outlined with the goals

S.P. Lajoie (✉) · T. Doleck · A. Jarrell
Department of Educational and Counselling Psychology, McGill University,
3700 McTavish Street, Montreal, QC H3A 1Y2, Canada
e-mail: susanne.lajoie@mcgill.ca

T. Doleck
e-mail: tenzin.doleck@mail.mcgill.ca

A. Jarrell
e-mail: amanda.jarrell@mail.mcgill.ca

E.G. Poitras
Advanced Instructional Systems and Technologies Laboratory, University of Utah
Educational Psychology, 1721 Campus Center Drive SAEC 3220, Salt Lake City,
UT 84112-8914, USA
e-mail: ASSISTlaboratory@gmail.com

© Springer International Publishing Switzerland 2015
A. Peña-Ayala (ed.), *Metacognition: Fundaments, Applications, and Trends*,
Intelligent Systems Reference Library 76, DOI 10.1007/978-3-319-11062-2_13

of capturing metacognitive activities as they unfold throughout problem solving, and guiding the design of scaffolding tools in order to promote higher levels of competence in novices.

Keywords Tools · Scaffolding   approaches · ITS · Metacognition · Problem-solving · Bioworld · Medical education · Novice-expert overlay · Help-seeking

Abbreviations

ANN   Artificial Neural Networks
HMM   Hidden Markov Models
MNB   Multinomial Naïve Bayes
NB   Naïve Bayes
SMO   Sequential Minimal Optimization
TRE   Technology-Rich Learning Environment

13.1 Modeling Metacognitive Activities in Medical Problem-Solving with BioWorld

Medical diagnostic reasoning is complex and ill-defined in that there is no single problem solving sequence for obtaining the correct answer. There are many routes to solving the problem and one medical problem may lead to a new set of medical issues that need to be resolved. Well-defined problems, on the other hand, often have clear procedures and outcomes. Ill-defined problems are more difficult to solve since there is no set of rules that will lead to the right answer [1].

Consequently, novice physicians must learn to monitor and control their problem-solving efforts by executing actions that will help them progress towards a correct diagnosis. Self-regulation is critical for effective medical problem-solving in that physicians must orient their actions and evaluate the consequences of such actions before planning new ones.

BioWorld is a technology-rich learning environment (TRE) designed to help medical students regulate their learning about medical reasoning by providing feedback in the context of learning to solve virtual patient cases [2, 3]. BioWorld provides tools that scaffold the learner's requisite cognitive and metacognitive activities.

Novices attain higher levels of competence through deliberate practice [4] as the system dynamically assesses their performance against expert solution paths and provides the necessary feedback.

Dynamic assessment in this system relies on a novice-expert overlay. This overlay is used to develop feedback when novices request help. When help-seeking occurs, help is provided by the tutoring module which applies a set of pre-defined rules based on the context of the learner's activity. The system also provides cumulative feedback by comparing the novice solution with an expert solution following the completion of the case. Our contention is that each individual has a different learning trajectory within specific problem solving contexts [5]. This trajectory can be identified by designing a learner model within a computer-based learning environment that captures the learner's competence and performance within a domain of study.

BioWorld assesses the learning model against an expert model of competence and performance and provides scaffolding that fosters cognitive and metacognitive activities within the domain of medical problem-solving. This allows the system to assess novice performance along the path towards competence and enables the system to deliver support and feedback tailored to the individual needs of different novices; a key factor in successfully fostering the development of metacognitive skills and knowledge. In this way BioWorld captures and assesses learners' trajectories towards expertise in medical reasoning. However, there are several challenges involved in fostering metacognitive activities while solving problems in BioWorld using an expert model.

This chapter covers the essential design guidelines for scaffolding metacognitive activities in problem-solving within the domain of medical education. Specifically, we review recent advances in modeling metacognition by outlining analytical techniques to design, evaluate, and develop expert models by capturing metacognitive activities in problem-solving. We demonstrate the use of on-line measures, including concurrent think-aloud protocols as well as video-screen captures, and log-file traces of user interactions. Educational data mining techniques are outlined with the goal of capturing metacognitive activities as they unfold throughout problem solving.

These trace methodologies are used to model self-regulatory processes along the trajectory towards competency in diagnostic reasoning. We summarize three studies that examine help-seeking activities in the context of BioWorld. These findings lead to insights with respect to designing appropriate scaffolding tools in order to promote higher levels of competence in novices. Future directions for expert-driven models of metacognition are outlined. We commence our chapter with a detailed discussion of how BioWorld is designed to scaffold medical problem solving and metacognition. Some of the principles for designing metacognitive scaffolding tools for BioWorld can be generalized to designing metacognitive scaffolding tools for other computer-based learning environments.

13.2 A Model of Metacognitive Activities in Problem-Solving

The study of self-regulation within domains requires consideration of the task that is performed by a learner as well as the strategic processing demands that are inherent to the domain [6]. In the medical field, self-regulation has been studied in numerous ways but from different lenses, such as: self-assessment in the context of professional development [7]; examining the interaction between personal attributes and environmental affordances [8]; and examining the developmental phases that occur through clinical practice [9].

For the purposes of this chapter, we outline a model that synthesizes existing accounts of self-regulation in problem-solving and situates the underlying activities in diagnosing patient cases in the medical domain [10–12]. We model cognitive and metacognitive activities in the context of BioWorld, a TRE that serves as a platform to support novices in solving problems within the medical domain.

We conceptualize self-regulation as a super-ordinate construct that encompasses metacognition, namely the ability to orient oneself in the problem space, plan and execute actions, monitor outcomes, as well as evaluate and elaborate a solution [3]. Solving such problems requires more than clinical experience, it requires the ability to regulate problem-solving by adapting one's approach to solving the problem.

Social cognitive models of self-regulation characterize metacognitive activities as occurring as part of a recursive and iterative process involving forethought, performance and reflection, where adjustments to the solution are made on the basis of progressively refining the problem space [13, 14].

In the forethought phase, novices orient themselves in the problem space, at the same time, formulating a plan to solve the problem. The performance phase is characterized by the novice's efforts to solve the problem by executing the planned steps and monitoring the outcomes. The self-reflection phase involves the novice's evaluations of the overall progress and elaborations about the problem space, resulting in conclusions about the case. The problem-solving process is recursive in that the outcomes of prior steps inform the next ones that are taken to solve the problem (as shown in Fig. 13.1).

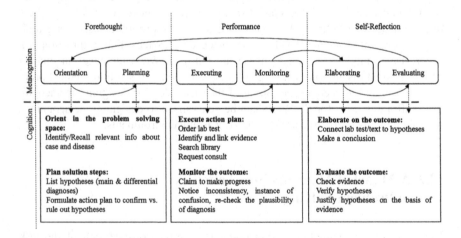

Fig. 13.1 Model of metacognitive activities in problem-solving

Metacognitive activities determine a learner's progress through each phase of diagnostic reasoning. For example, diagnostic hypotheses are refined progressively by engaging in strategic processing until a final diagnosis is reached. We distinguish amongst several types of metacognitive activities; namely, orienting, planning, executing, monitoring, evaluating, and elaborating.

Self-regulated learners orient themselves to a problem space by identifying or recalling information that is relevant to their efforts to outline tentative hypotheses (diagnoses) for the patient condition. In an effort to test these hypotheses, learners formulate plans that involve ordering lab tests, searching for specific symptoms or information about the disease, and asking for consults.

Once the plan is executed, learners make efforts to monitor the outcomes by evaluating their own progress or noticing unexpected or conflicting information. This appraisal may lead learners to revisit their own diagnosis by reviewing all the available evidence and hypotheses, leading to changes in the final diagnosis.

On the other hand, learners might elaborate further their own diagnosis by making conclusions on the basis of the evidence. As an example, a proficient learner may notice pertinent vital signs, such as the patient heart rate exceeds the normal range, which could be caused by a tumour of the adrenal glands. To test this assumption the learner formulates a plan to test for pheochromocytoma by verifying the serum levels of the catecholamines, adrenalin and noradrenalin.

The results indicate that the serum levels are elevated, which is pertinent to a diagnosis of pheochromocytoma. As such, the learner order a series of tests to rule out known alternatives to the diagnosis, while concluding that pheochromocytoma is a likely explanation for the patient condition.

Although the above example indicates a linear solution path, learners experience challenges to regulating their progress toward reaching a solution. Learners may fail to identify patient symptoms and vital signs that are indicative of the correct disease due in part to the low prevalence of this disease in the broader population.

Successful problem-solving requires that the learner makes an appropriate judgement that the serum-levels are in fact elevated, which implies that the learner is knowledgeable of the normal range of serum-level values.

Finally, learners may vary considerably in their levels of confidence in the final solution, depending on their ability to rule out their other hypotheses and to provide evidence that their final diagnosis is correct. BioWorld allows learners to practice their medical diagnostic reasoning by providing them with virtual patient cases.

BioWorld provides learners with feedback that highlights the similarities and differences between learners' approaches to solving the problem and the expert solution path. In the next section, we review the design guidelines of BioWorld, and how it supports learners in regulating their own progress in solving patient cases.

13.3 BioWorld: A Deliberate Practice Environment
for Diagnostic Reasoning with Virtual Patients

BioWorld is designed using the principles of a cognitive apprenticeship approach
to instruction [15, 16]. Cognitive apprenticeships provide learners with opportunities to link abstract knowledge to real world experiences. In this case students
apply their knowledge to medical patient cases. When diagnosing virtual patient
cases is too difficult for learners to do alone, they can still appropriate knowledge
from skilled experts and mentors.

BioWorld provides mentorship in the form of computer coaching and cognitive tools [17] embedded in the learning environment, which help structure the
learning experience for novice learners. Medical students test their knowledge by
formulating a diagnostic plan to solve a case, gather pertinent information about
the patient, and obtain feedback in relation to the diagnostic process, as shown in
Fig. 13.2.

In doing so, the software serves as a training platform for novices to practice
regulating the diagnostic process and become more proficient. The path taken by
expert physicians in solving the problem is modeled and made explicit to novices, thereby supporting them in diagnosing rare diseases that would be otherwise
beyond their reach [2]. Forethought processes include framing the patient's problem prior to formulating a diagnosis about a patient case.

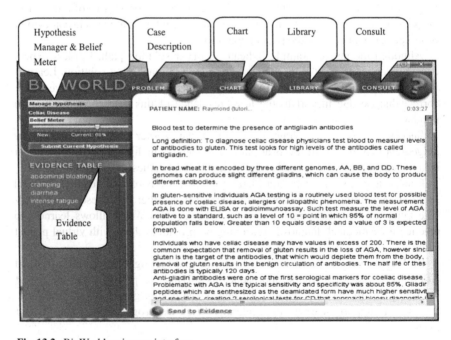

Fig. 13.2 BioWorld main user interface

Novices begin by gathering evidence (i.e., patient symptoms, history, etc.) from the case description and highlighting the evidence they consider important to the patient case. They send their evidence to an Evidence Table that is visible throughout their problem solving activity. Medical students use the Hypothesis Manager dropdown menu to indicate their differential hypotheses.

The menu displays a comprehensive list of diseases organized by the biological system that is afflicted. Physicians formulate differential diagnoses where they consider more than one disease at the same time. They can pick up to 10 diseases at the same time; however, they must indicate the level of confidence they have in each hypothesis by using the Belief Meter and select which disease they believe to be the most likely.

During the performance phase, novices search for evidence pertaining to a disease, conduct diagnostic tests to confirm or disconfirm their hypotheses, and search the online Library for additional information regarding the typical symptoms and transmission routes of a disease, definitions of standard medical terminology, and range of normal values for a specific laboratory procedure. Learners monitor their progress in solving the problem by evaluating their diagnostic test results and evaluating the patient's vital signs in the context of the Chart interface.

The Consult allows novices to request hints that are delivered in increasing order of specificity. After learners have submitted their final diagnosis, BioWorld supports them with additional tools that foster reflection on the diagnostic process.

The Categorization panel allows them to categorize their own evidence items stating which items confirm, refute, or are irrelevant to their final diagnosis.

Following categorization students use the Prioritization panel to rank the evidence items in terms of their relative importance to the final diagnosis. In the Case Summary panel, novices write a brief justification for their final diagnosis on the basis of the evidence items that were gathered throughout the case.

This summary is written for the next hypothetical physician who would see the patient. Finally, the Student Report provides formative feedback to each student by highlighting the similarities and differences between their solution steps and that of a validated expert.

The expert solution path also provides a case summary, written by an expert, which outlines in detail the steps that were taken to solve the case, and how each step contributed to formulating the final diagnosis. In the next section we focus on the importance of metacognitive scaffolding within BioWorld using expert models.

13.4 Metacognitive Scaffolding with Expert Models: Novice-Expert Overlay Component of the Expert Model

Developing expertise requires practice with appropriate levels of scaffolding. BioWorld [2] provides opportunities for deliberate practice [4] by making "expert models of performance and competency more visible to learners in the context of the problem solving" [18: 805]. BioWorld was developed using a cognitive apprenticeship framework [15] where medical students learn clinical diagnostic reasoning by practicing realistic diagnostic tasks and are scaffolded in the context of their learning with expert models.

In BioWorld diagnostic reasoning is assessed using a novice-expert overlay system [19, 20]. An overlay model highlights differences between the solution paths of learners and experts, often revealing learner misconceptions. It is important to identify differences between novices and experts to support the novice along a learning trajectory that will lead to more expert levels of performance.

Variations of novice-expert models have been used in TREs to serve different purposes. In BioWorld, novice-expert models have been used to examine clinical reasoning dynamically, with a particular focus on process models as well as outcome comparisons [21]. The BioWorld user-model detects relevant patient symptoms and patient history, diagnostic laboratory tests ordered, and library information accessed.

Evidence items characterize the learner's path to solve the problem, which are analysed to identify similarities and differences with the expert's approach. Visualizations of the novice and expert models provide learners with the opportunity to review a representation of their problem solving steps and become cognizant of where their solution path differed from an expert's solution path.

In addition, this comparison can help learners attend to evidence they missed or evidence that is not part of the expert solution. Identifying these differences in the reasoning path or the decision-making procedure is particularly useful for revealing learners' misconceptions and incorrect reasoning strategies.

BioWorld captures user interactions and compares them to an expert model for the purpose of adapting instruction to the specific needs of the learner. The expert model fosters metacognition as learners are supported in formulating plans, monitoring their progress, and adaptively engaging in strategic actions while solving problems. The novice-expert overlay model individualizes feedback by highlighting similarities and differences between their respective solution paths.

Another example of this methodology is demonstrated by [22] who uses learning trajectories to model metacognition as learners plan, monitor, and evaluate their cognitive behaviors in a problem space.

The Interactive MultiMedia Exercises platform allows learners to practice solving simulated chemical analyses problems. Learning trajectories are computed based on problem outcomes, using item response theory estimates of solution frequencies.

The solution process is then analyzed in terms of strategic activities using Artificial Neural Networks (ANNs) and Hidden Markov Models (HMMs) [23]. BioWorld analyzes strategic activities while solving problems by providing learners with an external representation of the trajectory towards competency in diagnostic reasoning.

This representation is composed of evidentiary items that justify the case solution obtained through a cognitive task analysis of several experts, wherein the symptoms and laboratory tests pertinent in solving a case are identified.

The feedback provided in BioWorld highlights similarities and differences between the expert's and novice's approaches. Expert-driven modeling in BioWorld assists in supporting the help-seeking behaviors of learners through the use of the novice-expert overlay.

13.5 Developing an Expert Model to Trace Metacognitive Activities in Problem-Solving

Developing an expert model requires knowledge extraction from experts. We developed CaseBuilder (see Fig. 13.3), a case authoring tool for expert physicians, to create virtual patient cases for medical students to solve in BioWorld [24]. The experts that we work with are also medical instructors.

The expert must externalize his or her knowledge about a case by filling in all of the requisite elements needed to solve the case and thereby defining the problem space by identifying the elements used in the novice-expert overlay model that are needed to individualize feedback. CaseBuilder is also used to create context-specific hints that are delivered through the help-seeking model.

The physician enters each case through the fields that are provided. Currently, patient information is linked to vital signs, disease categories, diagnostic tests, and

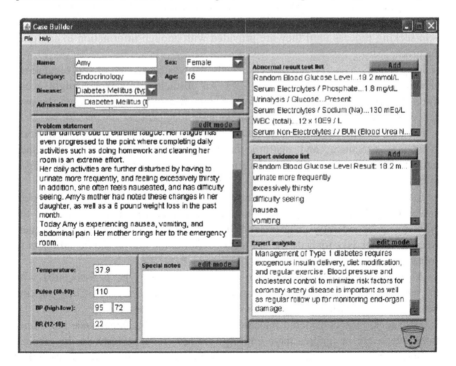

Fig. 13.3 CaseBuilder

expert arguments for solving the cases. Once the expert enters the data, the case is stored and it is usable by the BioWorld engine to present the new case to medical students. The CaseBuilder is an excellent tool for extracting expertise from the medical instructor.

In the section below we examine the use of the expert model to foster metacognition. In particular, we examine help-seeking behavior to determine when novices are at a learning impasse. The expert overlay model is used in the context of help-seeking to foster metacognitive and cognitive activities.

13.6 Examining Help-Seeking as an Indicator of Metacognition

Help-seeking is considered a metacognitive process since it indicates that learners monitor their problem solving and identify when they lack prior knowledge or the competency to continue the task independently. Help-seeking is particularly important in the context of solving problems since it is an indicator of obstacles and learning impasses [25].

To overcome these obstacles the learner can ask for help from a more knowledgeable other or from the TRE's help system. Once the learner has received help they must evaluate if the help was useful and if it was sufficient to continue solving the problem. If the help provided was not sufficient the learner can repeat the help-seeking process until they are provided with enough information to continue the task.

Help-seeking is a cyclical process which involves every step of self-regulation: planning, monitoring and reflection [26]. An obstacle in BioWorld might occur when the laboratory tests ordered do not support a hypothesis or when the evidence provided does not fit a specific hypothesis.

Karabenick [27] distinguishes between three types of help-seeking behaviors, avoidance, executive, and instrumental help-seeking. Help-seeking avoidance refers to instances when a learner fails to request help, despite the fact that help is in fact required.

Executive help-seeking involves a request for help when it is in fact needed; however, the learner requests final answers and makes no attempt to solve the problem independently. Instrumental help-seeking consists of a help-request where a learner asks for only the amount of information that would be sufficient to solve the problem independently.

As such, the distinction between each help-seeking type is associated with the necessity of the relevant information [28]. These three types of help-seeking behaviors can be further categorized as maladaptive or adaptive.

Help-seeking is considered maladaptive when a learner fails to use help functions effectively or ignores them entirely (see [29] for a review). In the medical domain, learners avoid the use of help functions such as glossaries, hyperlinked lectures, and expert advice, despite acknowledging that they lack the prior knowledge necessary for diagnosing the case on their own. Thus metacognitive awareness of what is known or not known does not lead to the execution of appropriate actions. In one study, fourth year medical students avoided or ignored help more than adaptively incorporating the help provided into to their diagnosis [30]. Maladaptive help-seeking behaviors were significantly correlated with poor quality solutions. This example demonstrates that even advanced medical students avoid or ignore help and this is problematic because maladaptive help behaviors are associated with poor quality diagnoses.

Help-seeking is considered to be adaptive when it enables students to continue the learning task independently [31] and this process is described as an exchange between self-regulation and other-regulation [32]. The support provided by more knowledgeable others scaffold learners to continue the task independently enabling them to complete a task that they would otherwise be unable to complete [33]. This exchange between self-regulation and other-regulation is necessary in order for the individual to become an autonomous learner.

Learners can obtain help from TREs as well, and scaffolding is provided to learners based on their current level of performance. The hints provided by the system supply the learner with enough information to continue the learning task independently without providing the final solution. In the same study mentioned above, adaptive help-seeking, although less prevalent than maladaptive help-seeking, was significantly correlated with better quality diagnoses [30].

It was suggested that for medical students to use help functions effectively, the help provided must be contextualized so that learners can apply the additional information offered by the system directly to the problem [30]. In BioWorld, learners have access to two help-seeking tools: the Consult Tool and the Library Tool. The Consult Tool provides context specific on demand hints that are delivered in increasing order of specificity and the Library Tool provides a glossary of medical terminology, diagnostic testing procedures, and typical symptoms and transmission routes of a disease [2].

Both of these tools have been designed with the purpose of scaffolding learners throughout diagnostic reasoning to foster adaptive help-seeking behaviors. Rule-based approaches to learner modeling have been used to study help-seeking episodes that result from self-monitoring while solving problems [34].

A series of decision rules determine how help-seeking behaviors are classified as effective or ineffective with the aim of providing appropriate feedback [35, 36]. BioWorld uses a help-seeking model to determine the type and level of hint to deliver a learner when they request a consult based on the expert model. Learners are supported through the analysis of previous help-seeking and problem-solving behaviors.

A series of rules allow the system to analyse user interactions and deliver hints in increasing order of specificity with the aim of gradually supporting learners to engage in the correct path to solving the case. Another aspect of the help-seeking model is to provide learners with supplementary knowledge in relation to diseases, lab-test procedures, and so on. The model encompasses search behaviors in terms of topics searched and pages viewed. We provide an overview of our research findings on help-seeking below.

13.7 Overview of Empirical Evidence of BioWorld's Role in Fostering Help-Seeking

Adaptive help-seeking behaviors are conducive for learning and lead to accurate diagnoses. Learners are able to ask for help at any time during their performance using BioWorld to solve patient cases. The following empirical studies address how and when learners ask for help to facilitate their diagnostic reasoning. We determine how students help-seek with the goal of encouraging learners to use help options more effectively.

13.7.1 Study 1: Using Process Data to Examine Self-Regulatory Behaviors During Clinical Problem Solving Using Technology

This study consisted of 30 students (28 medical and 2 dental students) who were registered at a Canadian University. All students had passed the same basic science course. Sequential pattern mining techniques were used to describe participant self-regulatory processes in BioWorld.

Self-regulation, for the purpose of this study, was defined as help-seeking behavior that was indicated by using the Consult Tool to receive help in the context of solving a case. Sequential pattern mining is a data mining technique that can be used to identify regularly occurring patterns in learning activities and behaviors [37].

In this study the sequential pattern mining technique classifies help requests according to groups (or clusters) of help requests whose sequence of activities occur prior to asking for help. We used this method to interpret the reasons why novices request help by identifying patterns in how they regulate diagnostic reasoning before asking for help from the Consult Tool.

A consult request was defined as clicking on the Consult Tool button with the goal of receiving a hint. For the purposes of this analysis no hints were provided when a student asked for help to observe how students naturally regulate their learning before and after requesting a consult.

Log-files were use to identify the behaviors that occurred before and after requesting help and these behaviors served as the boundaries of our unit of analysis for transcribing and coding the concurrent think-aloud protocols.

Data analyses suggest that students ask for help during the later stages of solving the problem and the amount of consult requests varied across patient cases. On average 83 % of the time taken to solve the case had elapsed (SD = 18.0 %) prior to asking for help. More consult requests made while diagnosing a rare disease such as Pheochromocytoma (52 %) and less requests were made while solving more common diseases, such as Diabetes mellitus Type 1 and Grave's disease (i.e., 28 and 21 %, respectively).

Consult requests were often preceded by ordering a lab test (72 %) and were followed by either: (a) submitting the final diagnosis (28 %), (b) changing their conviction in regards to their hypotheses (21 %), or (c) reading a topic in the library (14 %).

The results of the sequential pattern mining technique supported 5 distinct categories in the taxonomy of self-regulatory processes. However, the most interesting help-seeking pattern occurred before and after conducting diagnostic tests where the test results were unexpected. Help-seeking also occurred more frequently when reasoning about a rare disease rather than a common one.

These results have important implications for creating more effective forms of adaptive instruction by anticipating when students experience difficulty during reasoning and how to promote adaptive help-seeking in these instances.

For example, targeted prompts can be designed to encourage the appropriate use of help functions. This will help circumvent help-seeking avoidance behaviors. In study 2 below we examine another form of help-seeking, which pertains to knowledge acquisition that is gained by looking up information in the on-line-library.

13.7.2 Study 2: Supporting Diagnostic Reasoning by Modeling Help-Seeking in BioWorld

It is expected that if learners effectively monitor their learning then they will identify gaps in their knowledge and ask for help to improve performance. In addition to the Consult Tool in BioWorld, learners can address knowledge gaps by visiting the on-line library.

In study 2 we explored search behaviors in the Library Tool in relation to final solution accuracy using data from the same sample described in study 1. We hypothesized that if participants recognised a lacked of prior knowledge, then he or she would conduct a library search, leading to an improved final diagnosis.

In order to analyse learners' library search behaviors the RapidMiner C4.5 decision tree algorithm [38, 39] was used to split the data set of search behaviors into a tree-like network (Fig. 13.4). The nodes of the decision tree depict topics searched in the library derived from student log file data while solving three patient cases, Amy, Cynthia and Susan. The topics stated in the model are comprised of topics critical for solving the case and topics that highlight learner misconceptions about the nature of the unknown disease.

The results indicate which search behaviors were predictive of selecting a correct diagnosis. For example, in solving the case of Susan, learners who read about hyperthyroidism had a 100 % (5/5) chance of selecting the correct diagnosis: hyperthyroidism. However, learners who did not read about hyperthyroidism had only a 39 % (25/64) chance of selecting the correct diagnosis.

In other cases, topics searched by learners lead to diagnostic errors, which indicate learner misconceptions. In solving the case of Amy, learners who read about

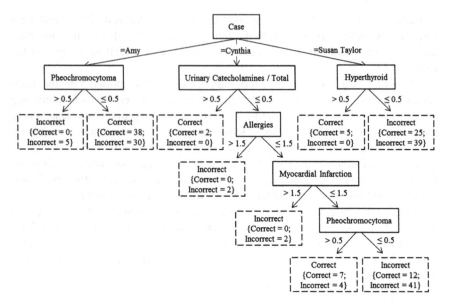

Fig. 13.4 BioWorld library search behavior model

pheochromocytoma had a 0 % (0/5) chance of selecting the correct diagnosis: phe-ochromocytoma. But, learners who did not read about pheochromocytoma had a 56 % (38/68) chance of selecting the correct diagnosis. Thus reading about pheo-chromocytoma decreased the likelihood that the student would arrive at the correct diagnosis. These results suggests that learners engage in both effective and inef-fective help-seeking by the Library Tool and this can lead to correct or incorrect diagnoses.

The results indicate which search behaviors were predictive of selecting a cor-rect diagnosis. For example, in solving the case of Susan Taylor, learners who read about hyperthyroidism had a 100 % (5/5) chance of selecting the correct diagno-sis: hyperthyroidism. However, learners who did not read about hyperthyroidism had only a 39 % (25/64) chance of selecting the correct diagnosis.

In other cases, topics searched by learners lead to diagnostic errors, which indi-cate learner misconceptions. In solving the case of Amy, learners who read about pheochromocytoma had a 0 % (0/5) chance of selecting the correct diagnosis: phe-ochromocytoma. However, learners who did not read about pheochromocytoma had a 56 % (38/68) chance of selecting the correct diagnosis.

Thus reading about pheochromocytoma decreased the likelihood that the stu-dent would arrive at the correct diagnosis. These results suggests that learners engage in both effective and ineffective help-seeking by the Library Tool and this can lead to correct or incorrect diagnoses, respectively.

Taken together, studies 1 and 2 indicate that learners engage in help-seek-ing and the nature of this behavior influences diagnostic reasoning and task performance.

Analyzing the patterns of behaviors that precede and proceed help-seeking suggest that help-seeking during diagnostic reasoning can be adaptive or maladaptive such that some of the search behaviors are more likely to yield a correct diagnosis while others are more likely to yield an incorrect diagnosis. The pattern of these behaviors and their respective outcomes are predictable, making it possible to develop individualised support to prompt adaptive help-seeking behaviors and to correct for maladaptive help-seeking thus promoting positive learning outcomes.

Studies 1 and 2 were based on comparing learner models that collected dynamically while problem solving, consisting of process data, and comparing to an expert overlay. Study 3 examines the expert case summary, which can be considered the final outcome of the problem solving, where learners summarize how they diagnosed the patient case.

13.7.3 Study 3: Case Summary Data-Diagnostic Learning Outcomes

In this study we compare how novice case summaries compare to written case summaries of experts who solve the case in BioWorld. Kellogg [40] highlights the importance of effective written communication in both academic and professional settings. As such, the written case summaries constitute an important exercise for learners. In BioWorld, a typical written case summary highlights the vital signs, relevant symptoms, and lab-tests that were germane to solving the case.

The written case summary is a written justification of a learner's solution and it is a unique organization of thoughts, actions, and plans that reflect the learner's knowledge after completing a particular case.

An important challenge towards developing the novice-expert overlay model in BioWorld is the analysis of unstructured data, such as the written case summaries. The unstructured text-based data collected from the written case summaries can provide novel insights into learners' reasoning after problem-solving that complement user interactions that are analysed in the current version of the novice-expert overlay model.

Thus, the case summaries provide an additional comparative model to guide learners. We have taken first steps for developing a robust novice-expert overlay model using the case summaries written by learners and experts.

Toward this end, we examine the accuracy of commonly used text-mining algorithms in terms of differentiating case summaries written by novices and experts. The resulting text classification model will highlight the key linguistic features that characterize expert knowledge and performance on this task.

The findings stand to inform design guidelines of the revised user model that would be capable of assessing novice case summaries in order to guide instruction necessary to support expertise development in clinical reasoning.

13.7.4 Study 3: Text Mining Algorithms

Text classification is used in a number of domains [41], including news filtering and organization, document organization and retrieval, opinion mining, and email classification and spam filtering [42]. The advantage of automatic text classification is that it can significantly reduce the cost and time involved in manual categorization.

Text classification algorithms are commonly used in intelligent tutoring systems for assessment purposes [43]. The linguistic features that characterize highly proficient texts are of particular interest to instructional designers [44].

These quality indices can ascertain the aspects of text that are most germane to problem solving, and can provide designers with guidance on a response mechanism when the system detects their deficiency or absence. The text classification problem addressed in this study is the recognition (classification) of case summaries written by novices and experts in BioWorld.

For the purposes of classifying novice and expert written case summaries, the following algorithms were selected on the basis of their frequent use in text classification problems and the ease of implementation in the revised user model [45, 46]: Naïve Bayes (NB), Multinomial Naïve Bayes (MNB), and Sequential Minimal Optimization (SMO). The WEKA [47] toolkit was used to train and compare the text classification algorithms.

WEKA is a comprehensive workbench for machine learning algorithms for data mining tasks, comprising of a myriad of tools for data pre-processing, classification, clustering, and feature selection.

13.7.5 Study 3: Dataset

The dataset used to train the text classifiers included a total of 74 case summaries written by both novices and experts. The case summaries were labeled as written by either a novice (n = 60) or an expert (n = 14). A sample of a case summary written by a novice is given next: "16 year old girl, previously active and with no significant family history with onset of extreme fatigue, polyurea, polydipsea, difficulty concentrating and 6 lbs weight loss. Lab showed +FBS, and glucose in urine."

A sample of a case summary written by an expert is shown as follows: "This previously well 16 year old female presents to the ER with abdominal pain and nausea that started today. She complains of 6 months of fatigue limiting her activities of daily life, frequent urination, thirstiness, blurred vision, and weight loss. On exam she was in shock with a blood pressure of 95 systolic and a tachycardia of 100/min, and was mildly tachypneic at 22/min. Her temperature was normal. Investigation revealed a random blood glucose of 18.2, ketones in her blood, serum K of 5.8 with a normal ECG, elevated anion gap and a WBC count of 12. Further investigations for a precipitant of the ketoacidosis (CXR, Urine leukoyte esterase, and abdominal ultrasound) were negative."

In order to evaluate the efficacy of the classifiers, five datasets were generated to compare several text pre-processing approaches. We manipulated the way in which case summaries were indexed (IDFT, TFTT) and transformed (Lower case, Stemmer, Stopwords) for training the text classifiers (see Table 13.1). In doing so, our evaluation will take into account several alternative approaches to addressing the classification problem.

13.7.6 Study 3: Results

After pre-processing the data for experimentation, we performed the classifications using the WEKA toolkit. A 10-fold cross-validation was used to evaluate the performance (classification accuracy) of the NB, MNB, and SMO algorithms for each dataset. The results (overall comparison between the classifiers and the breakdown of the classification accuracies) obtained are shown in Table 13.2.

The three classifiers: Naïve Bayes, Multinomial Naïve Bayes, and Sequential Minimal Optimization that we employed in our experiments displayed high accuracies in the classification tasks. The SMO algorithm, which provided the best accuracy of 93.42 %, was found to be the most accurate in terms of distinguishing between the novice and expert case summaries. These findings warrant the use of SMO to revise the current version of the novice-expert overlay system used in BioWorld.

Table 13.1 Dataset pre-processing steps

Data set	IDF transform	TFT transform	Lower case tokens	Output word counts	Stemmer	Stop words
1	False	False	False	False	False	False
2	False	False	False	True	False	False
3	False	False	True	False	True	True
4	False	False	True	True	True	True
5	True	True	True	True	True	True

Table 13.2 Text classification accuracy

Data set	NB (%)	MNB (%)	SMO (%)
1	86.84	88.16	92.11
2	88.16	86.84	89.47
3	86.84	89.47	93.42
4	88.16	88.16	93.42
5	86.84	82.89	93.42

The findings from study 3 suggests that the classifiers employed perform well in the context of differentiating novice-expert written case summaries and supports the idea of leveraging text classification in developing a novice-expert overlay model.

The findings in this study highlight that the SMO algorithm is the most accurate classifier (with the highest accuracy reaching 93.42 %) of novice-expert differences in written case summaries.

The findings could be improved in the future with the collection of more case summaries. Future research will explore the key linguistic features in written case summaries that should serve as quality indices by identifying similar and different terms mentioned in novice and expert case summaries.

In doing so, the findings will apprise domain expert and instructional designers by specifying the remedial steps in the response mechanism that should be taken by the intelligent tutoring system when case summaries are found to be deficient of or lacking the identified quality indices.

This study represents a first step towards developing a novice-expert overlay model component of the expert model, which will help to promote a more expert-like approach to diagnostic reasoning amongst learners.

13.8 Conclusion

A key dimension of expertise is metacognition, knowing what one knows and does not know [5]. Consequently, our expectation is that competent physicians know what they know and do not know. BioWorld is a TRE designed to foster metacognitive activities of physicians-in-training (novice medical students).

Novices need to know when to ask for help and we need to identify when they reach a learning impasse, when they need more information, and when they have misconceptions. BioWorld was designed to support help-seeking activities in the context of solving virtual patient cases. We presented an underlying model of the cognitive and metacognitive activities that occur in the context of diagnostic/clinical reasoning.

In particular, we documented the three phases of metacognition, forethought, performance and reflection that occur during clinical reasoning with BioWorld.

Theoretical definitions were provided of these important constructs in the context of medical reasoning. Furthermore, we discussed the importance of designing computer-based learning environments that provide cognitive tools to support metacognitive activities throughout the problem solving process.

BioWorld provides a cognitive apprenticeship for novice physicians to deliberately practice their diagnostic reasoning skills with scaffolding. The use of expert overlay models were described in terms of how they support adaptive help seeking and how they support the analysis of help-seeking while learning to diagnose a patient's disease.

The expert model also was used to examine the process and products of the diagnostic process. Specific advances were described in terms of modeling meta-cognition through analytical techniques to design, evaluate, and develop expert models by capturing metacognitive activities in problem-solving.

Educational data mining techniques are outlined with the aims of capturing metacognitive activities as they unfold throughout problem solving. These trace methodologies were used to model self-regulatory processes along the trajectory towards competency in diagnostic reasoning. In particular, sequential data mining techniques were used to identify patterns in help seeking behavior in the context of solving cases.

This method revealed that the most comment antecedent to help seeking was receiving a diagnostic test result that did not support a hypothesis about a disease. An analyses of help-seeking in the form of library usage indicated that those students that looked up a disease in the library were more likely to solve the case. Finally, text mining techniques were used to compare expert and novice case summaries in an attempt to assess differences in case solutions.

These techniques were highly accurate and can be used in future studies to assess learner trajectories. These findings lead to insights with respect to designing appropriate scaffolding tools in order to promote higher levels of competence in novices. The empirical findings will be used to inform our future work in the design and delivery of appropriate feedback. We anticipate that the principles for designing metacognitive scaffolding tools in BioWorld can generalize to designing metacognitive scaffolding tools for other computer-based learning environments.

References

1. Lesgold, A.M.: Problem solving. In: Sternberg, R.J., Smith, E.E. (eds.) The Psychology of Human Thought. Cambridge University Press, Cambridge (1988)
2. Lajoie, S.P.: Developing professional expertise with a cognitive apprenticeship model: Examples from avionics and medicine. In: Ericsson, K.A. (ed.) Development of Professional Expertise: Toward Measurement of Expert Performance and Design of Optimal Learning Environments, pp. 61–83. Cambridge University Press, Cambridge (2009)
3. Lajoie, S., Naismith, L., Poitras, E., Hong, Y., Panesso-Cruz, I., Ranelluci, J., Wiseman, J.: Technology rich tools to support self-regulated learning and performance in medicine. In: Azevedo, R., Aleven, V. (eds.) International Handbook of Metacognition and Learning Technologies. Springer, Amsterdam (2013)
4. Ericsson, K.A., Krampe, RTh, Tesch-Romer, C.: The role of deliberate practice in the acquisition of expert performance. Psychol. Rev. **100**(3), 363–406 (1993)
5. Lajoie, S.P.: Transitions and trajectories for studies of expertise. Educ. Researcher **32**, 21–25 (2003)
6. Alexander, P.A., Dinsmore, D.L., Parkinson, M.M., Winters, F.I.: Self-regulated learning in academic domains. In: Zimmerman, B., Schunk, D. (eds.) Handbook of Self-Regulation of Learning and Performance. Routledge, New York (2011)
7. White, C.B., Gruppen, L.D.: Self-regulated learning in medical education. In: Swanwick, T. (ed.) Understanding Medical Education. Wiley-Blackwell, Sussex (2010)

8. Evensen, D.H., Salisbury-Glennon, J.D., Glenn, J.: A qualitative study of six medical students in a problem-based curriculum: Toward a situated model of self-regulation. J. Educ. Psychol. **93**, 76–659 (2001)
9. Brydges, R., Butler, D.L.: A reflective analysis of medical education research on self-regulation in learning and practice. Med. Educ. **46**, 71–79 (2012)
10. Meijer, J., Veenman, M.V.J., Van Hout-Wolters, B.H.A.M.: Metacognitive activities in text-studying and problem-solving: Development of a taxonomy. Educ. Res. Eval. **12**(3), 209–237 (2006)
11. Lu, J., Lajoie, S.P.: Supporting medical decision making with argumentation tools. Contemp. Educ. Psychol. **33**, 425–442 (2008)
12. Lajoie, S.P., Lu, J.: Supporting collaboration with technology: Does shared cognition lead to co-regulation in medicine? Metacogn. Learn. **7**, 45–62 (2012)
13. Zimmerman, B.J.: Self-regulated learning and academic achievement: An overview. Educ. Psychol. **25**(1), 3–17 (1990)
14. Zimmerman, B.J., Campillo, M.: Motivating self-regulated problem solvers. In: Davidson, J.E., Sternberg, R. (eds.) The Nature of Problem Solving pp. 233–262. Cambridge University Press, New York (2003)
15. Collins, A.: Cognitive apprenticeship. In: Sawyer, K. (ed.) Cambridge Handbook of the Learning Sciences pp. 47–60. Cambridge University Press, New York (2006)
16. Lajoie, S.P.: Aligning theories with technology innovations in education. Br. J. Educ. Psychol.—Monogr. Ser. II **(5)** Learning through Digital Technologies, 27–38 (2007)
17. Lajoie, S.P.: Cognitive tools for the mind: The promises of technology: Cognitive amplifiers or bionic prosthetics? In: Sternberg, R.J., Preiss, D. (eds.) Intelligence and Technology: Impact of Tools on the Nature and Development of Human Skills, pp. 87–102. Erlbaum, Mahwah (2005)
18. Lajoie, S.P., Azevedo, R.: Teaching and learning in technology-rich environments. In: Alexander, P.A., Winne, P.H. (2nd ed.) Handbook of Educational Psychology pp. 803–821. Lawrence Erlbaum Associates, Mahwah (2006)
19. Goldstein, I.P.: The genetic graph: a representation for the evolution of procedural knowledge. In: Sleeman, D., Brown, J.S. (eds.) Intelligent Tutoring Systems pp. 51–77. Academic Press, London (1982)
20. Shute, V.J., Zapata-Rivera, D.: Adaptive educational systems. In: Adaptive Technologies for Training and Education, pp. 7–27 (2012)
21. Naismith, L., Lajoie, S.P.: Using expert models to provide feedback on clinical reasoning skills. In: Aleven, V., Kay, J., Mostow, J. (eds.) 10th International Conference on Intelligent Tutoring Systems, LNCS, vol. 6095, pp. 44–242. Springer, Berlin (2010)
22. Stevens, R.: Machine Learning Assessment Systems for Modeling Patterns of Student Learning, pp. 349–365. Games and Simulation in Online, Learning (2007)
23. Stevens, R., Beal, C.R., Sprang, M.: Assessing students' problem solving ability and cognitive regulation with learning trajectories. In: International Handbook of Metacognition and Learning Technologies pp. 409–423. Springer, New York (2013)
24. Lajoie, S.P., Faremo, S., Wiseman, J.: A knowledge-based approach to designing authoring tools: From tutor to author. In: Moore, J.D., Redfield, C., Johnson, L.W. (eds.) Artificial Intelligence in Education: AI-ED in the Wired and Wireless future pp. 77–86. IOS Press, Amsterdam (2001)
25. Järvelä, S.: How does help seeking help?–New prospects in a variety of contexts. Learn. Instr. **21**(2), 297–299 (2011)
26. Newman, R.S.: Adaptive help-seeking: a strategy of self-regulated learning. In: Schunk, D.H., Zimmerman, B.J. (eds.) Self-Regulation of Learning and Performance: Issues and Educational Applications pp. 283–301. Erlbaum, Hillsdale (1994)
27. Karabenick, S.A.: Strategic Help Seeking: Implications for Learning and Teaching. Erlbaum, Mahwah (1998)
28. Huet, N., Escribe, C., Dupeyrat, C., Sakdavong, J.-C.: The influence of achievement goals and perceptions of online help on its actual use in an interactive learning environment. Comput. Hum. Behav. **27**, 413–420 (2011)

29. Aleven, V., Stahl, E., Schworm, S., Fischer, F., Wallace, R.: Help-seeking and help design in interactive learning environments. Rev. Educ. Res. **73**(3), 277–320 (2003)
30. Gräsel, C., Fischer, F., Mandl, H.: The use of additional information in problem oriented learning environments. Learn. Environ. Res. **3**, 287–305 (2000)
31. Newman, R.S.: Children's help-seeking in the classroom: the role of motivational factors and attitudes. J. Educ. Psychol. **82**, 71–80 (1990)
32. Newman, R.S.: The motivational role of adaptive help seeking in self-regulated learning. In: Motivation and Self-Regulated Learning: Theory, Research, and Applications, 315–337 (2008)
33. Vygotsky, L.S.: Mind in Society: The Development of Higher Psychological Processes. Harvard University Press, Cambridge (1978)
34. Aleven, V.: Help seeking and intelligent tutoring systems: theoretical perspectives and a step towards theoretical integration. In: International Handbook of Metacognition and Learning Technologies pp. 311–335. Springer, New York (2013)
35. Aleven, V., McLaren, B., Roll, I., Koedinger, K.: Toward meta-cognitive tutoring: a model of help seeking with a cognitive tutor. Int. J. Artif. Intell. Educ. **16**, 101–128 (2006)
36. Aleven, V., Roll, I., McLaren, B.M., Koedinger, K.R.: Automated, unobtrusive, action-by-action assessment of self-regulation during learning with an intelligent tutoring system. Educ. Psychol. **45**(4), 224–233 (2010)
37. Kinnebrew, J.S., Mack, D.L.C., Biswas, G.: Mining temporally-interesting learning behavior patterns. In: 6th International Conference on Educational Data Mining, Memphis (2013)
38. Quinlan, J.R.: Improved use of continuous attributes in c4.5. J. Artif. Intell. Res. **4**(1), 77–90 (1996)
39. Mierswa, I., Wurst, M., Klinkenberg, R., Scholz, M., Euler, T.: YALE: rapid prototyping for complex data mining tasks. In: Ungar, L., Craven, M., Gunopulos, D., Eliassi-Rad, T. (eds.) 12th ACM SIGKDD International Conference on Knowledge Discovery and Data Mining KDD-06. ACM, New York (2006)
40. Kellogg, R.T.: Professional writing expertise. In: Ericsson, K.A., Charness, N., Feltovich, P.J., Hoffman, R.R. (eds.) The Cambridge Handbook of Expertise and Expert Performance. Cambridge University Press, New York (2006)
41. Sebastiani, F.: Machine learning in automated text categorization. ACM Comput. Surv. **34**(1), 1–47 (2002)
42. Aggarwal, C.C., Zhai, C.: A survey of text classification algorithms. In: Aggarwal, C.C., Zhai, C. (eds.) Mining Text Data pp. 163–222 Springer (2012)
43. McNamara, D.S.: IIS: A marriage of computational linguistics, psychology, and educational technologies. In: Wilson D., Sutcliffe G. (eds.) 20th International Florida Artificial Intelligence Research Society Conference pp. 15–20. The AAAI Press, Menlo Park (2007)
44. McNamara, D.S., Crossley, S.A., McCarthy, P.M.: Linguistic features of writing quality. Written Communic. **27**(1), 57–86 (2010)
45. Kibriya, A.M., Frank, E., Pfahringer, B., Holmes, G.: Multinomial naïve Bayes for text categorization revisited. In: Webb, G.I., Yu, X. (eds.) Advances in Artificial Intelligence pp. 488–499. Springer, Heidelberg (2004)
46. Platt, J.C.: A fast algorithm for training support vector machines. Technical Report MSR-TR-98-14 (1998)
47. Hall, M., Frank, E., Holmes, G., Pfahringer, B., Reutemann, P., Witten, I.H.: The WEKA data mining software: an update. SIGKDD Explor. **11**(1), 10–18 (2009)

Chapter 14
The MetaHistoReasoning Tool: Studying Domain-Specific Metacognitive Activities in an Intelligent Tutoring System for History

Eric G. Poitras

Abstract This chapter reviews empirical research on the MetaHistoReasoning (MHRt) tool, an intelligent tutoring system that aims to support students in regulating their own understanding of historical events in accordance with disciplinary-based practices. The design of the system is guided by a domain-specific account of the metacognitive activities involved in learning while performing inquiries into the causes of historical events. The system relies on modularization as a mechanism for delivering instruction and promoting the development of metacognitive skills. The Training Module supports skill acquisition from examples, while the Inquiry Module facilitates skill practice and refinement through problem-solving. Both modules fulfill complementary roles in skill development, since the learning outcomes for a module determines subsequent learning processes. The modular nature of the system also allows flexibility in implementing novel approaches for instruction and testing that impact towards several aspects of skill development. A pedagogical agent interacts with the learner to facilitate the transition across each module as skills become increasingly sophisticated. The aim of our research program is to improve the interactive capabilities of the agent by building assessment mechanisms that target critical aspects along this transition as a means to intervene and foster skill development. As such, we provide an overview of trace measures and analyses that are used to study how learners set goals, use strategies, and monitor the outcomes in the context of their investigations. We will review recent advances in building assessment mechanisms that target these disciplinary-based activities in order to recommend pedagogical strategies for the virtual agent embedded in the MHRt tool.

E.G. Poitras (✉)
Advanced Instructional Systems and Technologies Laboratory,
University of Utah Educational Psychology, 1721 Campus Center Drive SAEC 3220,
Salt Lake City, UT 84112-8914, USA
e-mail: ASSISTlaboratory@gmail.com

© Springer International Publishing Switzerland 2015
A. Peña-Ayala (ed.), *Metacognition: Fundaments, Applications, and Trends*,
Intelligent Systems Reference Library 76, DOI 10.1007/978-3-319-11062-2_14

Keywords MetaHistoReasoning tool · Confusion · Pedagogical agent · Problem-solving · Metacognitive tool · Domain-specific · Metacognition

Abbreviation

MHRt   MetaHistoReasoning tool

14.1 Introduction

Technology-rich learning environments refer to a learning environment where any application software is used in supporting learners to achieve instructional goals [1]. A fundamental characteristic of this type of learning environment is that the design and evaluation of technology is guided by theories of learning and instruction. A case in point is the design of computers as cognitive tools [2–6], a metaphor that conceptualizes the design process as the creation of external representations aligned with the cognitive activities that are involved in learning. In doing so, the application software may perform several functions, namely, to support logic, memory, or any other activities that would be out of the learner's reach, and to direct attentional resources to higher-order processes by automating lower-order thinking skills [7].

The use of computers as metacognitive tools [8–11] emerged from this long-standing research tradition. This development led to emphasize learners' efforts to regulate their own learning during the design process. Self-regulation requires a learner to set goals, use strategies to achieve these goals, and monitor their own progress [12–16]. It involves motivation and awareness as well as the capacity to adjust by evaluating one's own learning. Self-regulation raises an important design challenge for metacognitive tools since the learners' efforts to regulate their own learning involve latent and unobservable processes, which should be captured and analyzed by the software application in an unobtrusive manner [17, 18]. A considerable amount of literature has been published during the last decade on the adaptivity of metacognitive tools and how this type of assessment can improve instruction for learners that have difficulties regulating their own learning [19–21].

As a matter of fact, there is a growing body of empirical evidence showing learners' difficulties to regulate their own learning of complex topics in the basic sciences [22] and social sciences [23, 24]. Researchers have documented these different classes of failures that lead to minimal learning, referring to them as instances of dysregulated learning [25]. In studying historical texts, for instance, dysregulated learning may consist of insufficient amounts of activities related to planning and monitoring, in spite of the fact that setting goals, in particular, is predictive of declarative knowledge gains [23]. In addition, although learners often summarize texts and take notes, these strategies are traditionally less effective as compared to engaging in elaborative and inferential activities.

The structure of historical texts is an important antecedent to instances of dys-regulated learning while studying historical texts. Learners often fail to notice instances of confusion and offer plausible explanations while reading historical texts that do not mention the causes of events [24]. In doing so, self-regulatory knowledge acts as compensatory processing to infer the most likely causes that led to the occurrence of the event under investigation. Self-regulated learners are able to search across multiple text documents and recall prior knowledge in an effort to build a coherent mental representation of a chain of events.

Given that learners may lack the requisite knowledge, researchers have outlined principled methods to revise the causal structure of historical texts with the aim of facilitating comprehension [26, 27]. This approach assumes that uncertainty is undesirable, and should be minimized by providing coherent explanations. This line of research hypothesizes that more coherent texts require fewer inferential processing; therefore, learners should demonstrate better learning outcomes when texts are revised in order to make them more coherent. This effect is mediated by the interaction of several factors, including the source of the incoherence, the amount of prior knowledge of the reader, and whether learning is assessed in terms of the ability to recall or understand the relevant material.

On the other hand, others have maintained that confusion can be conducive to learning if appropriately induced and resolved while providing the necessary assistance [28, 29]. This alternative approach maintains that technology-rich learning environments can intentionally induce confusion to promote deep inquiry that can benefit learning, albeit if learners engage in the requisite activities with the help of software features. This research tradition states that confusion is beneficial to learning given the occurrence of activities that are associated with the search for a solution, namely, causal reasoning and effortful elaboration. Rather than eliminating potential sources of confusion that may arise in future learning situations, learners should be scaffolded in terms of resolving these issues, which increases the likelihood that learners will apply the relevant skills to other situations.

This chapter examines the latter approach by describing the MHRt tool, a computer-based learning environment designed to induce confusion to benefit learning through problem-solving within the domain of history [30]. The MHRt induces confusion by failing to mention any information pertaining to the causes of an event. Learners are expected to attain a coherent understanding of the event by searching and transforming information obtained from authentic source documents in accordance with disciplinary-based practices. Modules embedded in the MHRt target the requisite skills that are involved in regulating one's own investigation into the causes of the event. The scope of this chapter is limited to comparing and contrasting assessment mechanisms with respect to different stages of skill development. To do so, an illustrative case study is reviewed to exemplify how the assessment mechanisms adapt instruction to the specific needs of different learners. The next section provides a brief review of the three-phase model of cognitive and metacognitive activities in historical inquiry, the theoretical framework that is used to define the aforementioned skills.

14.2 The Three-Phase Model of Cognitive and Metacognitive Activities in Historical Inquiry

The existing models of self-regulated learning share several basic assumptions [31, 32]. First, learners are actively involved in making sense of information given the resources that originate from their own cognitive system or the external environment. Second, the notion of phases characterizes learners' efforts to plan, monitor, control, and evaluate their own learning in an iterative manner. Third, conditions that are inherent to the learner and a situation constrain the self-regulation of learning, including relevant cognitive, affective, behavioral, and contextual factors. The fourth assumption is related to knowledge about self-regulation determines skill deployment in response to obstacles and challenges to learning. The fifth concerns to the deployment of these skills mediate learning outcomes. Although self-regulated learning theorists have outlined detailed accounts of these mechanisms [13–16], researchers have recently called for further clarification of the domain-generality or –specificity of the relevant constructs [17, 33, 34].

With regard to the domain of history, the three-phase model of cognitive and metacognitive activities in historical inquiry provides a domain-specific account of self-regulated learning [34]. According to the model, history learners regulate their own search for the causes of historical events. The search process is characterized by several phases, spanning from an initial lack of knowledge about the causes of the event to the reinstatement and attainment of a coherent understanding. Theoretical constructs from models of historical reasoning [35–37] and self-regulated learning [15, 38–40] are synthesized in order to account for the regulatory mechanisms that facilitate the learners' transition across each phase. These mechanisms consist of metacognitive activities that are adaptively and iteratively deployed while investigating the causes of historical events.

Metacognitive monitoring activities involve the comparison of one's own comprehension of an event against standards for causal coherence. Causation constrains the inquiry process through the need to interpret information obtained from sources in terms of events that logically follow from their antecedents [35, 41]. However, the causal structure of a narrative text is not necessarily conducive to comprehension since relevant information may be missing from the account of an event [27]. Self-regulated learners continually evaluate their understanding of the causes of historical events and take remedial actions when the explanation is unknown or uncertain.

Planning related activities refer to setting goals that define the desired result of an inquiry into the causes of an event. In the early stages of an investigation, when the exact causes of the event under investigation are still unknown, self-regulated learners search for evidence to confirm a potential cause. However, as the learners' understanding of the causes gradually becomes more certain, learners attempt to weigh the likelihood of other potential causes or to anticipate counter-arguments against their own account of the event. In doing so, self-regulated learners reinstate coherence in understanding the causes of an event by building an increasingly sophisticated argument.

Metacognitive control activities refer to the disciplinary-based strategies that are involved during the learners' inquiries. These strategies, also known as historical thinking skills [37], stipulate how to evaluate the trustworthiness of a source document, gather and situate evidence within the time and place of its creation, find corroborating information across other sources, and use substantive concepts pertaining to the event under investigation. Self-regulated learners are able to choose and deploy the strategies appropriately and evaluate the certainty of the resulting argument.

As an example, a typical learner may notice that a text does not explain why an event occurred. As an example, the causal factors that led to the occurrence of the 2008 world financial crisis were not mentioned in the circumstances stated in the text. Confused as to why investors were pulling their money from banks, the learner may set the goal of investigating further by attempting to find information that would confirm that financial institutions were highly levered. To reach this goal, the learner first formulates a question: "What is the degree of financial leverage of a major financial institution, in particular, Lehman Brothers Holdings Inc, during the end of 2007?" Using a credible source of information, the annual report of the firm, the learner finds a leverage ratio of 31 to 1, suggesting that Lehman was at considerable risk. As such, the learner argues that investor panic was partially attributed to levered financial institutions, a claim that is corroborated by the fact that shares for Lehman plummeted sharply during the same time period. The learner may also contextualize this information by recalling that investor confidence was lowered by the near collapse of another firm, Bear Stearns, at the beginning of the following year. The learner may then engage in an additional line of inquiry in order to answer a follow-up question: "Did Lehman Brothers and Bear Stearns share a similar investment portfolio?" The example described here illustrates how the activities involved in regulating one's own investigation are recursive as the outcome of the previous search determines the direction of the next.

The three-phase model of cognitive and metacognitive activities in historical inquiry guides the development of the MHRt by decomposing the relevant activities into skill components. These skill components serve as the instructional goals of the MHRt as modules embedded within the system are designed to facilitate skill development. The Training Module implements example-based skill acquisition as an instructional approach, allowing learners to study examples of the requisite skills and to receive help in the form of hints and prompts [42, 43]. The Inquiry Module allows the learner to practice and refine the skills that were acquired in the previous module by performing a structured inquiry-based learning task [44–47]. The following sections describe the design of both modules, and how the system assesses the learners' progress through each stage of skill development.

14.3 The MetaHistoReasoning Tool Training Module

14.3.1 The Design Guidelines of the Training Module

The Training Module supports skill acquisition by providing learners with a set of examples and prompting them to analyze and differentiate each skill. The module is organized according to a series of phases, referred to as the training, categorization, and self-explanation phase. The training phase is completed by the learner at the beginning of the session, where an instructional video introduces the topic under investigation, the relevant skills, and the interface features of the module. The categorization phase requires that the learner analyze a series of examples by identifying the corresponding skill among a list of options, which include the correct response. Learners make as many attempts as necessary to choose the correct option. The self-explanation phase starts at pre-determined intervals, where the learner explains how the skills shown in a set of examples contributes to the investigation of the topic. The examples are displayed on the lower left corner of the screen, as shown in Fig. 14.1 an example consists of a brief verbal utterance that resembles a historian talking aloud while analyzing a historical document.

Although each example demonstrates a specific skill through a unique utterance, the learner is also provided with sets of examples in order to illustrate how skills are interrelated with each other in the context of an investigation. Each set is delivered in

Pedagogical agent panel

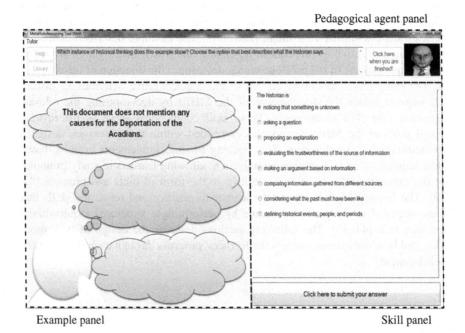

Example panel   Skill panel

Fig. 14.1 The main interface of the training module

increasing order of complexity, including one, three, and five examples in a given set. The order of each set is determined on the basis of learners' performance during the categorization phase as four sets of examples, containing both three and five examples each, are used to establish a baseline. If the baseline performance is greater than the 70 % accuracy threshold, then the learner skips the following sets of examples to either solve more complex sets of examples or investigate the event in the Inquiry Module.

The artificial pedagogical agent is located in the upper right corner of the screen. The agent interacts with the learner by providing definitions, prompts, and feedback. The definition of each skill was provided by the agent at the beginning of the session, when the learner solved the first sets of examples, each one demonstrating a single skill (e.g., "This example shows an historian asking a question. In doing so, the historian begins to search for the most important cause of the Acadian Deportation.").

The prompts were meant to encourage the learner to either categorize an example (e.g., "Which instance of historical thinking does this example show? Choose the option that best describes what the historian says.") or write a self-explanation regarding the skills that are shown in a set of examples (e.g., Explain how each instance of historical thinking relates to the historian's goal, which is to explain why the Acadian Deportation occurred.). The feedback was provided by the agent immediately after the learner categorized an example, and was either positive (e.g., "Your answer is correct.") or negative (e.g., "Your answer is incorrect, try again.").

14.3.2 Modeling Skill Acquisition in the Training Module

The skill acquisition model allows the MHRt to generate learning curves, a representation of the increasing rate of skill acquisition as a function of exposure to several examples of different skills in the context of the Training Module. The rate of skill acquisition is inferred on the basis of performance on the categorization task. A learning curve can be decomposed according to several performance metrics. These metrics include the observed and predicted cumulative percentage of correct attempts, the error ratio, and the time taken to categorize an example.

The cumulative percentage of correct attempts illustrates the rate of correct categorizations for each opportunity. Researchers have outlined several methods to model the rate of skill acquisition on the basis of user interactions with interface features [48, 49]. The skill acquisition model relies on a logistic function to predict whether a categorization attempt is correct or incorrect. The following parameters are included in the model: (1) the elapsed time duration in seconds; (2) the number of attempts; (3) the amount of exposure to examples of a particular skill; (4) the type of skill illustrated by the example.

On the one hand, the benefits of practice can be ascertained by comparing the observed and predicted performance as a function of the increasing amount of opportunities to categorize examples. Figure 14.2 shows the cumulative average percentage of correct categorizations obtained by a learner and predicted by the model. The predicted probability value is also plotted across each opportunity to categorize an example. The slope of the learning curve has a good fit to the

Fig. 14.2 The cumulative percentage of correct attempts to categorize examples as a function of the number of opportunities

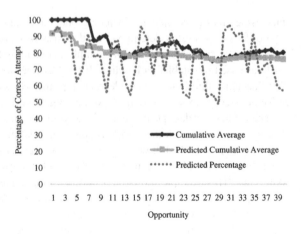

Fig. 14.3 The predicted percentage of a correct attempt to categorize an example on a specific opportunity

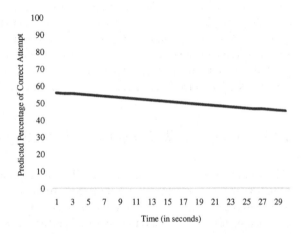

predictions of the model, as the average values range consistently above the 70 % threshold, suggesting that the learner's rate of skill acquisition is satisfactory.

On the other hand, the predicted performance can be plotted for a specific opportunity in order to determine when the system should intervene and provide assistance to the learner. Figure 14.3 shows the predicted percentage of correct categorization obtained on the learner's 10th opportunity, plotted across the elapsed time taken to categorize the example. The downward slope of the learning curve suggests that the probability of correctly categorizing the example decreases as a function of the elapsed time. The system can rely on this information in order to provide remedial instruction when the learning curve reaches predetermined thresholds. To do so, the pedagogical agent could deliver prompts to elaborate that specific type of skill or provide the learner with a hint.

The error ratio consists of the probability of an incorrect categorization on a first attempt, relative to the probability of a correct categorization. The bar chart shown in Fig. 14.4 shows the error ratios corresponding to each skill, calculated

Fig. 14.4 The error ratios for each type of skill exemplified in the training module

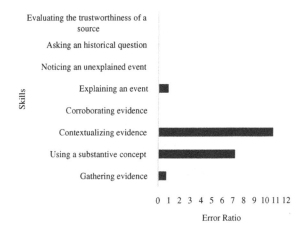

for the learners' entire session with the Training Module. As an example, the learner correctly categorized a total of 25 examples, and one of these correct categorizations corresponded to the skill of contextualization (i.e., 1:25 = 0.04). The learner incorrectly categorized a total of 7 examples, and 3 of those examples corresponded to the aforementioned skill (i.e., 3:7 = 0.4286). Therefore, the error ratio related to contextualizing evidence is 10.714 (i.e., 0.4286/0.04 = 10.714), suggesting that the learner is more likely to incorrectly categorize an example that shows this particular type of skill. The error ratio is a useful metric for ordering the sequence of examples that are shown to the learner since greater amounts of examples can be delivered as a means to target specific deficiencies in skill acquisition.

The elapsed time taken to study an example can be calculated in seconds by adding all the time durations for each opportunity that the learner made to categorize an example. The time duration can also be plotted for examples where the first attempt was correct or incorrect, as shown in Fig. 14.5. The stacked barchart indicates that although the skill of contextualization was associated with the highest error ratio, the learner nonetheless spent on average less time to study and categorize the relevant examples. The average elapsed time to incorrectly categorize an example of contextualizing evidence was 7.33 s, suggesting that the system should intervene by encouraging the learner to further analyze such examples.

14.4 The MetaHistoReasoning Tool Inquiry Module

14.4.1 The Design Guidelines of the Inquiry Module

The Inquiry Module supports the application and refinement of skills by allowing learners to inquiry into the causes of historical events. The module facilitates a learner's investigation through a digital collection of primary and secondary source documents with the help of embedded investigative tools. These tools are

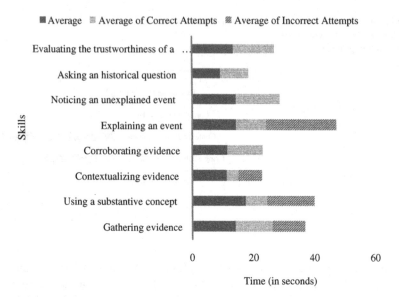

Fig. 14.5 The average elapsed time taken to categorize each type of skill exemplified in the training module

both dynamic and interactive (i.e., pedagogical agent) or static (i.e., series of instructional videos, the annotation tool, a digital library, as well as the explanation and evidence palette), as shown in Fig. 14.6. Using these tools, a learner is able to iteratively revise their explanation in light of new evidence and improve his understanding of the event.

A new session with the Inquiry Module begins with the learner reading a short narrative text that describes the circumstances surrounding the event. However, the text purposely makes no mention of any causes that would allow the learner to explain why the event occurred. Once the learner is done reading the text, the agent prompts them to monitor their own understanding by highlighting the missing information (i.e., "Read this text and you will notice that it does not explain why Charles Lawrence made the decision to deport the Acadians.") and asking an appropriate question (i.e., "What was the most important cause of the Acadian Deportation?").

The task of the learner is to search across the digital collection of source documents in order to answer the question. The system interface is designed to structure the learner's investigation into a series of steps, each step involving the use of a specific skill. For instance, the learner first evaluated the trustworthiness of a source, then gathered evidence from this source, searched across other sources for similar or contradictory information, and situated the evidence within that time period. During the initial line of inquiry, the pedagogical agent guides the learner through each step. As a result of each line of inquiry, the learners' explanation is revised in light of the new evidence that is obtained.

Fig. 14.6   The main interface of the inquiry module

The static tools embedded in the module are tailored to support learners in using specific skills. Instructional videos are made available to the learner in order to explain how to use each skill within the context of the module. The annotation tool allows the learner to write notes and select listbox items that are constrained to facilitate the following skills: evaluating the credibility of sources, gathering, corroborating, and contextualizing evidence. A digital library allows the learner to use a wide range of substantive concepts corresponding to the time period, including the relevant historical figures (e.g., Governor Charles Lawrence), the broader societal and political context (e.g., the Seven Year's War, and the governmental policies (e.g., Treaty of Utrecht). The explanation palette enables the learner to formulate an explanation by ranking the likelihood of several causal factors while investigating the event. The evidence palette serves as an external memory aid, allowing the learner to review a record of their own annotations.

14.4.2 Modeling Skill Practice and Refinement in the Inquiry Module

The skill practice and refinement model allows detecting states that are indicative of proficiency while the learner performs inquiries into the causes of historical events in the Inquiry Module. Learner states are classified through a series of decision rules applied on the items selected in the annotation tool and the causes ranked in the explanation palette. This argument-driven approach classifies learner states in terms of the type of goal pursued by the learner and whether strategies are appropriately used to achieve the goal, as shown in Figs. 14.7 and 14.8.

As an example, the learner investigated the causes of the Acadian Deportation, the forceful removal of the French inhabitants of Nova Scotia by the British authorities during the Seven Years' War. The explanation palette allows the learner to rank the likelihood of five plausible causes at the beginning and end of each line of inquiry. The event may be due to the influence and intentions of political figures, referring to British Governor Charles Lawrence's discontent towards the Acadians. The deportation might be attributed to the political situation as the Acadian deputies and communities refused to swear the unconditional oath of allegiance. An alternative is the economic situation at the time, which may

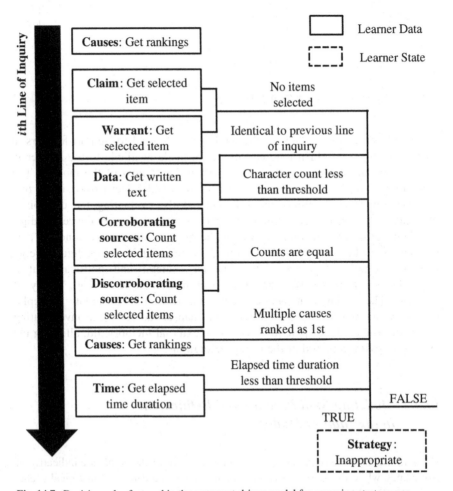

Fig. 14.7 Decision rules featured in the argument-driven model for assessing strategy use

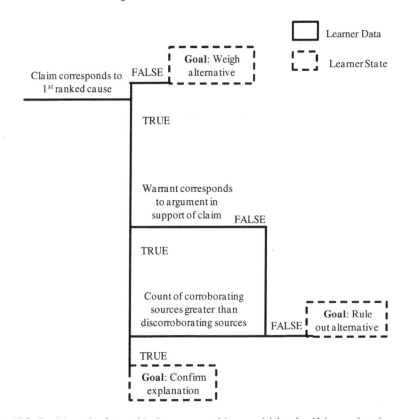

Fig. 14.8 Decision rules featured in the argument-driven model for classifying goal-setting

have motivated Charles Lawrence to seize Acadians' land, property, and livestock. The deportation may have been ordered for ideological reasons, as the Acadians would likely become loyal British subjects if they could be assimilated across the colonies. Charles Lawrence however may have wanted to prevent the Acadians from joining with their enemies in the conflict between the French and British empires.

At the beginning of the learner's first line of inquiry, the assimilation of the Acadians and the need to avoid a military conflict were both ranked as the most probable causes of the deportation. However, the learner annotated a source document that was found to support the claim that the deportation was due to Charles Lawrence's discontent towards the Acadians. This claim was supported by a quote taken from the source document that described an attack on the French army at Fort Beauséjour, which was ordered by Charles Lawrence. Therefore, "it is reasonable to infer that he displays general discontent for their presence and/ or refusal to swear oaths and loyalty". The learner corroborated this piece of evidence, noting that five other source documents mentioned similar information, whereas only three sources refuted the evidence. As a result, the learner's

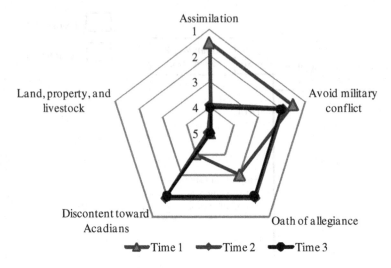

Fig. 14.9 Timeline of changes in the explanation for the event

explanation changed in favour of the Governor's discontent, the Acadian's refusal to swear the oath of allegiance, and the need to avoid a military conflict, as shown in Fig. 14.9.

In the following line of inquiry, the learner argued in favour of the Acadian's refusal to swear the oath of allegiance as the most likely cause for the event. The annotation referred to two sources, wherein the "Acadians were content with the first treaty with Philipps [past Governor of Nova Scotia Richard Philipps] evident by their letter to Cornwallis", but that the "British clearly weren't as per the source provided by the library". This evidence suggests that the refusal to swear an unconditional oath of allegiance meant "a cultural threat against British dominance [...] thus it is reasonable to assume that this was the final stimulus amongst many others that finally drove the British to expel the French in order to ensure dominance". The learner later indicated that the majority of sources agreed with this notion, referring to a total of six source documents that corroborated the evidence.

The skill practice and refinement model classifies both lines of inquiry as inappropriate in terms of achieving the different types of goals stated in the model. At the end of both lines of inquiry, the causal rankings suggest that the learner considered multiple causes in their explanation for the event under investigation. In doing so, the model cannot classify either lines of inquiry as an attempt to confirm an explanation, weigh an alternative cause, or rule out an alternative. To address the learner's uncertainty, the pedagogical agent should support the learner to monitor their own understanding of the event. The agent may challenge the learner's beliefs by highlighting information obtained from other source documents that either confirm or refute a piece of evidence, thereby prompting the learner to re-evaluate their own explanation for the event.

14.5 Discussion

Modularization enables the MHRt to capture and analyze user interactions at several stages of skill development. On the one hand, the Training Module generates learning curves to track skill acquisition as the learner categorizes illustrative examples of these skills and explains their underlying purpose. On the other hand, the Inquiry Module relies on an argument-driven model to characterize how the learner practices and refines the use of skills to investigate the causes of historical events. The learners' progress is assessed along a trajectory towards competency that is particular to the domain [50]. The modules complement each other as the learning outcomes at a previous stage dictate the learners' progress through the next stage.

The pedagogical agent is thus capable of facilitating transitions along this trajectory by selecting and delivering the instructional content that is most suitable to the needs of different learners. As evident in the review of the case study, the challenge is to identify the critical moments along the different trajectories of individual learners. First and foremost, the rate of skill acquisition varies greatly from one skill to another, depending on the complexity of the procedure that is applied and the variability of the information that is transformed. As a result, the agent should have an active role in selecting the examples that are delivered to the learner, providing just-in-time hints and prompts, as well as engaging learners in elaborative and evaluative processes. Furthermore, the agent should provide better guidance in relation to learners' efforts to plan their investigations and evaluate the outcomes of their inquiries into the causes of the event. When a learner is unsure of the most important cause for the event under investigation, the agent could challenge learners' beliefs by outlining a rebuttal argument or facilitate their search by referring to corroborating evidence obtained from other source documents.

There are several issues to consider in order improving the adaptive capabilities of the MHRt. One of the most important issues is to enhance both the quantity and quality of the self-regulation of learning. Although quantity can be strictly defined as the amount of lines of inquiry performed by the learner, each line also differs in terms of the amount of sources that were consulted and the pieces of evidence that were found to warrant or corroborate a particular claim.

The quality of these activities, however, reflects the depth of processing involved in each line of inquiry. For instance, learners who ruled out alternative explanations in addition to attempting to confirm an explanation built a more persuasive argument. In contextualizing evidence, the amount of elaborated information makes an argument more comprehensive to an audience, but the diversity of aspects that are considered is critical, such as whether the location of the event was described, a timeline established, and the values of the characters explained, These examples illustrate the importance of improving the assessment capabilities of the system in terms of targeting both the quantity and quality of self-regulation.

The main limitation to the modularization approach is that the quantity and quality of processing during the early stages of skill development determines the level of performance at the later stages. As a case example, the system detects that learners are engaging in inappropriate strategy use given a particular goal. The learners are restricted to the affordances of the interface elements embedded in this module. The system does not allow learners to repeat the Training Module and differentiate examples of appropriate and inappropriate strategy use. This is due to the fact that the modularization approach used in the MHRt relies on static interface elements to structure the self-regulation of learning.

After discussing the benefits and limitations of modularization, the following paragraphs of this paper now moves on to consider ways to improve this approach to the design of metacognitive tools. Improvements are proposed with respect to several areas, including the development of external representations, assessment mechanisms, and pedagogical agents. These are discussed in the context of a computer-based learning environment called the MetaEnquirer, a system under development at the University of Utah by the Advanced Instructional Systems and Technologies laboratory.

14.5.1 The Role of External Representations

An open-learner model may be defined as a representation that is made visible to a learner and that displays acquired knowledge during task performance [51]. In other words, the content of the learner model is continually updated with the aim of fostering self-reflection. The progress made in knowledge acquisition is inferred on the basis of user interactions logged by the system. What we know about open-learner models is largely based upon empirical studies that have compared the impacts of several characteristics of these representations in order to establish evidence-based design guidelines that are generalizable across systems [52]. Together these studies provide insights into the manner in which the MetaEnquirer should illustrate the learners' progress in investigating the causes of historical events.

The MetaEnquirer should represent how learners change their arguments as a result of searching for evidence across source documents. The benefit of this approach is that learners become more aware of the outcomes of each line of inquiry, which is hypothesized to improve learners' planning of their investigation. Besides highlighting how the outcomes of each investigation inform the next, the pedagogical agent embedded in the MetaEnquirer could challenge learners by critiquing weak points in their arguments, only to support them later in searching for clues.

14.5.2 The Role of Assessment Mechanisms

A novice-expert overlay model is an assessment approach whereby learners' steps that were taken to solve a problem are compared to an ideal solution, which is typically validated from several domain experts [53]. As such, computer-based learning environments allow learners to visualize the similarities and differences between both solution paths. Lajoie [7] implemented novice-expert overlay models in BioWorld, a computer-based learning environment that allows novices to practice medical diagnostic reasoning. The findings obtained show that experts pursue different paths in solving a case; however, their reasoning can be modeled in the form of commonly identified evidence items that are pertinent to obtaining the correct diagnosis. The model allows the system to track user interactions and compare the evidence items identified as relevant by the novices to the ones of the experts. Individualized reports are provided to the learners that highlight these similarities and differences as well as explain the correct approach to managing and treating the patient.

The MetaEnquirer stands to improve the quality of the feedback that is delivered to learners by assessing the self-regulation of learning in accordance with how experts perform inquiries into the causes of historical events. This method builds on previous work with argument-driven models in the context of the MHRt since the user interactions are appraised not only in terms of the requirements of achieving a particular goal, but also for the correctness of written annotations. An expert model of annotations that relies on decision rules to appraise the quality of learners' investigation into the causes of historical events allows the MetaEnquirer to individualize instruction. Feedback is delivered to the learner in order to distinguish between pieces of evidence that were similar and different from the ones obtained by the experts. The same model can also be used by the system to recommend source documents that may corroborate or discorroborate certain explanations for the event under investigation.

14.5.3 The Role of Pedagogical Agents

Multi-agent intelligent tutoring systems rely on several pedagogical agents that emulate different roles with the aim of achieving an instructional objective. As an example, Betty's Brain allows learners to teach Betty, an artificial pedagogical agent, and evaluate her understanding of river-ecosystem processes [54]. Mr. Davis supports learners by delivering quiz results, guiding them in their search of the library, and scaffolding learners' efforts to regulate their own learning. MetaTutor assigns each pedagogical agent to support a construct from the information processing theory of self-regulated learning in order to scaffold them in using the relevant skills [55]. These agents include Mary the monitor, Sam the strategizer, and Pam the planner.

The use of multiple agents embedded in a modular system such as the MetaEnquirer stands to address the issue in relation to the current design of the MHRt. Modularization as distinct configurations of interface elements, each set designed for an instructional objective, is limited in terms of its flexibility. However, the dialogue that occurs between different agents and the learner can be tailored by the system to compensate for this lack of flexibility, while also guiding the learner across each module.

For instance, the role of the first agent may be to support skill acquisition by asking learners to differentiate between examples of appropriate and inappropriate strategy use. The second agent coaches learners in practicing and refining the use of these strategies by suggesting relevant content and highlighting deviations from the experts' written annotations. Both agents could intervene at appropriate moments depending on how the system appraises the quality of learners' investigations. The benefit of this approach is that the revised version of the system would be capable of targeting the specific needs of learners at different stages of skill development.

14.6 Conclusion

In summary, this chapter compared and contrasted two assessment mechanisms that each targeted different stages of skill acquisition. An illustrative case study of one learner was reviewed as an example of the use of learning curves to model skill acquisition and an argument-driven model to assess skill practice and refinement. The role of these assessment mechanisms was explained in terms of their capacity to adapt instruction in the context of the MHRt. In doing so, the MHRt modules facilitate the regulation of learning while performing inquiries into the causes of historical events in accordance with disciplinary-based practices.

Difficulties arise, however, when modules are completed in a linear manner, as the early stages of skill development are critical to ensuring consolidation during the later stages. Since learners are not allowed to go backward along the trajectory to develop the targeted skills, modularization assumes that prerequisite knowledge and skills have been gained for future learning to be successful. In reviewing recent advances pertaining to the roles of external representations, assessment mechanisms, and pedagogical agents in the context of metacognitive tools, a set of design principles were outlined to guide the development of the MetaEnquirer, which will address this issue by redefining modules as dynamic components that are delivered to the learner when necessary. Considerably more work will need to be done to determine whether modularization as a mechanism to deliver instruction within metacognitive tools is generalizable.

Acknowledgments I would like to thank Dr. Susanne Lajoie, my graduate research supervisor, for her guidance throughout the completion of this research project. I also would like to acknowledge contributions from the members of my doctoral thesis committee, Drs. Roger Azevedo and Nathan Hall.

References

1. Lajoie, S.P., Azevedo, R.: Teaching and learning in technology-rich learning environments. In: Alexander, P., Winne, P. (eds.) Handbook of Educational Psychology, 2nd edn, pp. 803–821. Erlbaum, Mahwah, NJ (2006)
2. Pea, R.D.: Beyond amplification: using the computer to reorganize mental functioning. Educ. Psychol. **20**(4), 167–182 (1985)

3. Perkins, P.R.: The fingertip effect: how information processing technology shapes thinking. Educ. Res. **14**, 11–17 (1985)
4. Salomon, G., Perkins, D., Globerson, T.: Partners in cognition: Extending human intelligence with intelligent technologies. Educ. Res. **20**, 10–16 (1991)
5. Lajoie, S.P. (ed.): Computers as Cognitive Tools vol 2: No More Walls. Erlbaum, Mahwah, NJ (2000)
6. Jonassen, D.H.: Using cognitive tools to represent problems. J. Res. Technol. Educ. **35**(3), 362–381 (2003)
7. Lajoie, S.P.: Developing professional expertise with a cognitive apprenticeship model: Examples from avionics and medicine. In: Ericsson, K.A. (ed.) Development of Professional Expertise: Toward Measurement of Expert Performance and Design of Optimal Learning Environments, pp. 61–83. Cambridge University Press, Cambridge, UK (2009)
8. Azevedo, R.: Computer environments as metacognitive tools for enhancing learning. Educ. Psychol. **40**, 193–198 (2005)
9. Azevedo, R.: Using hypermedia as a metacognitive tool for enhancing student learning? The role of self-regulated learning. Educ. Psychol. **40**(4), 199–209 (2005)
10. Azevedo, R.: The role of self-regulation in learning about science with hypermedia. In: Robinson, D., Schraw, G. (eds.) Recent Innovations in Educational Technology that Facilitate Student Learning, pp. 127–156. Information Age Publishing, Charlotte, NC (2008)
11. Azevedo, R., Aleven, V. (eds.): International Handbook of Metacognition and Learning Technologies. Springer, Berlin (2013)
12. Bjork, R., Dunlosky, J., Kornell, N.: Self-regulated learning: Beliefs, techniques, and illusions. Annu. Rev. Psychol. **64**, 417–444 (2013)
13. Pintrich, P.R.: A conceptual framework for assessing motivation and self-regulated learning in college students. Educ. Psychol. Rev. **16**(4), 385–407 (2004)
14. Schunk, D.H.: Self-regulated learning: the educational legacy of Paul R Pintrich. Educ. Psychol. **40**, 85–94 (2005)
15. Winne, P.H., Hadwin, A.F.: The weave of motivation and self-regulated learning. In: Schunk, D.H., Zimmerman, B.J. (eds.) Motivation and Self-Regulated Learning: Theory, Research, and Applications, pp. 297–314. Lawrence Erlbaum Associates, Mahwah, NJ (2008)
16. Zimmerman, B.J.: Investigating self-regulation and motivation: historical background, methodological developments, and future prospects. Am. Educ. Res. J. **45**(1), 166–183 (2008)
17. Schraw, G.: Measuring self-regulation in computer-based learning environments. Educ. Psychol. **45**(4), 258–266 (2010)
18. Veenman, M.V.J.: Assessing metacognitive skills in computerized learning environments. In: Azevedo, R., Aleven, V. (eds.) International Handbook of Metacognition and Learning Technologies, pp. 157–168. Springer, New York, NY (2013)
19. Winters, F.I., Greene, J.A., Costich, C.M.: Self-regulation of learning within computer-based learning environments: A critical analysis. Educ. Psychol. Rev. **20**(4), 429–444 (2008)
20. Tsai, C.-W., Shen, P.-D., Fan, Y.-T.: Research trends in self-regulated learning research in online learning environments: a review of studies published in selected journals from 2003 to 2012. Br. J. Educ. Technol. **44**(5), E107–E110 (2013)
21. Devolder, A., van Braak, J., Tondeur, J.: Supporting self-regulated learning in computer-based learning environments: systemic review of effects of scaffolding in the domain of education. J. Comput. Assist. Learn. **28**(6), 557–573 (2012)
22. Azevedo, R., Moos, D.C., Greene, J.A., Winters, F.I., Cromley, J.G.: Why is externally-facilitated regulated learning more effective than self-regulated learning with hypermedia? Educ. Tech. Res. Dev. **56**, 45–72 (2008)
23. Greene, J.A., Bolick, C.M., Robertson, J.: Fostering historical knowledge and thinking skills using hypermedia learning environments: the role of self-regulated learning. Comput. Educ. **54**, 230–243 (2010)
24. Poitras, E.P., Lajoie, S.P., Hong, Y.J.: The design of technology-rich learning environments as metacognitive tools in history education. Instr. Sci. **40**, 1033–1061 (2012)

25. Azevedo, R., Feyzi-Behnagh, R.: Dysregulated learning with advanced learning technologies. J. e-Learn. Know. Soc. **7**(2), 9–18 (2011)
26. Linderholm, T., Everson, M., van den Broek, P., Mischinski, M., Crittenden, A., Samuels, J.: Effects of causal text revisions on more- and less-skilled readers' comprehension of easy and difficult texts. Cogn. Instr. **18**, 525–556 (2000)
27. Gilabert, R., Martinez, G., Vidal-Abarca, E.: Some good texts are always better: Text revision to foster inferences of readers with high and low prior background knowledge. Learn. Instr. **15**(1), 45–68 (2005)
28. D'Mello, S., Lehman, B., Pekrun, R., Graesser, A.C.: Confusion can be beneficial for learning. Learn. Instr. **29**, 153–170 (2014)
29. Lehman, B., D'Mello, S.K., Graesser, A.C.: Confusion and complex learning during interactions with computer learning environments. Int High. Educ. **15**, 184–194 (2012)
30. Poitras, E., Lajoie, S.P.: A three-pronged approach to the design of technology-rich learning environments. In: Atkinson, R. (ed.) Learning Environments: Technologies, Challenges and Impact Assessment. Nova Science Publishers Inc., Hauppauge, NY (2013)
31. Pintrich, P.R.: Multiple goals, multiple pathways: the role of goal orientations in learning and achievement. J. Educ. Psychol. **92**, 544–555 (2000)
32. Zimmerman, B.J.: Theories of self-regulated learning and academic achievement: An overview and analysis. In: Zimmerman, B.J., Schunk, D.H. (eds.) Self-Regulated Learning and Academic Achievement: Theoretical Perspectives, 2nd edn, pp. 1–37. Erlbaum, Mahwah, NJ (2001)
33. Alexander, P., Dinsmore, D., Parkinson, M., Winters, F.: Self-regulated learning in academic domains. In: Zimmerman, B.J., Schunk, D.H. (eds.) Handbook of Self-Regulation of Learning and Performance, pp. 393–407. Routledge, New York, NY (2011)
34. Poitras, E., Lajoie, S.P.: A domain-specific account of self-regulated learning: the cognitive and metacognitive activities involved in learning through historical inquiry. Metacogn. & Learn. **8**(3), 213–234 (2013)
35. Carretero, M., López-Manjón, A., Jacott, L.: Explanining historical events. Int. J. Educ. Res. **27**(3), 245–253 (1997)
36. Nokes, J.D., Dole, J.A., Hacker, D.J.: Teaching high school students to use heuristics while reading historical texts. J. Educ. Psychol. **99**(3), 492–504 (2007)
37. van Drie, J., van Boxtel, C.: Historical reasoning: Towards a framework for analyzing students' reasoning about the past. Educ. Psychology Rev. **20**(2), 87–110 (2008)
38. Winne, P.H.: A perspective on state-of-the-art research on self-regulated learning. Instr. Sci. **33**, 559–565 (2005)
39. Winne, P.H.: A cognitive and metacognitive analysis of self-regulated learning. In: Zimmerman, B.J., Schunk, D.H. (eds.) Handbook of Self-Regulation of Learning and Performance, pp. 15–32. Routledge, New York, NY (2011)
40. Winne, P.H., Hadwin, A.F.: Self-regulated learning and sociocognitive theory. In: Peterson, P., Baker, E., McGraw, B. (eds.) International Encyclopedia of Education, vol. 5, pp. 503–508. Elsevier, Amsterdam (2010)
41. Voss, J.F., Wiley, J.: Developing understanding while writing essays in history. Int. J. Educ. Res. **27**(3), 255–265 (1997)
42. Renkl, A.: Instruction based on examples. In: Mayer, R.E., Alexander, P.A. (eds.) Handbook of research on learning and instruction. Routledge, New York, NY (2010)
43. Renkl, A., Hilbert, T., Schworm, S.: Example-based learning in heuristic domains: A cognitive load theory account. Educ. Psychol. Rev. **21**(1), 67–78 (2009)
44. Hmelo-Silver, C.E., Duncan, R.G., Chinn, C.A.: Scaffolding and achievement in problem-based learning and inquiry learning: a response to Kirschner, Sweller, and Clark (2006). Educ. Psychol. **42**(2), 99–107 (2007)
45. Krajcik, J.S., Blumenfeld, P.: Project-based learning. In: Sawyer, R.K. (ed.) The Cambridge Handbook of the Learning Sciences, pp. 317–334. New York, NY, Cambridge (2006)
46. Levstik, L.S.: Learning history. In: Mayer, R.E., Alexander, P.A. (eds.) Handbook of Research on Learning and Instruction, pp. 108–126. Routledge, New York, NY (2011)

47. Loyens, S.M.M., Rikers, R.M.J.P.: Instruction based on inquiry. In: Mayer, R., Alexander, P. (eds.) Handbook of Research on Learning and Instruction. Routledge, New York, NY (2011)

48. Gong, Y., Beck, J.E., Heffernan, N.T.: Comparing knowledge tracing and performance factor analysis by using multiple model fitting procedure. In: Aleven, V., Kay, J., Mostow, J. (eds.) Intelligent Tutoring Systems, pp. 35–44. Springer, Berlin Heidelberg (2010)

49. Martin, B., Mitrovic, T., Mathan, S., Koedinger, K.R.: Evaluating and improving adaptive educational systems with learning curves. User Model. User-Adap. Inter. **21**, 249–283 (2011)

50. Lajoie, S.P.: Transitions and trajectories for studies of expertise. Educ. Res. **32**(8), 21–25 (2003)

51. Bull, S., Kay, J.: Student models that invite the learner in: the SMILI:() open learner modelling framework. Int. J. Arti. Intell. Educ., **17**(2), 89–120 (2007)

52. Dimitrova, V., McCalla, G., Bull, S.: Open learner models: future research directions. Special Issue of the IJAAIED (Part 2). Int. J. Arti. Intell. Educ. **17**(3), 217–226 (2007)

53. Shute, V.J., Zapata-Rivera, D.: Adaptive educational systems. In: Durlach, P. (ed.) Adaptive technologies for training and education, pp. 7–27. Cambridge University Press, New York NY (2012)

54. Segedy, J.R., Biswas, G., Sulcer, B.: A model-based behavior analysis approach for open-ended environments. J. Educ. Technol. Soc. **17**(1), 272–282 (2014)

55. Azevedo, R., Johnson, A., Chauncey, A., Burkett, C.: Self-regulated learning with MetaTutor: advancing the science of learning with MetaCognitive tools. In: Khine, M., Saleh, I. (eds.) New Science of Learning: Computers, Cognition, and Collaboration in Education, pp. 225–247. Springer, Amsterdam (2010)

Author Index

© Springer International Publishing Switzerland 2015   367
A. Peña-Ayala (ed.), *Metacognition: Fundaments, Applications, and Trends*,
Intelligent Systems Reference Library 76, DOI 10.1007/978-3-319-11062-2

Printed in the United States
By Bookmasters